Sustainability and Economic Development in Hill Agriculture

Sustainability and Economic Development in Hill Agriculture

— Editors —
Durga Charan Kalita
and
Birendra Kumar Mishra

Department of Rural Development and Agriculture Production
North-Eastern Hill University
Tura Campus – 794002
Meghalaya (India)

BIOTECH
2012
BIOTECH BOOKS®

ISBN 978-81-7622-260-0

Published by: **BIOTECH BOOKS®**
4762-63/23, Ansari Road, Darya Ganj,
New Delhi - 110 002
Phone: +91-011-23262132
E-mail: biotechbooks@yahoo.co.in

Printed at: **Chawla Offset Printers**
Delhi - 110 052

PRINTED IN INDIA

Acknowledgements

We are at loss of words to bring out our deep sense of veneration and gratitude to Prof. E. Thomas, PVC, North-Eastern Hill University, Tura campus for his valuable encouragement and suggestion at the initial stage of our carrier and also in pursuit the present work.

The academic encouragement also received from Sri Jitendra Nath Pathak, Retired Joint Director, State Department of Agriculture, Government of Assam for his message and encouragement for this book.

We are highly indebted to the family members of North-Eastern Hill University, Tura Campus and to our colleagues of the Department and our friends of the campus for encouraging us to publish this edited volume.

We are very thankful to all the authors for their moral support and continuous encouragement to publish this book.

We wish to record a word of appreciation to Biotech Book Publishers, New Delhi for accepting the manuscript and bringing out this book in a graceful manner.

Last but not least, it is our pleasure to record and acknowledge the cheerful cooperation of our family members.

Durga Charan Kalita

Birendra Kumar Mishra

Foreword

North Eastern Region is predominantly an agricultural region with over 90 per cent of its population living in rural areas and dependant on agriculture. The region possesses some features which are peculiar in nature. Its geographical location, physiographic, climatic features, ethnic diversities, socio-cultural sphere are unique by themselves. Being a natural greenery, it is also ravaged by natural calamities. Agriculture has been practised, on one end in the most primitive ways like jhuming to the other end in well organised modern ways practised in the tea estates. By and large crop production and animal husbandry have been ways of life rather than any commercial enterprise. The progress in agriculture made during the last half a century by most of other states in the country is not seen in the region. This sluggishness is ascribed to natural phenomena, infrastructural deficiencies, inadequacies in input supply and in extension support.

Hill agriculture is facing the twin challenges of eco-environmental hazards and loss in productivity. The development of agriculture in hilly areas has been constrained by several factors which include lack of scientific land use according to capability, lack of economic and social institutions, lack of incentives, proper training and extension education, loss of support area for the forest dwellers and risk and uncertainty in hill farming system. Farming systems in hills offer vast scope for adoption of a diverse mix of farm enterprises including crop cultivation and vegetable farming, horticulture and livestock. The production potential of hill agriculture has not been exploited fully due to several constraints. The thirty five papers included in this volume, apart from a few are mainly based on field studies, conducted in the hill areas of North Eastern Region, identify the constraints in the development of hill agriculture and suggest measure to overcome them. Few papers are from the hill areas of other states of the country. Most of the papers deal with agribusiness opportunities, role of NGO's, resource conservation, homestead agro-forestry, shifting cultivation which seems to have a greater impact on sustainability of hill agriculture.

Apart from that one paper deals with the concept, need and scope of sustainable agriculture in North Eastern region.

In recent times agricultural development in the region is seen all around, even though not very spectacular. It was very necessary to document the issues and I congratulate Dr. D.C. Kalita and Dr. B.K. Mishra for bridging this deficiency. These types of documents are invaluable reference material for educational institutions and field workers connected to agriculture in hill region. It is hoped that the volume will promote further thinking on the subject and a better understanding of the constraints faced in the development of hill agriculture.

I think that the authors can be confident that there will be many grateful readers who will have gained a broader perspective of the disciplines of sustainability issues relating to hill agriculture as a result of their efforts.

I am thankful to the editors and hope they as well as others would strive to enrich the agricultural community with similar publications in future. Finally, I hope this compilation will be of good use to the readers.

C. Hazarika

Professor,

Department of Agricultural Economics,

Assam Agricultural University, Jorhat, Assam

Prof. Eugene D. Thomas,
Pro-Vice-Chancellor,
North-Eastern Hill University,
Tura Campus, Tura 794002,
Meghalaya.

Foreword

It gives me great pleasure to write a Forward to this book entitled *"Sustainability and Economic Development in Hill Agriculture."* The present publication is a collection of articles on sustainable development vis-à-vis agriculture and allied activities in the hills.

The Brundtland Report of 1987 defines sustainable development as one "that meets the needs of the present without compromising the ability of future generations to meet their own needs." Underlining this understanding of sustainable development requires one to view our activities in this earth that we live in as a system that connects both space and time. It is also accepted that the quality of life too needs to be viewed as a system in sustainable development.

Globally, dramatic changes have been witnessed since the latter pan of the twentieth century in the social and industrial sectors with advancements in agriculture, health care and medicine, leading to an unparallel economic growth. However, this development has not been without a price. Wider disparities in income and inequality have been seen coupled with million of farmers still using traditional implements and low-productivity techniques. There has been unsystematic and unplanned harvesting of physical resources due to the greed of a few and allowed to occur by an equally corrupt government administration; harnessing of renewal energy has not made an appreciable impact. The increased production and consumption has led to wastes and by-products which have polluted our waters and air, and gradually destroying the flora and fauna. The increased population has led to encroachment on agricultural land and forests by farmers and forest dwellers. This type of development cannot be sustained!

Agriculture in the hills has a strong correlation with environmental changes as it affects the latter. It also is a catalyst to develop and enlarge non- agricultural economic activity.

Farming in the hills is characterized with different constraints. The topography calls for different methods of cultivation. Land holdings are scattered and small. Most of the land is inaccessible and infrastructure inadequate. Communications is poor or non-existent and thus transportation of the product is greatly dependant on the climate. The fanning is dependant, in most cases, on rainfall. Monsoons are erratic. The land is prone to soil erosion. This is also worsened by migratory grazing. Fanners are vulnerable to natural disasters like floods and hailstorms. Animal-man conflict adds to the fanner's woes. Pressure on usable land for non-agricultural and housing purposes has reduced areas of cultivable land.

Yet, the hills have a vast wealth of natural resources, whether it is in fertility of the soil, the rich flora and fauna, the major and minor forest produce, the minerals, and the natural pastures. There is richness in the bio-diversity. Hills are also the source of water catchment areas that so badly need to be protected and preserved.

Balancing today's agricultural needs without destroying the natural wealth in the hills for the future generations calls for more extensive, in-depth and meaningful research in hill agriculture. Sustainable agricultural development also calls for those in authority to invoke ethics in politics and economic activity.

I extend my appreciation to Dr. D.C. Kalita and Dr. B.K. Mishra for their endeavor in bringing together researchers in this area of study culminating in this volume. I wish it receives a wide readership among the academia, educationists and policy makers.

Eugene D. Thomas

Sri Jitendra Nath Pathak
Retired Joint Director
State Department of Agriculture
Govt of Assam

Message

I am happy to know that Dr. D.C. Kalita and Dr. B.K. Mishra, Department of RDAP, NEHU, Tura Campus, Meghalaya are going to publish the edited book on "*Sustainability and Economic Development in Hill Agriculture*". Agriculture is the backbone of the nation. After the civilization, man takes different ways for living depending upon the nature of soil, topography, ecological situation for cultivation, Originally the hill people were not settled permanently in one place and their system of cultivation were also shifting cultivation which is known as jhumming. This system of cultivation affects the soil health, ecology system, and creates pollution to the nearby areas by so many ways. When time passed, they now settled permanently in one place and cultivation is converted to settled cultivation where the working force is also less. Farmers are adopting different cropping systems and crop mixtures in different situation to get minimum production for their livelihood. Rice is the main crop in hill as well as in plain areas. With rice different farming systems with goattery, poultry, piggery, horticultural crops, bee keeping and dairy can adopted in hill agriculture which in turn raise the standard of living of people and contribute to the economic development, Therefore, diversification is more suitable answer for food security and economic development.

I firmly believed that this book will help the researchers, academicians, planners and policy makers to take appropriate measure for sustainability and economic development in hill agriculture.

J.N. Pathak

J.N. Pathak

Contents

List of Contributors

Aditya, A.K., Formality Professor, School of Life Science, Visva-Bharati, Santiniketan

Ahmed, Jaynal Ud-din, Department of Management, North Eastern Hill University, Tura Campus, Tura – 794 002, Meghalaya

Ao, R., Research Scholar, Department of RDAP, NEHU, Tura Campus, Meghalaya

Arengh, Akaibel D., Research Scholar, Department of RDAP, NEHU, Tura Campus, Meghalaya

Bahuguna, Abhishek, Senior Research Officer and Senior Research Fellow, GBPUA&T, Hill Campus, Ranichauri, Tehri Garhwal – 249 199, Uttarakhand

Barbhuiya, Anjam Hussain, Department of Zoology, Don Bosco College, Tura, Meghalaya

Bhagat, Deepak, Assistant Professor, Department of Management, North-Eastern Hill University, Tura Campus, Tura – 794 002, Meghalaya

Bhowmick, B.C., Department of Agricultural Economics, AAU, Jorhat – 13, Assam

Bordoloi, Bhriguraj, Jaynal Ud-din Ahmed, Department of Management, North Eastern Hill University, Tura Campus, Tura – 794 002, Meghalaya

Das, Amit, Research Associate (NAIP) KVK, ICAR, Tura

Das, Anup, Senior Scientist, ICAR Research Complex for NEH Region, Umiam, Meghalaya

Das, Apu, Ph.D. Scholar, CIFE, Mumbai – 61

Das, Tarun Kumar, Subject Matter Specialist (Agri. Extension), KVK, ICAR, Tura, Meghalaya

Debnath, Biswajit, SMS (Fisheries), KVK (ICAR), South Tripura, SMS (Plant Protection), Tripura

Dev, Rahul, College of Forestry and Hill Agriculture, G.B. Pant University of Agriculture and Technology, Hill Campus, Ranichauri, Tehri Garhwal – 249 199, Uttarakhand

Devi, Ch. Victoria, Ph.D. Scholar, Dept. of Agril. Extension, BCKV, Mohanpur, District Nadia –741 252, West Bengal,
E-mail: ch_venus2005@yahoo.com

Dhar, U.R., Professor, Department of Business Administration, Guwahati University, Assam

Doley, Sunil, Senior Scientist, Division of Poultry Science, ICAR, NEH Region, Shillong

Ghosh, R.C., Professor, Pathology Department, College of Veterinary Science and A.H., Durg – 491 001, Chhattisgarh

Hazarika, C., Department of Agricultural Economics, AAU, Jorhat – 13, Assam

Islam, M., Subject Matter Specialist (Agronomy), KVK (ICAR), Tura, Megahalaya

Kalita, D.C., Reader, Department of RDAP, NEHU, Tura Campus, Meghalaya

Kar, Devashish, Department of Life Science, Assam University, Silchar, Assam

Kumar, Sunil, Assistant Professor (Animal Science), College of Forestry and Hill Agriculture, G.B. Pant University of Agriculture & Technology, Hill Campus, Ranichauri, Tehri Garhwal – 249 199, Uttarakhand

Laskar, Sahnaj Rahmatulla, Department of Management, North Eastern Hill University, Tura Campus, Tura – 794 002, Meghalaya

Makwana, A.K., Assistant Professor, SMC College of Dairy Science, Anand Agricultural University, Anand, Gujarat – 388 110

Mandal, Lalmohan, Professor, Department of Animal Science and, Ex-Dean, Faculty of Agriculture, Bidhan Chandra Krishi Viswavidyalaya, Mohanpur – 741252, Nadia, W.B.

Mishra, Anjani Kumar, Assistant Professor, M.P. Pashu Chikitsa Vigyan Vishwa Vidyalaya, College of Veterinary Sciences &A.H, Jabalpur – 482 001, M.P.

Mishra, B.K., Department of RDAP, North Eastern Hill University, Tura Campus, Tura –794 002, Meghalaya

Momin, J.W., Department of RDAP, NEHU, Tura Campus, Meghalaya

Mondal, Sagar, Associate Professor, Dept. of Agril. Extension, BCKV, District Nadia, West Bengal
E-mail: sagarmondal.bckv@gmail.com

Mugaokar, Pankaj K., Ph.D. Scholar, CIFE, Mumbai – 61

Munda, G.C., Principal Scientist, ICAR Research Complex for NEH Region, Umiam, Meghalaya

Nakhro, R., Department of Agricultural Economics, Nagaland University, SASRD, Medziphema Campus

Nath, L.K., Programme Assistant, KVK, Tura, Meghalaya

Ngachan, S.V., Director, ICAR Research Complex for NEH Region, Umiam, Meghalaya

Panda, B.B., Senior scientist, Central Rice Research Institute, Cuttack – 753 006, Odisha

Patel, D.P., Senior Scientist, ICAR Research Complex for NEH Region, Umiam, Meghalaya

Prasad, Birendra, Senior Research Officer and Senior Research Fellow, GBPUA&T, Hill Campus, Ranichauri, Tehri Garhwal – 249 199, Uttarakhand

Purohit, V., Research scholar, Department of Management, NEHU, Tura Campus – 794 002, Meghalaya

Pushpajali, Guest Faculty, SRT Campus, Badshaithaul, HNB Garhwal Central University, Srinagar – 249 199, Uttarakhand

Pyne, S.K., Professor in Animal Science and Head of the Department, ASEPAN, PSB, Visva-Bharati, W.B.

Roy, Kishor Kr., SMS, Assam Agricultural University, Akbarpur, Karimganj – 788 710, Assam

Sah, V.K., College of Forestry and Hill Agriculture, G.B. Pant University of Agriculture and Technology, Hill Campus, Ranichauri, Tehri Garhwal – 249 199, Uttarakhand

Saikia, R.S., Department of Agricultural Economics, AAU, Jorhat – 13, Assam

Samajdar, Tanmay, Programme Coordinator, KVK, Tura, Meghalaya

Sangtam, L.T., Department of Agricultural Economics, Nagaland University, SASRD, Medziphema Campus

Sarma, G.K., Farm Manager, Krishi Vigyan Kendra, Darrang, Assam

Sharmah, A.K., Technical Consultant (Hort.), NGO, Dishari, Tripura

Sharmah, D., Tripura Subject Matter Specialist (Fisheries), KVK (ICAR), South Tripura, Tripura

Sharmah, Diganta, Ph.D. Scholar, CIFE, Mumbai – 61
E-mail: ch_venus2005@yahoo.com

Singaiah, G., Professor, Department of Management, NEHU, Tura Campus – 794 002, Meghalaya

Srivastava, N. Kumar, Research scholar, Department of Management, NEHU, Tura Campus – 794 002, Megahalaya

Tripathi, Shailesh, College of Forestry and Hill Agriculture, G.B. Pant University of Agriculture and Technology, Hill Campus, Ranichauri, Tehri Garhwal – 249 199, Uttarakhand

Vanlalruali, M.Sc., Agril. Extension, Dept. of Agril. Extension, BCKV, Mohanpur, District Nadia – 741 252, West Bengal

Yadav, Rashmi, Senior Scientist, NBPGR, New Delhi

Yadav, V.K., College of Forestry and Hill Agriculture, G.B. Pant University of Agriculture and Technology, Hill Campus, Ranichauri, Tehri Garhwal – 249 199, Uttarakhand

Yadav, V.K., Principal Scientist, IGFRI, Jhanshi, Uttar Pradesh

Introduction

North-Eastern region of India consists of the contiguous seven sisters (Assam, Meghalaya, Manipur, Nagaland, Tripura, Mizoram and Arunanchal Pradesh) states and Sikkim. It is ethnically distinct from the other states of the country. More than 2000 Km. of boundary is shared with other countries. These states comprised of 72 district; 39,946 villages and 245 towns. The region is underdeveloped in terms of livestock production than rest of the country. Very few organized industries are available in this region. This region separates from other regions for its unique culture, handicrafts, martial arts and scenic value. This region is rich in natural resources despite of having so many problems like lack of infrastructure, poor planning and offers. Since beginning of the economic liberalization in 1990,s studies have shown that this region is lagging behind the others in terms of development.

Flanked by hills and with the mighty Brahmaputra river slashing a central path between its north and south, the North East is bounded by the States of Manipur, Meghalaya, Mizoram, Nagaland, Tripura with Assam comprising the heart. Arunachal Pradesh lies to its north and Sikkim a little away in the North West bordering China and Bhutan. Bangladesh and Myanmar lie to its southwest and east. The torrential Brahmaputra deposits its rich alluvial silt along the banks of the plains of Assam. Tropical rain forest, rich in flora and fauna, spread their arms across Arunachal Pradesh into Assam. It is a land where tea is an industry, handicrafts a major occupation and martial arts a favoured sport.

The economy of Northeast India has got its definite identity due to its peculiar physical, economic and socio-cultural characteristics. This region consists of eight states *viz.*, Assam, Arunachal Pradesh, Manipur, Meghalaya, Mizoram, Nagaland, Tripura, Sikkim. The NER of India covers an area of 2.62 lakh sq. km. It accounts for 7.9 per cent of total geographical area of the country. With a total population of 39 million (2001), it accounts for 3.8 per cent of total population of India. NER as a whole is placed amongst the poorest regions in the country. The per capita income at Rs. 12407/- is less than the national average of Rs.17978/- (2001-02).

The pattern of agricultural growth has however remained uneven across regions and crops. The NER continues to be a net importer of food grains even for its own consumption. In spite of covering 7.7 per cent of the country's total geographical area, NER produces only 1.5 per cent of the country's total food grain production. Agriculture provides livelihood support to 70 per cent of the population of NER.

The agricultural production systems in the hilly areas differ from the plough cultivation in the plain areas. In one side highly modern agricultural cultivation practices flourishes while on other side age old agricultural practices dominates. Thus dualism in agriculture still persists in the hilly areas. This system of dualism determines the standard of living of people and economic development. Further, there are two distinct agricultural production systems in the hilly areas, *viz.*, settled and shifting cultivation. The tribal people practice the shifting cultivation which is locally known as Jhumming and this system is the first step of transition from hunting to food production system. Usually the settled cultivation is practiced in foothills and terraces' in gentle slopes. The production behaviour in both the system is more or less similar as crop mixtures are grown in both the production system. However, the productivity under settled cultivation is higher than that of shifting cultivation. These productivity differences lead to differences in livelihood standard of people and economic development. The productivity in these two production systems differs greatly and thereby causes a serious problem of food availability. Therefore, it is necessary as well as important to know the economics under both the production systems.

The global food security will be a cause of concern in the coming decades if environmental foundations for sustainable agriculture are not preserved. In this light the need of redoubling efforts in enhancing agricultural production and promoting agrarian prosperity needs no emphasis. To attain this goal promotion of sustainable agriculture is the only solution. Sustainable agriculture is that form of farming which produces sufficient food to meet the needs of the present generation without eroding the ecological assets and the productivity of the life supporting system of future generation (Thakur, 1995). The study of the system will provide a broad basis for developing farm plans which in turn help the farmer to utilize his resources optimally for achieving sustainability to ensure a minimum standard of living. In the light of increase population in the country in general and the north east states in particular, promotion of sustainable agriculture is the only solution to feed the people. Attainment of sustainability in terms of food and fuels without impairing the soil fertility status and distributing the minimum forests coverage is the main objective for developing plans for future. Keeping view the above facts an attempt is made in this edited book to know the sustainability and economic development in hill agriculture.

As Indian economy is mainly based on agriculture, 70 per cent of the population is engaged in it directly or indirectly. But agriculture alone is unable to provide necessary employment and income to the farmers-small and marginal in general and the landless labourers in particular. The All India Rural Credit Review Committee has also emphasized the need for providing subsidiary occupations to the peasants. Hence, the government through the Departments of Agriculture and Animal Husbandry has to encourage subsidiary and allied occupations to agriculture like dairy, fishery, poultry, sheep-rearing, etc.The rearing of cattle and dairying go hand in hand with agriculture and help improve the socio-economic conditions of the rural folk. It also provides continuous and regular employment to all the members of the family of the farmer, as dairying requires a lot of manual labour for its various kinds of operations like maintenance of animals, feeding, milking and marketing of milk, etc. Besides, the dairy industry requires skilled and unskilled labour in procurement, processing and distribution of milk. Thus, the employment potential in the country can tremendously be increased through dairy-farming. The importance of animal husbandry is recognized in India since times immemorial. Bull was scared to the Indus people. The water-buffalo was known in Harappa, long before it appeared anywhere in the world. The cow gained prominence with the advent of the Aryans who worshipped it. The Aryans who settled in India during 2000-1500 B.C. were rearing cattle on large scale as the mainstay of their profession. When they switches-over to agriculture in the vast Genetic plains, the cattle were used in farming.

Dairying has grown with human history over years into art, science and technology. It has provided much needed nutritional support and food fantasy. This has enabled dairying to carve a niche in commerce as the oldest branch of global agriculture segment in practice and the first activity to assume status of processed food industry.

Technology involved in some areas of dairy processing are simple but have limitations to transcend commercial operations and aspirations. Commercial growth of dairying needs rational application of science, technology and management. This needs considerable support of qualified professionals, physical and financial resources. Obviously commercial dairying as an individual's enterprise will have constraints. Consortium of dairy-farmers located over a large geographic area can hold promise of well tooled professional enterprise. Most of the developing countries are in the process of transformation from the mediaeval social relations and modes of production to those of advanced countries of modern times. India is no exception to this. Though India achieved political freedom some sixty years ago, the nation has yet to free her from the shackles of socio-economic inequalities. Keeping this in view, the development of agriculture and cottage industries was given priority in the First Five Year Plan (1951-56). India with its vast arable lands and sufficient scope of irrigation potential has to depend on agriculture and allied industries to absorb the millions of uneducated un-employed of rural India.

In achieving these aims, special livestock improvement programmes to benefit small and marginal farmers and landless agriculture labourers play an important role. Thus, dairying as a subsidiary to agriculture occupies a significant position in the rural economy. Dairying has played a predominant role in strengthening India's rural economy. It has been recognized as an instrument to bring about socio-economic transformation. A symbiotic relationship exists between agriculture and dairy farming. The agriculture byproducts provide feed and fodder for the cattle, whereas cattle provide necessary drought power for various agricultural operation, besides nutritional security and ready cash to farming families from sale of surplus milk.

According to the 1992 livestock census the cattle population was 204.58 and buffalo population was 84.2 and the total Livestock population was 470.86 which in

2004 census stands at 186.18 cattle population and 97.92 and the livestock population was 485. The following Tables 1 and 2 below show the Total Bovine and Livestock Population in India:

Table 1: Livestock Population In India, 1992-2003 (in Million Numbers)

Species	1992	1997	2003
Cattle	204.58	198.88	185.18
Adult female Cattle	64.36	64.42	64.51
Buffalo	84.21	89.92	97.92
Adult Female Buffalo	43.81	46.77	50.97
Total Bovines	289.00	288.80	283.10
Sheep	50.78	57.49	61.47
Goat	115.28	122.72	124.36
Horses and Ponies	0.82	0.83	0.75
Camels	1.03	0.91	0.63
Pigs	12.79	13.29	13.52
Mules	0.19	0.22	0.18
Donkeys	0.97	0.88	0.65
Yak	0.06	0.06	0.07
Total Livestock	470.86	485.39	485.00

Table 2: Number of Milch Animals–1992, 1997 and 2003 State-wise (in Thousands)

State/UTS	Milch Cows						Milch Buffaloes		
	Crossbred			Indigenous					
	1992	1997	2003	1992	1997	2003	1992	1997	2003
Andhra pradesh	205	294	475	2,171	2,234	1,934	4,462	4,731	5,227
Arunachal Pradesh	5	3	5	73	95	111	2	4	5
Assam	112	132	161	2,558	2,006	2,035	263	180	199
Bihar	59	68	569	4,954	5,541	2,840	2,423	2,662	2,678
Chattisgarh	–	40	67	–	2,482	2,313	–	345	280
Goa	4	4	6	24	21	17	19	16	16
Gujarat	121	170	318	1,878	2,000	2,120	2,983	3,533	3,938
Haryana	138	260	262	477	431	309	2,007	2,040	2,726
Himachal Pradesh	113	147	334	536	515	479	436	464	466
Jammu and Kashmir	216	328	543	681	650	598	380	397	543
Jharkhand	–	–	70	–	–	1,830	–	–	390
Karnataka	258	690	903	3,396	2,978	2,500	2,205	2,335	2,216
Kerala	842	1,050	791	759	271	152	105	27	18
Madhya Pradesh	87	83	136	8,429	5,643	5,704	2,940	5,345	4,055

Contd...

Table 1–Contd...

State/UTS	Milch Cows						Milch Buffaloes		
	Crossbred			Indigenous					
	1992	1997	2003	1992	1997	2003	1992	1997	2003
Maharashtra	827	1,132	1,390	4,442	4,219	3,532	3,448	3,318	3,478
Manipur	25	22	23	147	88	78	33	24	22
Meghalaya	8	10	15	154	213	215	7	4	5
Mizoram	3	3	4	15	7	7	2	2	2
Nagaland	35	41	87	51	53	50	6	6	7
Orissa	238	354	385	3,677	3,399	3,235	388	355	358
Punjab	642	866	741	386	287	124	3,168	3,398	3,106
Rajasthan	41	78	235	4,189	4,557	4,248	3,769	4,813	5,222
Sikkim	13	20	33	38	28	26	1	1	1
Tamil Nadu	725	1,376	2,353	2,228	1,796	1,341	1,420	1,282	792
Tripura	43	26	25	240	384	215	7	6	6
Uttar Pradesh	585	638	642	5,768	4,633	4,901	9,412	8,429	10,378
Uttaranchal	–	43	99	–	572	595	–	607	682
West Bengal	397	411	489	4,659	4,799	5,293	214	207	247
Union Territories									
A &N Island	1	2	5	17	19	15	5	5	6
Chandigarh	5	4	3	–	0	0	15	16	15
Dadra and Nagar Haveli	–	–	1	11	10	11	2	2	2
Daman and Diu	–	–	0	2	1	1	0	1	0
Delhi	8	36	35	17	22	20	150	124	139
Lakshadweep	–	–	1	1	1	0	0	0	0
Pondicherry	26	24	28	11	10	7	3	2	2
All India	5,792	8,355	11,231	52,019	49,875	46,855	40,275	42,730	47,224

Source: Livestock census 2003, Govt. of India.

This book contains 35 chapters. All the issues related sustainability and economic development in hill agriculture are included in this book. We hope this edited volume will help the students, researchers, planners and policy maker etc.

Chapter 1

Role of NGOs in Empowerment of Rural Women for Sustainable Livelihood Development

A.K. Sharmah[1], D. Sharmah[2], B. Debnath[3] and M. Islam[4]

[1]Technical Consultant (Hort.), NGO, Dishari, Tripura
[2]Subject Matter Specialist (Plant Protection)
[3]Subject Matter Specialist (Fisheries), KVK (ICAR), South Tripura, Tripura
[4]Subject Matter Specialist (Agronomy), KVK (ICAR), Tura

ABSTRACT

This paper reviews some of the roles and programs of NGOs, such as microfinance, and capacity building for empowerment of rural women. Microfinance programs helps in the up-liftmen of economic condition of rural women by job creation and income generation resulting in the sustainable development. Capacity building programmes of NGO's develop beneficiary's capacities such as ability, skill and knowledge of mobilizing resources and solving problems to gain the goals. It also helps rural women to improve quality of their lives. In this way, NGOs contribute in promoting empowerment of rural women for sustainable livelihood development. In short, this paper demonstrates that all these programs and functions of NGOs could contribute towards the promotion of sustainable livelihood development by empowering of rural women.

Keywords: Capacity Building, Microfinance, Rural women, Empowerment.

Introduction

Non-Governmental Organizations (NGOs) have been gaining much attention as a promoter of development process among rural women for sustainable livelihood development. NGOs act as a viable alternative to government as channels of development assistance, particularly in developing countries like India. In the late 1980s and early 1990s, non-governmental organizations (NGOs) have also taken on an increased role in the area of women's empowerment (Sadik, 1988). The provision of micro-finance by the NGO's to women SHG's has helped the groups to achieve a measure of economic and social empowerment. It can developed a sense of leadership, organizational skill, management of various activities of a business, right from acquiring finance, identifying raw material, market and suitable diversification and modernization (Manimekalai and Rajeshwari, 2001).

Women play a vital part in the household work and are very hard worker; they toil through the day, starting with the family works, nurturing children and livestock. In the midst of all these worries, a woman remains firmly devoted to her work, keeps her composure and is honest. Enhancing social security for these women, programmes like microfinance and capacity building can enhance employment opportunities to rural women. International organizations like the World Bank and United Nations have focused on women's issues especially the empowerment of poor women in rural areas.

Since the 1990's women have been identified as key agents of sustainable livelihood development and women's equality and empowerment are seen as central to a more holistic approach towards establishing new patterns and processes of development that are sustainable. The World Bank has suggested that empowerment of women should be a key aspect of all social development programs (World Bank, 2001). According to Anand (2001), empowerment is the process of building capacities of creating an atmosphere, which enables people to fully utilize their creative potential in pursuance of quality of life. Empowerment gives women the capacity to influence decision-making process, planning, implementation and evaluation.According to Gurumoorthy (2000), empowering women contributes to social development. Economic progress in any country whether developed or underdeveloped could be achieved through social development. The Self Help Group disburses micro credit to the rural women for the purpose of making them enterprising women and encouraging them to enter into entrepreneurial activities.

In this article two aspects will be discussed, first, the program undertaken by NGOs to empower rural women and another, how empowerment of rural women helps in sustainable livelihood development as a particular virtue of NGOs. Thus, keeping in view this paper will different NGOs' programmes would affect empowerment of rural women and finally contribute to sustainable livelihood development.

Results and Discussion

Role of NGOs in Women Empowerment

This section highlights NGOs, functions and the promotion of sustainable livelihood development. Specifically, the discussions are on the functions related to provision of microfinance and initiation of capacity building to rural women. Eventually, sustainable livelihood development will be achieved, particularly when rural women is empowered. According to Finger (1994), the bottom-up approach emphasizes community participation, grassroots movements and local decision making. It argues that community participation and grassroots initiatives promote participatory decision making and local self-reliance (Panda 2007).In bottom-up approach, rural women are able to define their own problems and having ability and capacity to solve it through organizing and participating themselves.

(*a*) Micro-Finance Programme of NGOs for Women's Empowerment

Micro-finance programmes not only give women access to savings and credit, but also bring them together regularly in organised groups. Although no 'magic bullet', they are potentially a very significant contribution to gender equality and women's empowerment, as well as pro-poor development and civil society strengthening. Through their contribution to women's ability to earn an income these programmes have potential to initiate a series of 'virtuous spirals' of economic empowerment, increased well-being for women and their families and wider social and political empowerment.

Microfinance is an effective means of poverty reduction (Rekha 1995; Cerven and Ghazanfar 1999; Pankhurst and Johnston 1999). Oxaal and Baden (1997). Mayoux (2000) and Cheston and Khan (2002) have pointed out the importance of microfinance in empowerment, particularly for women. Microfinance is defined as efforts to improve the access to loans and to saving services for poor people (Shreiner 2001). According to Cheston and Khan (2002), one of the most popular forms of economic empowerment for women is microfinance, which provides credit for poor women who are usually excluded from formal credit institutions. Since the 1990s, microfinance institutions have addressed the issues of sustainability, participation and empowerment.

The microfinance empowers women by putting capital in their hands and allowing them to earn an independent income and contribute financially to their households. This economic empowerment is expected to generate increased self-esteem, respect, and other forms of empowerment for women beneficiaries. Some evidence show that microfinance would empower women in some domains such as increased participation in decision making, more equitable status of women in the family and community, increased political power and rights, and increased self-esteem (Cheston and Kuhn 2002). Well-being as an out put of microfinance not only covers the economic indicators, but also other indicators such as community education, environment, recreation and accessibility to social services. It is related to the quality of life (Asnarulkhadi 2002). In order to gain economic sustainability, NGOs through microfinance help the communities to reduce poverty, create jobs, and promote income generation. In the developing countries, sustainability is linked more closely to issues of poverty and the gross inequalities of power and resources (Hamnett and Hassan 2003). This is due to the fact that in the Third World countries, the ecological system sometimes conflicts with the socio-economic needs of local people who depend on a local ecosystem for their survival (Nygren 2000). Manimekalai and Rajeshwari (2001) in their paper highlighted that the provision of micro-finance by the NGO's to women SHG's has helped the groups to achieve a measure of economic and social empowerment.

(*b*) Capacity Building Programme of NGOs for Women Empowerment

Capacity building is an approach to development that builds independence. Capacity building is another NGO's strategy that helps to bring about sustainable development. Before beginning to build capacity within programs, practitioners need to identify pre-existing capacities such as skills, structures, partnerships and resources. Frankish (2003) has counted a number of dimensions for community capacity including financial capacity (resources, opportunities and knowledge), human resources (skills, motivations, confidence, and relational abilities and trust) and social resources (networks, participation structures, shared trust and bonding). Capacity building is an approach to development not a set of pre-determined activities. NGOs, through the provision of education, skill and knowledge, develop the capacity of rural women towards achieving sustainable development. In fact, NGOs act as a capacity builder to help the rural women to develop the resources, building awareness, motivating to participation in project and finally improving the quality of social lives.

Empowerment is discussed at the level of individual empowerment (changes in skills, knowledge, consciousness and awareness, hope, action and beliefs in abilities to affect change) and changes in wider social structures and processes that result in increased resources and opportunities (Verity, 2007). Furthermore, with regards to sustainability, capacity building has been identified in much sustainable development policy as one of the key strategies for increasing the potential towards sustainable development. For example, in a program to train women to establish a small enterprise, a positive result will be that women have co-operated to start the enterprise, but a sustainable outcome will depend on whether women have the capacity to make it work and derive an income from it without external financial or technical assistance (Viswanath 1991). Local solutions to achieving sustainable and economic development are therefore seen as important outcomes of a capacity-building approach (Lennie 2005). Women's empowerment is not only in financial terms but also in attitudinal and motivational factors. There is no need for much inventions and innovation to empower rural women and what needed is reorientation, mobilization and realisation of women friendly environment in the rural areas, (Devasia, 2001).

Conclusion

NGOs through the micro-finance help rural women to access jobs, income-generation and improve economic condition. And then they would become empowered economically. Furthermore, capacity building programmes of NGOs such as skills, abilities, knowledge, assets and motivates to participate in the project to improve the quality of their lives. NGOs act as a capacity builder that helps the community to achieve the empowerment particularly individual empowerment. As empowerment is nothing but the greater control over livelihood resources. Therefore, the final outcome of women empowerment for sustainable livelihood development is the independence of the rural people from external agents in formulating its agenda and managing its affairs. This process involves capacity building, where rural women get involved in human capital training, transferring of authority from donor to recipient and receive supports from stakeholders (World Bank group 1999). When people become fully empowered, they are able to contribute toward sustainable development (Lyons *et al.*, 2001). This paper established the importance of NGOs in gaining sustainable livelihood development through empowering the rural women by facilitating micro-finance and capacity building programmes.

References

Anand, V.K. (2001), "Avenue for Sustainable Empowerment of Rural Women", Social Welfare, Vol.48, No.4, July, Pp.24-26

Baccaro L. (2001). Civil Society, NGOs, and Decent Work Policies: Sorting out the Issues. The Discussion Paper Series presents the preliminary results of research undertaken by the IILS in Geneva.

Bradshaw T, Winn K. (2000). Gleaners, Do-gooders, and Balers: Options for Linking Sustainability and Economic Development. Community Development Society, 31(1): 112-129.

Bridger JC, Luloff A. E. (1999). Toward an Interactional Approach to Sustainable Community Development, Journal of Rural Studies, 15(4): 377-387.

Cerven J, Ghazanfar S.M. (1999). Third World Microfinance: Challenges of Growth and Possibilities for Adaptation. The Journal of Social, Political and Economic Studies, 24(4): 445-462.

Clark J. (1999), Democratizing Development: The Role of Voluntary Organizations, West Hartford: Kumarian Press.

Desai V. (2005), NGOs, gender mainstreaming, and urban poor communities in Mumbai. *Gender and Development*, 13(2): 90-98.

Devasia, Leelamma. (2001). "Rural Women's Empowerment: A Grass Root Level Experiment". Social Welfare, Vol.48, No.4, July, Pp. 5-10

Fowler A. (1993), Non-governmental Organizations as Agents of Democratization: An African Perspective. Journal of International Development, 5(3): 325-339.

Frankish J. (2003), Conceptualization and Measurement of Community Capacity, Applied Research and Analysis Directorate (ARAD) Canada, (Executive Summary), Canada

Friedmann J. (1992). Empowerment: The Politics of Alternative Development. Cambridge

Gurumoorthy, T. R. (2000). "Self Help Groups Empower Rural Women", Kurushetra, Vol.48, No.5, February, p. 36-39

Hashemi Syed, Sydney Ruth Schuler and Ann Riley (1996): "Rural Credit Programs and Women's Empowerment in Bangladesh," *World Development* 1996, Volume 24, No. 4: 635-653

Kabeer N. (1999). Resources, Agency, Achievements: Reflections on the Measurement of Women's Empowerment. *Development and Change*, 30(3): 435-464.

Kishor, Sunita (2000). "Empowerment of Women in Egypt and Links to the Survival and Health of Their Infants" In *Women's Empowerment and Demographic Processes:*

Moving Beyond Cairo. Harriet Presser and Gita Sen, eds, New York: Oxford University Press. 2000a

Langran L.V. (2002). Empowerment and the Limits of Change: NGOs and Health Decentralization in the Philippine, Department of Political Science. Ph.D. Thesis, Toronto: University of Toronto

Lyons M, Smuts C, Stephens A. (2001): Participation, Empowerment and Sustainability: (How) Do the Links Work? *Urban Studies*, 38(8): 1233-1251

Malhotra, Anju and Mark Mather (1997). "Do Schooling and Work Empower Women in Developing Countries? Gender and Domestic Decisions in Sri Lanka." *Sociological Forum*, 12(4):599-630

McNamara Kerry (2003). "Information and Communication Technologies, Poverty and Development: Learning from Experience", *A Background Paper for the Information and Development Annual Symposium*, Geneva, Switzerland

Manimekalai, M. and Rajeshwari, G. (2001). Informal Self Help Groups – A Case From Tamil Nadu, Indian Journal of Agricultural Economics, Vol. 56 (3)

Nygren A (2000). Development Discourses and Peasant forest Relations: Natural Resource Utilization as Social Process. *Development and Change*, 31(1): 11-34

Panda B. (2007).Top Down or Bottom Up? A Study of Grassroots NGOs' Approach, *Journal of Health Management*, 9(257): 257–273

Rahman, A. (1986). "Consciousness raising efforts of Grameen Bank," *Grameen Bank Evaluation Project Working Paper No. 3.* Dhaka, Bangladesh Institute of Development Studies, Agriculture and Rural Development Division

Rekha M. (1995). *Credit for Women: Why is it so Important?* United Nations: International Research and Training Institute for the Advancement of Women.

Sadik, N (1988). "Women, the Centre of Development", *Development*, Vol. 1, 30-31.

Salamon L. (1994). The Rise of the Non-profit Sector, *Foreign Affairs*, 73(4): 109-122.

Sheraton M. (2004). *An Analysis of the Effectiveness of Microfinance: A Case Study in the Western Cape.* M.S. Thesis, University of the Western Cape

Sneddon C.S. (2000). Sustainability in Ecological Economics, Ecology and Livelihoods: A Review. *Progress in Human Geography*, 24(4): 521-549

Streeten P. (1997). Non-governmental Organizations and Development, *Annals of the American Academy of Political and Social Science*, 554: 193-210

Stromquist NP (2002). NGOs in a New Paradigm of Civil Society, *Current Issues in Comparative Education*, 1(1): 62-67

United Nations (2002). Report of the World Summit on Sustainable Development, plan of implementation, *A/CONF.199/20* New York: United Nations.

Chapter 2

Conservation and Genetic Improvement in Dairy Animals: Strategic Approach to Meet Future Challenges

Anjani Kumar Mishra

Assistant Professor, M.P. Pashu Chikitsa Vigyan Vishwa Vidyalaya College of Veterinary Sciences & A.H, Jabalpur – 482 001, M.P.

ABSTRACT

India is fast emerging as a leading country for milk production in world. The annual milk production of India has touched to 101 mMT in 2007 and is likely to reach 153mMT by 2010 (dairy India 2007). The per capita availability of milk was 124 g in 1966 which has increased to 230 g in 2007. Two factors for same are Operation Flood and strong rural based milk market network. The dairy sector has sustained annual growth rate 4 per cent -6 per cent in last decade. The aggregate price of milk production (2004) is higher than total price of wheat, rice and sugarcane. The planning commission has identified Livestock Sector as "Power House of Growth" to attain aggregate growth in agricultural sector at 4 per cent rate. The contribution of agricultural sector in national GDP has declined from 39 per cent in 1979 to 19.8 per cent in 2006-07, while contribution of livestock sector in agricultural sector has consistently increased from 19 per cent at 1950-51 to 26 per cent in 2005-06. In Gujarat it is close to 36 per cent. In large many villages of Gujarat income from milk is much larger than income from agriculture. The Gujarat state is a leading state in dairying with annual milk production 7.2 mMT in 2006-07. At present, state is procuring and processing more than 100 lac liters of milk per day under co-operative sector and purchase price of a milk ranges from Rs. 260-305/kg fat. The annual milk requirement of country (human population : 1.325-1.421 b) in 2020 has been pegged at 180-230 mMT. The demand of milk may rise further as consumption of milk is income elastic and Indian economy is poised to grow at 9 per cent -

10 per cent in coming decade. In view of future requirement, it is very much essential to strengthen and improve upon current growth rate of milk production through technological inputs, in which planned breeding strategy and organized breeding programmes covering large percentage of cattle and buffalo population will play vital role.

Keyowrds: Conservation, Genetic improvement, Cattle and buffalo.

Dairy Cattle and Buffalo Genetic Resources

India has rich genetic resources in form of 30 cattle breeds and 10 buffalo breeds. The cattle and buffalo population 2003 in India is 185.5 and 97.9 m respectively. The contribution of cattle and buffalo in total milk production is 39.7 per cent and 54.0 per cent respectively. The 20 per cent of dairy cattle population is under defined breed category while in buffalo 20 per cent of population is under defined breed, 20 per cent under graded group and rest under non descript category.

Status of Genetic Diversity and Productivity

The breed is a unit for breeding programmes and operation for all purpose. We must make strategic planning to use between and within breed genetic diversity through technological inputs to attain our target milk production. India is bestowed upon for great genetic diversity in form of 30 cattle breeds and 10 buffalo breeds (ICAR). Hammand and Leitch (1996) reviewed global animal genetic resources and concluded that within breed genetic variability accounts for 50-75 per cent of genetic variability in species. This one is sufficient to attain desired rate of genetic improvement. India and most developing countries are facing serious problem of genetic deterioration of indigenous germ plasm on account of a. intermixing b. infusion of exotic germ plasm c. absence of organized breeding programme d. absence of conservation programme e. geographic rearrangements.

Review of Ongoing Breeding Programmes

Success of any breed improvement programme depends on animal recording infrastructure, population size, extent of selection pressure, replacement of field animals and coverage of breedable population under A.I. On examining our past and present programmes we can realize that, we have to put large scale efforts in all aspects to attain meaningful results in breeding programmes. In post independence era Indian army established cattle and buffalo breeding farms which were mainly north India and for Murrah buffalo, Red Sindhi, Sahiwal, exotic and crossbreed cattle. Later on in sixties, key village schemes were implemented in different states to provide pedigreed breeding bulls at village level. Central Cattle Breeding Farms were established in 1967-1975 at Suratgarh (Tharparkar and its crosses with HF), Chiplima (Red sindhi and its crosses with Jersey), Sunabeda (Jersey), Andeshnagar (HF X Tharparkar), Hessargatta (HF), Dhamrod (Surti buffalo) and Alamadhi (Murrah buffalo) with objective to produce 10 progeny tested bulls/batch, which was never fulfilled. ICAR initiated AICRP on cattle at five locations (HAU, Hissar ; IVRI, Izatnagar ; JNKVV, Jabalpur; MPKVV,Rahuri and APAU, Lam) to produce grades of crossbreeds of HF, Jersey and Brownswiss using Gir and Haryana as indigenous breeds. At

NDRI crossbreeding programme was initiated to produce 'Karanwsiss' and 'Karanfries'. The AICRP on buffalo was initiated in 1970 at four station (NDRI, Karnal; PAU, Ludhiana; UAS, Dharvad and at University of Udaipur, Vallabhgadh). Ideas was to carry out Progeny Testing Programme for Murrah and Surti. It was operational up to 1980 and later on Network Project on for Murrah was started at NDRI, CIRB, PAU, HAU, IVRI and CBF by ICAR in 1993.

In addition to large scale coordinated breeding programmes, pure breeding and crossbreeding activities have been undertaken at more than 190 farms with state department, Indian army and SAUs. Many Gaushalas and trusts have also taken up pure breeding programme in Gir cattle, Kankrej cattle and many other cattle breeds. Few successful Field Progeny Testing Programmes are being implemented. These are Field Progeny Testing Programme for Mehsana buffalo by Dudhsagar Dairy, Mehsana, Gujarat. Field Progeny Testing Programme for Murrah buffalo and Crossbreed cattle under DIPA (Dairy Heard Improvement Programme Action) involving 4 districts (Panchmahal, Sabar, Surat and Vadodara). Similar programmes have been initiated in Karnataka, Maharashtra and Madhya Pradesh. BAIF has also introduced Field Progeny Testing Programme for Crossbred bulls in Gujarat, Maharashtra and other state. The development of 'Sunandini' breed from Field Progeny Testing Programme of CB bulls by Kerala Livestock Development Board is well known. The Intensive Cattle Development Projects (ICDPs) operated by many State Department of Animal Husbandry also undertake purebreeding and crossbreeding activities. Sabarmati Ashram Gaushala Project (Bidaj, Gujarat) is implementing ONBS programme since 1994 and is using MOET technology to produce large many genetically elite bulls. It has bulls for most of indigenous cattle and buffalo breeds, exotic breeds and crossbred grades. The NDDB and SAG, Bidaj has also undertaken MOET to produce genetically elite HF animals at AAU, Anand and Cattle Farm, Salon, Rai Barrielly. A joint venture company INDIAGEN has been established to market semen doses.

Recently NDDB has launched large scale conservation and breed improvement programme in Rathi, Kankrej, Gir cattle and Jaffrabadi buffalo. ICAR has recently launched breed improvement of Pandharpuri buffalo. NBAGR, Karnal has also initiated Network Project on Animal Genetic Resources with objective to characterize (Physical characteristics and molecular characteristics) indigenous breeds of livestock in IX FYP. It has successfully completed characterization work in most of breeds of livestock species. Gir Progeny Testing Programme at LRS, Junagadh was initiated in 1972. Through it was on limited scale (5-6 bulls/batch and 7-17 daughters/bull). It helped in identification of good bulls. The average performance of farm has improved to >1900 kg/lactation with elite animals producing >2200 kg/lactation. The average performance of daughters of best bulls ranged from 1806 to 2161kg (Anonymous, 2003) A joint venture company INDIAGEN has been established with a view to provide quality breeding services in rural area. The National Co-Operative Dairy Federation of India has been interested with marketing of semen dosage for various breeding activities.

Constrains in Genetic Improvement Programme

The information on success of various genetic improvement programme presents highly variable results and many of these have failed to achieve and significant results. The major constrains observed were small herd size, limited No. of bulls, lack of intensity of selection, frequent change in objectives, insufficient input. Under Field Progeny Testing Programmes major constrains observed were poor infrastructure for performance recording, poor AI coverage, poor AI services and competence, limited participation by farmers and lack of understanding about importance of programme by authority. Many NGOs operated in limited scale and in assortive manner.

Genetic Gain

The reports on genetic gain in indigenous cattle and buffalo breeds have indicated it to be of small magnitude. The genetic gains reported in first lactation yield ranged from -130.6 kg (-8.2 per cent) in Red Sindhi to + 5879 kg in Kankrej (Joshi *et al.*, 2003). The genetic gain in buffaloes for first lactation yield ranged from +4.5 kg (0.27 per cent) to +5.78 kg under different farms (Sethi, 2005). The most of results reported showed either no genetic gain or deteriorating trend in herds which might be due to variable factors.

Annual Growth Rate: Present and Desirable

The milk yield per annum has reached to nearly 101.0 mMT in 2007. At present rate we can attain nearly 158.0 mMT in 2019-20. This is less than targeted milk production. The targeted milk production growth in cows, crossbred cows and buffaloes numbers needs to be 0.15 per cent, 6.0 per cent and 2.3 per cent respectively. Further these animals should have increase in milk yield (kg/day) at rate of 3.2 per cent, 7.5 per cent and 2.5 per cent respectively. We must plan our breeding programmes in view of availability of inputs with farmer, agro climatic conditions, adaptability of genotype to local condition along with conservation efforts for indigenous breeds.

Breeding Strategies for Various Agro-climatic Conditions

It is well known fact that a genotype which is adapted to local agro climatic conditions can perform best. We must identify and scientifically breed our animals accordingly to suit agro climatic conditions and availability of feed resources along with socioeconomic conditions of farmers. We can divide our country in to different zones which can be identified as a) Comfortable climate with good resources b) Comfortable climate with poor resources c) Harsh climate with good resources d) Harsh climate with poor resources. The genetic improvement programmes must be planned for each zone in view of availability of different germ plasm, their production performance, percentage of crossbred population and types of indigenous breeds and or crossbred which can be promoted. This would involve selective breeding of indigenous germ plasm, upgrading of native non descript population with indigenous known breed, development of crossbred with 50 per cent and higher exotic inheritance as best suited to local agro climatic conditions. Further a strategy need to be planned and executed to cover larger percentage of breedable population under artificial insemination. We must also have planned strategy to convert low producing non-descript and well known breed animals in to high producing germ plasm.

Genetic Improvement Programme

The India has large variation and complexity in terms of cattle and buffalo germ plasm. The herd size of farmers is very low (1-2 breedable cattle and/buffalo per farmer only). Except small groups of nomadic tribes, farmers breed their animals through AI or natural service. In view of limited scope of purchase of genetically superior animals by farmers efforts need to be made by slow and steady transformation of farmer's animals through extensive use of genetically superior sires. For rest of breeds, genetically superior sires be identified from indigenous population and distributed in local areas under natural services/AI Programmes. Our main emphasis should be large scale production of genetically superior sires of different breeds/genetic groups and to cover large percentage (>50 per cent) of breedable females under selective breeding, crossbreeding and grading up programmes along with establishment of viable performance recording network for female animals.

Performance Recording and Field Progeny Testing Programme

The performance recording of animals under field condition is pre-requisite to success of any genetic improvement programme. Success of leading dairy countries (USA, Netherland, Israel, Newzeland) mainly depended on establishment of large scale field infrastructure for performance recording. This has facilitated in correct assessment of production potential of animals, identification of superior females as bull mothers and implementation of Progeny Testing Programmes. In USA 50 per cent of animals are milk recorded. 1600 bulls are progeny tested every year. More than 80 per cent of cows are inseminated using semen of genetically superior sires. In Newzeland 90 per cent cows are under milk recording and 600 bulls are progeny tested every year. Top 10 per cent bulls are used extensively and 96 per cent cows are inseminated using top breeding bulls.

In India we can successfully use infrastructure of Dist. Co-operative Milk Producer Unions, ICDPs, BAIF and other NGOs. Formation of breed society can also help. These breed societies and local NGOs, can identify elite females, collect their male calves, rear them and distribute as bulls in local areas. Such approaches have been successful in Gujarat for Gir and Kankrej breeds.

Further, well managed breeding farms of SAUs and army, well managed Gaushalas can play key role in adapting villages in radius of 10 km periphery and involving them in breed improvement programmes. The performance recording programmes in these villages along with nominated mating of elite females animals can help us in production of genetically superior pedigreed males for progeny testing and breeding purpose. The experience of Dudhsagar Dairy and DIPA programme can be utilized after necessary modification as per local conditions. It is envisaged that revolution in electronics, digitalized communication, software development, computerization, mobile technology etc can make performance recording easy, accurate and help in implementation of breeding programme. (Development of 'Personal Digital Assistance' concept – personal communication).

Use of Reproduction Technology

In country like India, where field recording network is poor, use of reproductive technology *i.e.* MOET, embryo/ova freezing, embryo splitting, sexing of semen can help us in production genetically superior breeding bulls and bull dams under organizational concept of ONBS. The farmers animals around MOET centers can be used as surrogate mothers and elite females can be used as donor. However, in buffaloes success rate of MOET is limited.

Development of Biotechnology

Large scale planned breeding programmes can put up threat for loss of valuable genes (experience in HF) and loss of small breeds which do not have apparent economic benefits due to changed agricultural production/management system. It would be most important to conserve these breeds/genes with help of developments in biotechnology. Ova freezing, Embryo freezing, DNA libraries can be great help. Development of microarray/chips can help in early scanning of young males and females calves for their genetic potential for large many qualitative and quantitative traits *i.e.* k-casein, growth hormone, growth hormone receptor, beta lactalbumin, prolactin, MHC, leptin, myostatin, milk synthesis genes etc. Further, microchips can be developed for marker assisted selection. With change in agricultural scenario and availability of alternative modes of successful transport in rural areas, demand for bullock is rapidly declining. At this time sexed semen can be very useful for large scale production of females or male calves as desired by farmers and breeders. For conservation of threatened breeds/endangered breeds 2000 semen doses from 25 bulls (randomly selected) and 20 ova/embryo for each of 200 female/elite female need to be preserved under frozen condition. Further, test/software's be developed on basis of battery of 12-20 micro satellites, to determine group of non descript animals for their closeness to any known breeds. Once they exhibit closeness to degree of 90 per cent and above, such animals/groups are included in breed improvement programme of that known breeds.

Other Supportive Strategies

Success of large scale genetic improvement programmes greatly depends on many factors *i.e.* effective involvement of farmers, use of local inputs, identification of elite animals, creation of data base, creation/establishment/strengthing and existing AI network, adaptation of AI, monitoring co-ordination and evaluation agency. This would require:

1. Active involvement of breed society and breed association members will be actively involved in conservation and propagation of breeds. Gaushala can be involved for rearing of elite male calves to be used as future breeding bulls.
2. Active involvement of local animal breeders, nomadic tribes who traditionally keep animals. Creation of directory of breeding animals with them.
3. Replacement of scrub bulls by registered pedigreed bulls.
4. Increased AI coverage, AI adaptation, distribution of quality semen and increased AI competence. Concept of door step AI be propagated.
5. Preparation and approval of draft document for breeding policy.
6. Development of notified areas (District and Talukas) for indigenous breeds, where in no body can introduce breeding males other than notified breed

for the area. These can be focal point/centre for intensive breeding activities for genetic improvement of breeds.

7. Preparation of web directory for elite females and maies for breed.
8. Planned strategy to convert non descript group of animals in to locally known breed (grading up) or cross bred as per availability of inputs and agro climatic conditions.
9. Establishment of monitoring and coordination cell at state level for success implementation of breeding programmes with effective co-ordination between many agencies operating a different areas of state.

References

Annonymus (2003). Annual report of cattle breeding farm Junagadh.

Joshi B.K., Singh A. and Mukherjee S. (2005). Genetic improvement of indigenous cattle for milk and draught: A review. Indian Journal of Animal Science 75 (3): 335-348.

Sethi (2005). Improvement of buffalo for milk, meat and draught: A reiew. Indian Journal of Animal Science 75 (3): 327-334

Hammond K. H. and Leitch H.W. (1996) FAO global programme for management of farm animal resources and application of biotechnology. Proc of Beltsville Symposium, USA, Am. Soc. Ani. Sci. Publication illinosis, USA

Nivsarkar A.E., Vij P.K. and Tantia M.S. (2000) Animal Genetic Resources of India: Cattle and Buffalo, ICAR publication.

Chapter 3

Plankton Diversity in the River Barak Valley of North East India

Anjam Hussain Barbhuiya[1] *and Devashish Kar*[2]

[1]*Department of Zoology, Don Bosco College, Tura, Meghalaya*
[2]*Department of Life Science, Assam University, Silchar, Assam*

ABSTRACT

The planktons are a group of floating plants and animals, an important component of aquatic ecosystem and play key role in maintaining proper equilibrium of aquatic ecosystem. They are primary producers and serve as food for macro-invertebrates and fish directly and indirectly. A total of 25 potamoplankton genera were recorded from the river Barak during the investigation period under 8 classes. Density of the total potamoplankton in the river varied from 401.4 u/l (Tipaimukh) to 576 u/l (Panchgram) with average of 473.4 u/l.

Keywords: Plankton, Diversity, Diversity Indices, Potamoplankton communities.

Introduction

Structural characteristics of the lotic environment are closely associated with the occurrence of fish species. Again importance of the habitat and the relationship between the fish and the habitat are major concerns of the fishery biologists, as the habitat features like depth, flow, substrate type, availability of food etc, are generally considered as major determinants in distribution and abundance of fishes from earlier times (Shelford, 1911).

The North Eastern region of India covering nearly 2, 62,379 sq. km. area has been divided into two bio-geographic zones – Eastern Himalaya and North East India, based on floristic composition, the naturalness of the flora and the local climate (Rodgers and Panwar, 1988). The North Eastern Region of India lies between 22°00′N and 29°5′N latitude and 88°00′E and 97°30′E longitudes, and shares international border with Bhutan, China, Myanmar and Bangladesh. The North-East region of India contains more than one-third of the country's total biodiversity. The region represents important part of Indo-Myanmar bio-diversity hotspot, one of 25 global biodiversity hotspots recognized.

This rich diversity of the region could be assigned to certain reasons, notably, the geo-morphology and the tectonics of this zone. The hills and undulating valleys of this area give rise to large number of torrential hill streams, which leads to big rivers and that finally, become part of the Ganga-Brahmaputra-Barak-Chindwin-Kolodyne-Gomati-Meghna system (Kar *et al.*, 2002a).

The Brahmaputra and Barak river system which drains the Brahmaputra Valley and Barak valley of Assam in North East India respectively have a rich fish fauna of torrential stream especially in the upper reaches. Assam is dominated by two river plains: the Brahmaputra plains (56,480 sq kms), drained by the Brahmaputra river and its 43 tributaries; and the Barak plains (6,962 sq kms), drained by the Barak river and its 11 tributaries.

The river Barak is one of the major river of North Eastern India and a part of the Surma-Meghna river system. It rises in the Manipur hills near Mao Song sang village of Nagaland-Manipur border (N 25°28′ E 94°17′), receives a lot of little hill streams and flows west through Manipur state, then takes U-turn in Karong first towards north-east direction and enters Nagaland, takes a round and moves towards south-

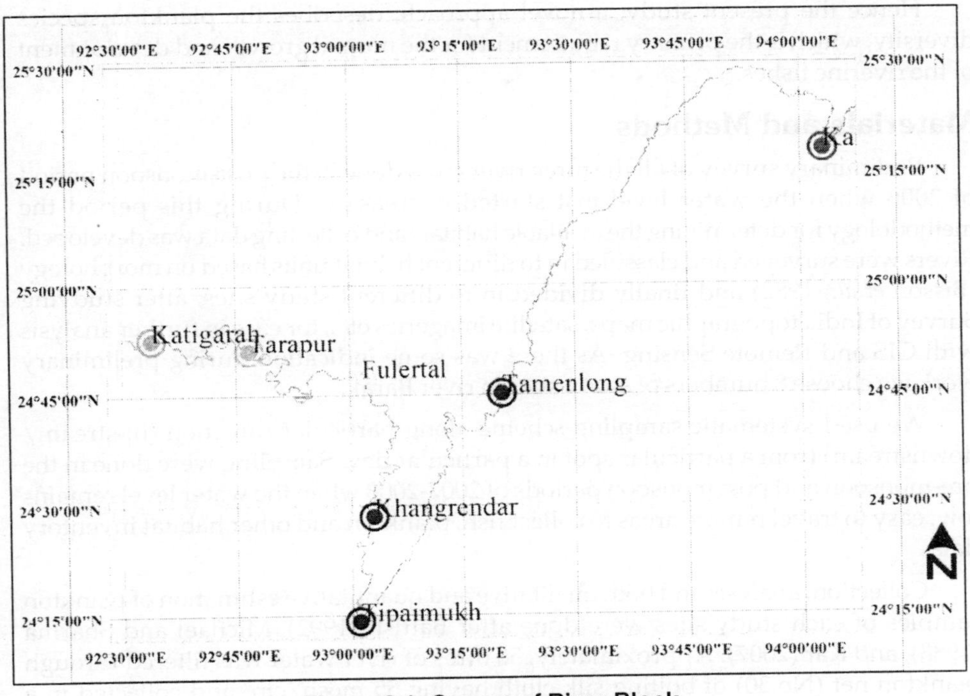

Figure 3.1: Course of River Block

east direction entered Manipur till reaches Tipaimukh in Manipur-Mizoram border. In Tipaimukh after joining with the river Tuivai it again takes an U-turn entered Assam and moves north ward till it receive river Jiri near Jirimukh (N 24°42'44.0" E 93°4'53.0"). Then the river starts crawling east ward in the plains crossing Silchar city and receiving several other tributaries like Chiri, Madhura, Sonai etc. It then flows through the western part of Silchar where it joined by the river Jatinga, Dhaleswari along with several other tributaries, bifurcates into Surma and Kushiara and enters Bangladesh before joining the Bay of Bengal.

Phytoplankton, a group of floating plants, are an important component of aquatic ecosystem and play key role in maintaining proper equilibrium of aquatic ecosystem. They are primary producers and serve as food for zooplankton, macro-invertebrates and fish directly and indirectly. However, in case of hilly streams true plankton are absent especially in upper stretch of the river due to turbulent and rapid flow (Welch, 2003). In the Barak drainages, phytoplankton community was comprised of Chlorophyceae, Bacillariophyceae, and Cyanophyceae etc. But Chlorophyceae was a most important group in the river Barak and Dhaleswari and the number of species were more than the other groups. But Bacillariophyceae (diatoms) was most important group in the river Jatinga, contributing more than 50 per cent of total phytoplankton.

While zooplankton is a group of micro-organisms which float at their own energy and they are at mercy of water currents (Kar and Barbhuiya, 2004). In upper reaches they were less in number at most of the sampling sites. Zooplankton community of the river mainly comprise Rotifer, Cladocera, Copepoda and Protozoon's.

Hence the present study, a novel approach, describes the plankton species diversity, which is the primary requirement for the normal growth and development of the riverine fishes.

Materials and Methods

Preliminary survey of all the three rivers was done in the post-monsoon period of 2006 when the water level just started decreasing. During this period the methodology for determining the available habitats and collecting data was developed. Rivers were surveyed and classified in to different habitat units based on morphology (Bisson *et al.*, 1982) and finally divided in to different study sites, after studying Survey of India topographic maps, satellite imageries etc., for ease in further analysis with GIS and Remote Sensing. As there was some indication during preliminary work we choose 8 numbers of study sites in river Barak.

We used systematic sampling scheme along particular direction (upstream/ downstream) from a particular spot in a particular day. Sampling were done in the pre-monsoon and post monsoon periods of 2007-2008 when the water level remains low, easy to travel remote areas to collect fish, plankton and other habitat inventory data.

Collection, analysis and both qualitative and quantitative estimation of plankton samples of each study sites were done after Battish (1992), Michael and Sharma (1988), and Kar (2007). Approximately, 50 litres of river water was filtered through plankton net (No.30) of bolting silk cloth having 55 mesh/cm^2 and collected in a plankton tube of 45 ml capacity. The sample was then fixed with 1 per cent formalin and taken to the laboratory. Finally, the identification that is qualitative analysis was done by the Lackeys drop method under a Labomed made binocular light microscope. For quantitative analysis thoroughly mixed sample was taken with graduated dropper to fill the 'Sedgewick Rafter cell' having capacity of 1 ml and the number of planktons per ml was estimated.

Diversity Indices of Planktons and Fishes

Measurements of diversity have been of historical significance and their importance still remains today given the obvious declines in habitat quality in almost every ecological system. The Shannon-Wiener Diversity Index is one of the most widely used species diversity indices for examining overall community characteristics comparing two or more distinct habitats. It is derived from a function used in the field of information and has been adapted by ecologists to describe the average degree of uncertainty of predicting the species of an individual picked at random from the community. The uncertainty of occurrence increases both as the number of species increases and as the individuals are distributed more and more evenly among the species already present.

A diversity index is a mathematical measure of species diversity in a community. Diversity indices provide more information about community composition than simply species richness (*i.e.*, the number of species present); they also take the relative abundances of different species into account. It is not necessary to key all organisms to their specific species nomenclature (*i.e.* organisms not expediently identified may be assigned numeric values such as species 1, 2, 3). However, in order to derive accurate diversity values, all organisms should be keyed to the lowest possible like taxonomic level.

The Shannon-Weiner Species Diversity Index is calculated by taking the number of each species, the proportion each species is of the total number of individuals, and sums the proportion times the natural log of the proportion for each species. Since this is a negative number, we then take the negative of this sum. The higher the number, the higher is the species diversity. In the ideal situation, one should compare populations that are the same size in numbers of individuals.

The formula is as follows: $'H' = -\Sigma pi \ln pi$

It combines two quantifiable measures; 1. The *Species Richness S* (the number of species in the community) 2. *Abundance N* (is the total number of individuals in the sample). The index is termed 'H' = - Sum of Pi ln Pi with higher values indicating increased diversity. 3. Pi = S/N.

When properly manipulated, it will result in a diversity value (*H*) ranging between 0 (indicating low community complexity) and 4 (indicating high community complexity). Diversity indices provide important information about rarity and commonness of species in a community. The ability to quantify diversity in this way is an important tool for biologists trying to understand community structure.

Potamoplankton Communities of River Barak at Different Study Sites

Station wise incidence on the total potamoplankton biomass recorded in the river Barak during the study period at eight study sites are shown in the Tables 3.1 to 3.8.

Discussion and Conclusion

The density of phytoplankton was observed to increase from upper to lower stretches of the rivers in pre-monsoon and post-monsoon seasons. However, different trend was observed in the density of plankton during monsoon which might be due to physical and chemical characteristics like turbidity and current velocity, that varying greatly during monsoon months and reflect multimodal peaks in plankton (Acharjee

Table 3.1: Plankton Diversity at Katigarah of River Barak

Name of the Plankton Species	*Name of Panktons/Drops Studied*										*No. of Planktons/ 0.5 ml*	*No. cf Plank-tons/ml*	*Nos./Litre of Original Water*
	1	*2*	*3*	*4*	*5·*	*6*	*7*	*8*	*9*	*10*			
Ankistrodesmus sp.	2	–	2	1	1	–	3	1	–	1	11	22	19.8
Spirogyra sp.	2	3	–	1	1	–	1	–	2	3	13	26	23.4
Hydrodictyon sp.	1	2	1	–	1	2	3	–	1	2	13	26	23.4
Scendesmus sp.	2	–	2	3	1	–	1	1	–	2	12	24	21.6
Characium sp.	1	3	–	1	–	2	2	–	2	1	12	24	21.6
Ulothrix sp.	1	–	4	1	–	1	2	1	–	3	13	26	23.4
Cladophora sp.	2	1	–	2	–	3	–	1	4	1	14	28	25.2
Oedogonium sp.	4	–	3	–	2	–	2	1	2	–	14	28	25.2
Nostoc sp.	1	3	–	2	1	1	–	1	2	1	12	24	21.6
Anabaena sp.	2	1	2	2	–	1	2	–	3	2	15	30	27.0
Oscillatoria sp.	2	–	1	2	1	–	3	–	2	1	12	24	21.6
Polycystis sp.	2	1	3	1	1	–	3	–	2	1	14	28	25.2
Spirulina sp.	1	2	–	3	1	2	–	4	–	1	14	28	25.2
Fragilaria sp.	–	2	–	3	–	3	1	–	–	3	12	24	21.6
Synedra sp.	1	–	2	–	1	1	–	2	1	1	9	18	16.2
Desmidium sp.	1	2	–	–	3	–	2	1	3	–	12	24	21.6
Tabellaria sp.	–	1	2	1	–	3	2	–	–	1	10	20	18.0
Keratela sp.	3	–	1	1	2	–	–	3	2	3	15	30	27.0
Filinia sp.	1	2	–	1	–	1	1	–	1	–	7	14	12.6
Daphnia sp.	2	1	–	2	–	1	1	2	–	1	10	20	18.0
Bosmina sp.	2	–	3	–	2	1	–	2	–	2	12	24	21.6
Cyclops sp.	3	–	1	–	2	2	1	–	1	2	12	24	21.6
Diaptomous sp.	1	–	1	–	1	3	–	2	–	2	10	20	18.0
Nauplius larvae.	1	2	–	2	2	2	–	–	2	3	14	28	25.2
Difflugia sp.	3	1	2	1	–	1	–	1	–	1	10	20	18.0

Table 3.2: Plankton Diversity at Panchgram of River Barak

Name of the Plankton Species	Name of Panktons/Drops Studied										No. of Planktons/ 0.5 ml	No. of Plank-tons/ml	Nos./Litre of Original Water
	1	2	3	4	5	6	7	8	9	10			
Ankistrodesmus sp.	3	1	2	1	4	–	1	–	1	–	13	26	23.4
Spirogyra sp.	3	1	–	–	1	–	1	–	3	1	10	20	18.0
Hydrodictyon sp.	4	–	2	2	1	2	1	–	3	–	15	30	27.0
Scendesmus sp.	1	–	4	2	1	–	3	2	–	2	15	30	27.0
Characium sp.	5	1	2	1	3	1	1	1	–	1	16	32	28.8
Ulothrix sp.	4	2	1	1	1	2	3	–	1	1	16	32	28.8
Cladophora sp.	2	1	–	–	3	1	2	1	4	–	14	28	25.2
Oedogonium sp.	–	–	1	–	2	–	2	1	1	–	7	14	12.6
Nostoc sp.	3	2	–	2	1	2	–	1	2	1	14	28	25.2
Anabaena sp.	–	3	3	2	–	1	2	–	2	–	13	26	23.4
Oscillatoria sp.	2	–	2	1	1	–	2	–	–	1	9	18	16.2
Polycystis sp.	2	1	4	1	2	–	3	–	1	2	16	32	28.8
Spirulina sp.	4	1	2	1	–	3	2	1	1	–	15	30	27.0
Fragilaria sp.	–	2	1	2	–	2	1	–	–	2	10	20	18.0
Synedra sp.	–	–	2	–	2	2	–	2	1	–	9	18	16.2
Desmidium sp.	1	1	–	1	2	–	2	2	3	–	12	24	21.6
Tabellaria sp.	1	2	1	–	3	2	1	–	1	3	14	28	25.2
Keratela sp.	2	–	2	2	2	–	–	2	2	1	13	26	23.4
Filinia sp.	4	2	1	3	2	1	1	–	–	1	15	30	27.0
Daphnia sp.	1	–	–	2	–	2	1	2	–	1	9	18	16.2
Bosmina sp.	2	–	–	–	1	1	–	2	–	1	7	14	12.6
Cyclops sp.	3	–	1	–	3	2	1	–	4	1	15	30	27.0
Diaptomous sp.	3	–	1	–	2	1	–	4	–	1	12	24	21.6
Nauplius larvae	2	3	1	–	3	2	3	2	1	1	18	36	32.4
Difflugia sp.	2	1	3	1	4	–	1	1	–	–	13	26	23.4

Table 3.3: Plankton Diversity at Tarapur of River Barak

Name of the Plankton Species	Name of Panktons/Drops Studied										No. of Planktons/ 0.5 ml	No. of Plank-tons/ml	Nos./Litre of Original Water
	1	2	3	4	5	6	7	8	9	10			
Ankistrodesmus sp.	1	–	2	1	1	–	1	1	–	–	7	14	12.6
Spirogyra sp.	2	2	–	1	1	–	1	–	2	1	10	20	18.0
Hydrodictyon sp.	1	1	1	–	–	2	1	–	1	2	9	18	16.2
Scendesmus sp.	2	–	2	2	1	–	1	2	–	2	12	24	21.6
Characium sp.	1	1	–	1	–	2	2	–	2	1	10	20	18.0
Ulothrix sp.	1	–	4	1	–	1	2	1	–	1	11	22	19.8

Contd...

Table 3.3–Contd...

Name of the Plankton Species	*Name of Panktons/Drops Studied*										*No. of Planktons/ 0.5 ml*	*No. of Plank- tons/ml*	*Nos./Litre of Original Water*
	1	*2*	*3*	*4*	*5*	*6*	*7*	*8*	*9*	*10*			
Cladophora sp.	2	1	–	2	–	2	–	1	2	–	10	20	18.0
Oedogonium sp.	4	–	1	–	2	–	1	1	2	–	11	22	19.8
Nostoc sp.	1	2	–	2	1	2	–	1	2	1	12	24	21.6
Anabaena sp.	2	1	2	2	–	1	2	–	3	–	13	26	23.4
Oscillatoria sp.	2	–	1	1	1	–	3	–	–	1	9	18	16.2
Polycystis sp.	2	1	1	1	1	–	3	–	1	1	11	22	19.8
Spirulina sp.	1	–	–	–	1	1	–	–	–	1	4	8	7.2
Fragilaria sp.	–	2	–	3	–	1	1	–	–	3	10	20	18.0
Synedra sp.	–	–	2	–	2	1	–	2	1	1	9	18	16.2
Desmidium sp.	1	1	–	–	4	–	2	1	3	–	12	24	21.6
Tabellaria sp.	–	1	2	1	–	–	2	–	–	1	7	14	12.6
Keratela sp.	3	–	2	1	2	–	–	2	2	3	15	30	27.0
Filinia sp.	1	–	–	1	–	1	1	–	1	–	5	10	9.0
Daphnia sp.	1	1	–	2	–	1	1	2	–	2	10	20	18.0
Bosmina sp.	2	–	3	–	1	1	–	2	–	1	10	20	18.0
Cyclops sp.	3	–	1	–	1	2	2	–	1	2	12	24	21.6
Diaptomous sp.	1	–	1	–	1	1	–	1	–	1	6	12	10.8
Nauplius larvae	1	2	–	2	1	2	–	–	2	1	11	22	19.8
Difflugia sp.	2	1	2	1	–	2	–	1	–	1	10	20	18.0

Table 3.4: Plankton Diversity at Fulertal of River Barak

Name of the Plankton Species	*Name of Panktons/Drops Studied*										*No. of Planktons/ 0.5 ml*	*No. of Plank- tons/ml*	*Nos./Litre of Original Water*
	1	*2*	*3*	*4*	*5*	*6*	*7*	*8*	*9*	*10*			
Ankistrodesmus sp.	2	1	2	–	3	1	2	2	–	1	14	28	25.2
Spirogyra sp.	2	2	–	–	2	–	1	–	2	1	10	20	18.0
Hydrodictyon sp.	2	1	3	1	1	–	–	1	1	–	10	20	18.0
Scendesmus sp.	2	–	2	2	1	1	3	2	–	2	15	30	27.0
Characium sp.	3	2	1	–	1	–	2	1	–	2	12	24	21.6
Ulothrix sp.	1	2	1	3	–	1	–	1	2	2	13	26	23.4
Cladophora sp.	3	2	1	–	1	1	3	2	1	–	14	28	25.2
Oedogonium sp.	2	–	1	–	2	–	1	1	–	–	7	14	12.6
Nostoc sp.	2	2	–	2	2	–	–	1	2	1	12	24	21.6
Anabaena sp.	–	1	2	2	–	2	2	–	3	1	13	26	23.4
Oscillatoria sp.	–	–	3	1	1	–	3	–	–	1	9	18	16.2

Contd...

Table 3.4–Contd...

Name of the Plankton Species	Name of Panktons/Drops Studied										No. of Planktons/ 0.5 ml	No. of Plank-tons/ml	Nos./Litre of Original Water
	1	2	3	4	5	6	7	8	9	10			
Polycystis sp.	2	1	1	2	–	–	2	–	2	1	11	22	19.8
Spirulina sp.	3	1	4	1	2	–	1	–	2	1	15	30	27.0
Fragilaria sp.	–	2	–	2	–	2	1	–	–	3	10	20	18.0
Synedra sp.	–	–	3	–	2	1	–	2	1	–	9	18	16.2
Desmidium sp.	2	1	1	–	3	4	1	–	–	–	12	24	21.6
Tabellaria sp.	4	1	2	1	–	1	1	2	1	–	13	26	23.4
Keratela sp.	3	–	2	1	2	–	–	2	2	1	13	26	23.4
Filinia sp.	2	1	3	1	2	1	–	1	–	3	14	28	25.2
Daphnia sp.	1	–	–	2	–	1	1	2	–	2	9	18	16.2
Bosmina sp.	2	–	–	–	1	1	–	2	–	1	7	14	12.6
Cyclops sp.	2	1	1	–	1	2	1	–	1	1	10	20	18.0
Diaptomous sp.	1	–	1	–	2	1	–	–	–	1	6	12	10.8
Nauplius larvae	–	1	1	–	–	2	1	3	1	1	10	20	18.0
Difflugia sp.	–	–	–	–	–	–	–	–	–	–	0	0	0

Table 3.5: Plankton Diversity at Khangrendar of River Barak

Name of the Plankton Species	Name of Panktons/Drops Studied										No. of Planktons/ 0.5 ml	No. of Plank-tons/ml	Nos./Litre of Original Water
	1	2	3	4	5	6	7	8	9	10			
Ankistrodesmus sp.	–	–	2	2	1	–	1	1	–	–	7	14	12.6
Spirogyra sp.	2	2	–	–	1	–	1	–	3	1	10	20	18.0
Hydrodictyon sp.	1	1	1	–	–	2	1	–	1	2	9	18	16.2
Scendesmus sp.	3	–	2	2	1	–	3	2	–	2	15	30	27.0
Characium sp.	1	1	–	1	–	1	2	–	3	1	10	20	18.0
Ulothrix sp.	1	–	–	1	–	1	2	1	–	–	6	12	10.8
Cladophora sp.	3	1	–	2	–	1	–	1	2	–	10	20	18.0
Oedogonium sp.	–	–	1	–	2	–	–	1	1	–	5	10	9.0
Nostoc sp.	3	2	–	2	1	–	–	1	2	1	12	24	21.6
Anabaena sp.	–	3	2	2	–	1	2	–	3	–	13	26	23.4
Oscillatoria sp.	2	–	1	1	1	–	3	–	–	1	9	18	16.2
Polycystis sp.	2	1	1	1	1	–	3	–	1	1	11	22	19.8
Spirulina sp.	4	1	1	1	–	–	1	1	2	1	12	24	21.6
Fragilaria sp.	–	2	–	2	–	2	1	–	–	3	10	20	18.0
Synedra sp.	–	–	3	–	2	1	–	2	1	–	9	18	16.2
Desmidium sp.	1	1	–	–	3	–	2	2	3	–	12	24	21.6

Contd...

Table 3.5–Contd...

Name of the Plankton Species	Name of Panktons/Drops Studied										No. of Planktons/ 0.5 ml	No. of Plank-tons/ml	Nos./Litre of Original Water
	1	2	3	4	5	6	7	8	9	10			
Tabellaria sp.	–	1	2	1	–	–	1	1	–	1	7	14	12.6
Keratela sp.	3	–	2	1	2	–	–	2	2	1	13	26	23.4
Filinia sp.	1	–	–	1	–	1	2	–	–	–	5	10	9.0
Daphnia sp.	1	–	–	2	–	1	1	2	–	2	9	18	16.2
Bosmina sp.	2	–	–	–	1	1	–	2	–	1	7	14	12.6
Cyclops sp.	3	–	1	–	1	2	1	–	1	1	10	20	18.0
Diaptomous sp.	1	–	1	–	2	1	–	–	–	1	6	12	10.8
Nauplius larvae	1	3	–	2	1	1	–	–	2	1	11	22	19.8
Difflugia sp.	2	1	2	1	–	1	–	1	–	1	9	18	16.2

Table 3.6: Plankton Diversity at Tipaimukh of River Barak

Name of the Plankton Species	Name of Panktons/Drops Studied										No. of Planktons/ 0.5 ml	No. of Plank-tons/ml	Nos./Litre of Original Water
	1	2	3	4	5	6	7	8	9	10			
Ankistrodesmus sp.	1	–	–	–	2	1	1	1	–	–	6	12	10.8
Spirogyra sp.	2	–	1	1	1	1	–	3	1	1	11	22	19.8
Hydrodictyon sp.	2	1	1	1	–	–	2	1	1	–	9	18	16.2
Scendesmus sp.	–	3	–	2	1	–	3	2	2	–	13	26	23.4
Characium sp.	2	1	1	3	1	–	–	–	1	–	9	18	16.2
Ulothrix sp.	2	1	1	3	–	–	–	1	1	–	9	18	16.2
Cladophora sp.	1	–	–	–	1	1	1	2	1	1	8	16	14.4
Oedogonium sp.	1	1	–	–	3	–	1	–	–	1	7	14	12.6
Nostoc sp.	2	–	1	2	–	2	2	–	–	1	10	20	18.0
Anabaena sp.	–	–	3	2	–	1	3	–	–	2	11	22	19.8
Oscillatoria sp.	–	–	2	–	–	1	1	–	2	1	7	14	12.6
Polycystis sp.	1	2	1	–	–	2	1	3	2	–	12	24	21.6
Spirulina sp.	3	1	1	–	–	–	1	–	1	–	7	14	12.6
Fragilaria sp.	2	1	1	–	2	–	3	–	1	1	11	22	19.8
Synedra sp.	3	2	–	2	2	3	–	–	–	–	12	24	21.6
Desmidium sp.	2	–	1	1	1	–	–	–	1	1	7	14	12.6
Tabellaria sp.	1	2	–	1	3	–	–	1	2	1	11	22	19.8
Keratela sp.	2	–	2	–	1	1	–	3	–	2	11	22	19.8
Filinia sp.	2	1	1	1	–	–	–	1	1	–	7	14	12.6
Daphnia sp.	1	1	–	3	–	2	2	–	–	1	10	20	18.0
Bosmina sp.	1	–	–	–	2	–	1	–	1	1	6	12	10.8

Contd...

Table 3.6–Contd...

Name of the Plankton Species	*Name of Panktons/Drops Studied*										*No. of Planktons/ 0.5 ml*	*No. of Plank-tons/ml*	*Nos./Litre of Original Water*
	1	*2*	*3*	*4*	*5*	*6*	*7*	*8*	*9*	*10*			
Cyclops sp.	2	–	1	1	–	1	–	1	1	–	7	14	12.6
Diaptomous sp.	1	–	2	1	–	1	1	–	1	1	8	16	14.4
Nauplius larvae	1	–	1	1	1	–	2	–	3	–	9	18	16.2
Difflugia sp.	1	–	–	–	–	1	1	2	–	–	5	10	9.0

Table 3.7: Plankton Diversity at Tamenlong of River Barak

Name of the Plankton Species	*Name of Panktons/Drops Studied*										*No. of Planktons/ 0.5 ml*	*No. of Plank-tons/ml*	*Nos./Litre of Original Water*
	1	*2*	*3*	*4*	*5*	*6*	*7*	*8*	*9*	*10*			
Ankistrodesmus sp.	1	1	2	1	3	–	–	–	1	1	10	20	18.0
Spirogyra sp.	1	2	1	–	–	3	–	3	–	1	10	20	18.0
Hydrodictyon sp.	3	2	1	1	–	–	–	1	1	–	9	18	16.2
Scendesmus sp.	1	2	–	–	2	1	–	3	–	1	10	20	18.0
Characium sp.	1	2	1	2	1	1	1	–	–	–	9	18	16.2
Ulothrix sp.	3	2	2	1	1	1	–	–	–	–	10	20	18.0
Cladophora sp.	2	1	1	1	1	1	2	2	–	–	11	22	19.8
Oedogonium sp.	–	1	–	–	1	1	–	2	–	1	6	12	10.8
Nostoc sp.	3	2	–	–	2	1	2	–	3	1	14	28	25.2
Anabaena sp.	2	2	3	–	–	3	2	–	2	1	15	30	27.0
Oscillatoria sp.	1	1	3	–	2	3	–	–	1	–	11	22	19.8
Polycystis sp.	2	–	–	1	–	–	2	2	–	2	9	9	16.2
Spirulina sp.	1	1	1	3	2	2	–	–	–	1	11	22	19.8
Fragilaria sp.	3	1	1	–	2	2	–	–	2	–	11	22	19.8
Synedra sp.	2	–	–	1	–	2	–	1	2	–	8	16	14.4
Desmidium sp.	1	2	–	1	3	2	–	1	–	–	12	20	18.0
Tabellaria sp.	1	1	1	–	–	–	1	1	–	1	6	12	10.8
Keratela sp.	1	2	1	1	–	–	–	1	1	1	8	16	14.4
Filinia sp.	2	3	3	1	1	1	–	–	–	1	12	24	21.6
Daphnia sp.	1	1	1	2	1	1	1	2	2	1	13	26	23.4
Bosmina sp.	2	2	1	1	1	1	–	–	–	2	10	20	18.0
Cyclops sp.	1	–	–	–	2	2	2	3	1	1	12	24	21.6
Diaptomous sp.	–	–	2	2	2	1	1	–	–	–	8	16	14.4
Nauplius larvae	1	1	1	2	1	3	1	–	–	–	10	20	18.0
Difflugia sp.	1	1	–	2	1	1	–	–	–	1	7	14	12.6

Table 3.8: Plankton Diversity at Karong of River Barak

Name of the Plankton Species	Name of Panktons/Drops Studied										No. of Planktons/ 0.5 ml	No. of Plank-tons/ml	Nos./Litre of Original Water
	1	2	3	4	5	6	7	8	9	10			
Ankistrodesmus sp.	1	–	–	2	1	1	–	–	1	1	7	14	12.6
Spirogyra sp.	2	2	–	1	3	1	–	1	–	–	10	20	18
Hydrodictyon sp.	2	1	1	–	–	–	2	1	–	3	10	20	18
Scendesmus sp.	2	3	–	2	–	2	–	1	–	1	11	22	19.8
Characium sp.	1	2	–	1	1	–	–	–	3	1	9	18	16.2
Ulothrix sp.	2	–	–	1	1	–	–	1	1	1	7	14	12.6
Cladophora sp.	1	3	1	–	–	–	3	2	1	–	11	22	19.8
Oedogonium sp.	1	—	2	–	–	1	1	–	–	1	4	8	7.2
Nostoc sp.	2	1	–	–	2	2	–	1	2	1	11	22	19.8
Anabaena sp.	2	3	2	–	–	2	3	–	–	3	15	30	27.0
Oscillatoria sp.	1	2	1	1	2	–	1	1	2	2	14	28	25.2
Polycystis sp.	2	1	2	2	–	2	–	–	1	2	12	24	21.6
Spirulina sp.	1	1	2	–	–	2	1	2	–	–	9	18	16.2
Fragilaria sp.	1	2	2	1	–	–	3	2	–	2	13	26	23.4
Synedra sp.	1	1	1	2	–	1	2	–	–	2	10	20	18.0
Desmidium sp.	1	1	1	2	2	–	–	1	1	1	10	20	18
Tabellaria sp.	2	–	–	–	–	2	1	1	1	1	8	16	14.4
Keratela sp.	1	–	–	–	2	2	2	1	1	1	10	20	18
Filinia sp.	2	2	1	1	1	3	–	–	1	1	12	24	21.6
Daphnia sp.	2	–	–	1	1	2	3	2	1	1	13	26	23.4
Bosmina sp.	2	1	1	3	2	–	–	1	1	1	12	24	21.6
Cyclops sp.	2	1	1	1	–	–	–	2	2	1	10	20	18
Diaptomous sp.	3	3	1	1	1	1	–	–	–	1	11	22	19.8
Nauplius larvae	3	1	1	1	–	–	1	1	–	–	8	16	14.4
Difflugia sp.	2	1	1	1	2	2	1	1	–	–	11	22	19.8

et al., 2003, Dey and Kar (1994), Reinhard, 1931,). The density was maximum at Panchgram of river Barak up to 576 u/l and a minimum at Tipaimukh up to 401.4 u/ l during. The plankton density decreased during monsoon season at all the sampling sites. Fluctuations in water discharge and turbidity seems to be the determining factor for the seasonal variation in the phytoplankton density while temperature and velocity determine the spatial variation (Berner, 1951).

Chlorophyceae was mainly represented by *Spirogyra* sp, *Chlamydomonus* sp, *Scenedesmus* sp, *Ankistrodesmus* sp and *Volvox* sp of which *Ankistrodesmus* sp and *Volvox* sp were consistently recorded throughout the year but poor in number in river Barak. Majority of the phytoplankton found are Diatoms specially *Fragillaria* sp,

Table 3.9: Enumeration of Diversity Index of the Potamoplankton Community of the River Barak

Name of the Plankton Species	*Total of 8 Sites*	*Avg (n1) in u/l*	*n_1/N=Pi*	*In Pi*	*Pi*InPi*
		Phytoplanktons			
Chlorophyceae					
Ankistrodesmus sp.	135	16.875	0.03564639	–3.3341075	–0.11885
Spirogyra sp.	151.2	18.9	0.03992395	–3.2207788	–0.12859
Hydrodictyon sp.	151.2	18.9	0.03992395	–3.2207788	–0.12859
Scendesmus sp.	185.4	23.175	0.04895437	–3.0168666	–0.14769
Characium sp.	156.6	19.575	0.04134981	–3.1856875	–0.13173
Ulothrix sp.	153	19.125	0.04039924	–3.2089443	–0.12964
Cladophora sp.	165.6	20.7	0.04372624	–3.129807	–0.13685
Oedogonium sp.	109.8	13.725	0.0289924	–3.5407217	–0.10265
Cyanophyceae					
Nostoc sp.	174.6	21.825	0.04610266	–3.0768846	–0.14185
Anabaena sp.	194.4	24.3	0.0513308	–2.9694643	–0.15242
Oscillatoria sp.	144	18	0.03802281	–3.2695689	–0.12432
Polycystis sp.	172.8	21.6	0.04562738	–3.0872474	–0.14086
Spirulina sp.	156.6	19.575	0.04134981	–3.1856875	–0.13173
Bacillariophyceae					
Fragilaria sp.	156.6	19.575	0.04134981	–3.1856875	–0.13173
Synedra sp.	135	16.875	0.03564639	–3.3341075	–0.11885
Desmidiaceae					
Desmidium sp.	156.6	19.575	0.04134981	–3.1856875	–0.13173
Tabellaria sp.	136.8	17.1	0.03612167	–3.3208622	–0.11996
		Zooplankton			
Rotifers					
Keratela sp.	176.4	22.05	0.04657795	–3.0666281	–0.14284
Filinia sp.	138.6	17.325	0.03659696	–3.3077902	–0.12106
Cladocera					
Daphnia sp.	149.4	18.675	0.03944867	–3.232755	–0.12753
Bosmina sp.	127.8	15.975	0.03374525	–3.3889157	–0.11436
Copepoda					
Diaptomous sp.	120.6	15.075	0.03184411	–3.446903	–0.10976
Cyclops sp.	158.4	19.8	0.0418251	–3.1742588	–0.13276
Nauplius larvae	163.8	20.475	0.04325095	–3.1407361	–0.13584
Protozoa					
Difflugia sp.	117	14.625	0.03089354	–3.4772083	–0.10742
	3787.2	473.4	1		–3.210
		Σ1=N			**H=3.210**

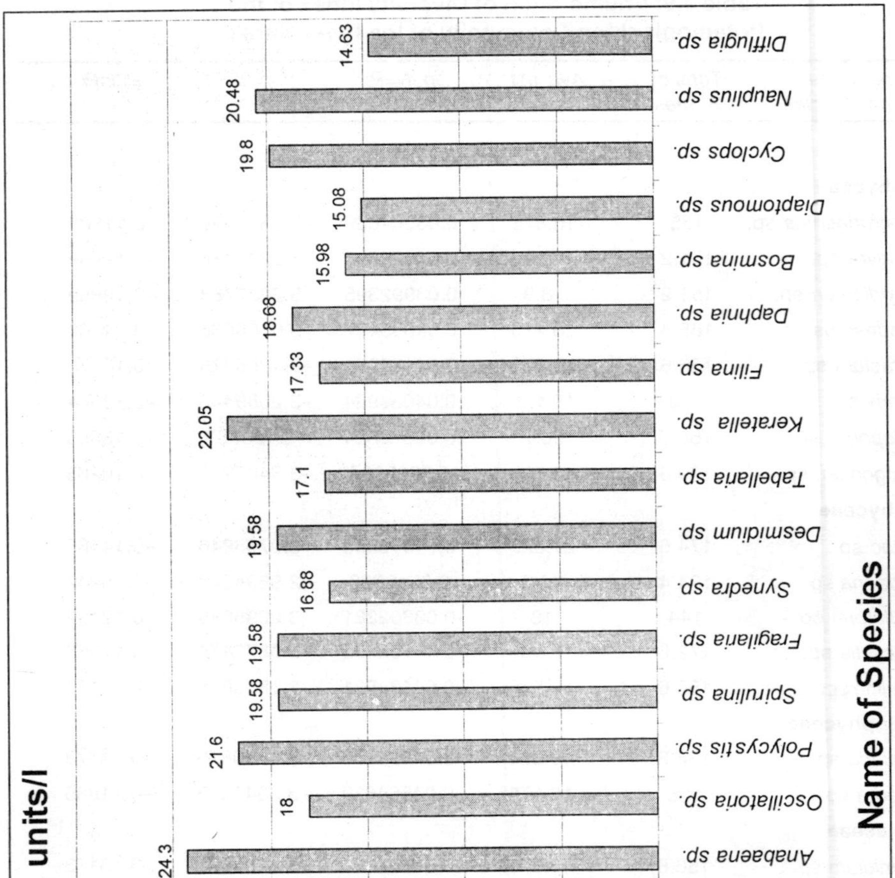

Figure 3.2: Average Records of Planktons in units/l of the River Barak

Synedra sp., *Anabaena* sp., *Polycystis* sp. The diversity index (H) for the river Barak is found to be 3.210.

The dominant zooplanktons of river Barak are Copepodans *viz.*, *Diaptomus* sp., *Cyclops* sp, *Nouplius* sp., *Cydorus* sp., occupy during the entire period of investigation. Generally the zooplanktons are fewer in number (8) than the phytoplankton (17).

It has been observed that fish density and plankton number are having relationship with water temperature which is in conformity with Chakravarty *et al.* (1959). Variation in bio chemical activity, velocity, volume, substrate, cover, water sources and a number of addition al features operating seasonally (Raid, 1961) may be possible factor for variation in temperature.

Occurrence of zooplankton in general is minimum in comparison to phytoplankton. Inverse relationship has been observed between phytoplankton and zooplankton which refuses the view of Riley and Bumpus (1946), Prasad (1956) and Chakravarty *et al.* (1959). An inverse relationship has also been obtained from the fish density and potamoplankton occurrence. During monsoon season, adverse effect on planktonic forms has been recorded particularly at certain gradient zones. It is apparent that sudden influx of water in the river during that season should be the causative factor for such phenomeno. The same view was also observed by Berner (1951). Occurrence of plankton in rapids and upstream of river Barak has been found to be very less. It has been observed that increase in water current and turbidity affects phyto and zooplankton adversely. A similar trend was also reported by Jana (1979).

The major controlling factor in the distribution of invertebrate is the nature of the river bed (Hynes, 1960). Ruggiero and Merchant (1979) reported that the distribution of benthic macro invertebrates was more closely related to the substratum of the stream then to the water quality. Hora (1936) considered substratum as an important factor in the ecology of torrential fauna in the Himalaya.

A total of 25 potamoplankton genera were recorded from the river Barak during the investigation period under 8 classes. Density of the total potamoplankton in the river varied from 401.4 u/l (Tipaimukh) to 576 u/l (Panchgram) with average of 473.4 u/l. Survey station wise distribution of total plankton showed maximum biomass in Panchgram (576 u/l) followed by Katigarah (544 u/l), Fulertal (482 u/l), Karong (464.4 u/l), Tamenlong (450.0 u/l), Tarapur (443 u/l), Khangrendar (426.6 u/l), Tipaimukh (401.4).

References

Acharjee, B. Choudhuury, M. and Dutta, A. (2003). On the abundance of phytoplankton and its relation to certain physico-chemical parameters of Dighali Beel in the lower Brahmaputra basin, Assam. In: Kar, D. Dey, S. C. and Dutta, N. C. (ed.). *Welfare Biology in the New Millennium*, Allied Publishers, pp. 97.

Battish, S.K. (1992). Freshwater Zooplankton of India, vi + 233, Oxford and IBH Publishing Co., Ltd. (New Delhi).

Berner, L. M. (1951). Limnology of the lower Missouri river, *Ecology*, 32: 1-12.

Bisson, P. A., Nielson, J. A., Palmason, R. A. and Grove, L. E. (1982). A system of naming habitat types in small streams, with example of habitat utilisation by Salmonids during low stream flow. pp. 62-73. In Armantrout, N.B. ed. Acquisition and utilisation of aquatic habitat inventory information. *American Fisheries Society*, Bethesda, Maryland.

Chakravarty, R. D., Roy, P. and Singh, S. D. (1959). A quantitative study of the plankton and physico-chemical condition of the river Jamuna at Allahabad in 1954-55, *Indian J. Fish.*, 6 (1): 186-203.

Dey, S. C. and Kar, D. (1994). Phytoplankton dynamics of Lake Sone in Assam. *Bull. Life Sciences*, 4: 49-54.

Hora, S. L. (1936). On a further collection of fish from the Naga Hills, *Rec. Indian Mus.*, XXXVIII: 317-331.

Hynes, H. B. N. (1960). *The Biology of Polluted Water*. Liverpool Univ. Press.

Jana, B. B. (1979). Tropical plankton succession and ecology of a tropical tank in West Bengal, *Int. Rev. org., Hydrobiologia*, 64(5): 661-671.

Kar, D. (2007). *Fundamentals of Limnology and Aquaculture Biotechnology*. Daya Publishing House (New Delhi), xiv+609pp.

Kar, D. and Barbhuiya, M.H. (2004). Abundance and Diversity of Zooplankton in Chatla Haor, a floodplain wetland in Cachar district of Assam. Environment and Ecology, 22 (1): 247-248.

Kar, D., Laskar, B.A, Nath, D., Mandal, M., and, Lalsiamliana (2002 a). *Tor progenius* (McClelland) under threat in River Jatinga, Assam. Science and Culture, 68 (7-8): 211.

Michael, R.G. and Sharma, B.K. (1988). Indian Cladocera: Fauna of India and Adjacent Countries Series, x + 261, Zoological Survey of India (Calcutta).

Prasad, R. R. (1956). Further studies on the plankton of the inshore waters of Mandapam, *Indian J. Fish.*, 3 (1): 1-41.

Raid, G. K. (1961). *Ecology of Inland waters and Estuaries*, Reinhold Book Corp., New York.

Reinhard, E. G. (1931). The plankton ecology of the Upper Mississippi, Mineapolis to Winona. *Ecol. Monogr.* 1: 395-464.

Riley, G. A. and Bumpus, D. P. (1946). Phytoplankton-zooplankton relationship on Gaorges Bank, *J. Mar. Res.*, 6: 33-46.

Rodgers, W. A. and Panwar, H. S. (1988). Planning a Wildlife Protected Area Network in India. Vol I and II. Wildlife Trust of India, Dehradun India Saala *et al.*, 2000. Global biodiversity scenarios for the year 2100. Science. Vol 287 pp 1770-1774.

Ruggiero, M. A. and Merchant, H. C. (1979). Water quality substrate and distribution of macro-invertebrates in the Patuxent river, Maryland, *Hydrobiologia.*, 65 (2): 183-189.

Shelford, V. E. (1911). Ecological Succession: Stream Fishes and the method of Physiographic analysis. *Biological Bulletin*, 21: 9-35.

Welch, P. S. (2003). *Limnological methods*. Narendra Publishing House. New Delhi, India. pp. xviii+381.

Chapter 4

Agribusiness Opportunities in Meghalaya: Some Facts and Emerging Issues

Deepak Bhagat[1] *and U.R. Dhar*[2]

[1]*Assistant Professor, Department of Management*
North-Eastern Hill University, Tura Campus, Tura – 794 002, Meghalaya
[2]*Professor, Department of Business Administration,*
Guwahati University, Assam

ABSTRACT

The diversification of consumption pattern towards high value agricultural products has been well established. Also there is an accelerated flow of export of high- value food commodities from developing to developed countries. In the mountainous region land utilization under high value agricultural commodities tend to yield higher level of both social and economic benefit and thus an attempt has been made in this paper to understand high value agricultural commodity scenario and agribusiness opportunity in one of the mountainous states of north east India i.e Meghalaya. The study clearly implies that farmers in the region are shifting their focus towards cash crops. However there are number of constraints affecting the sound development of agribusiness sector in the state. The issue of development of agribusiness sector in the region needs to be approached by fully exploring the potential of the region and strategizing the method of harnessing the same. It should be based on a holistic strategy to create a favourable environment by making available the infrastructural facilities and linkages necessary for economic productivity, proper and efficient supply chains, enhanced processing facilities, positioning of the product in domestic and export market and sustainable agribusiness development.

Keywords: Agribusiness, ACGR, CV, Agricultural commodities.

Introduction

The agricultural sector in most of the Asian countries is under transformation. Various studies have proved that there is a diversification of consumption pattern towards High- Value Agricultural Products (HVAPs). High value agricultural products are defined as products that are typically perishable, that are of specific high- value, and that are sold through specialized markets (CGIAR, 2004). HVAPs can include livestock, dairy products, fish, fruit and vegetables; products that are growing in importance, albeit at varying degree. Ravi and Roy (2006) projected the demand in India by 2020. Their study reveals that income growth and changes in other determinants such as urbanization is going to pronounce the diversification of consumption pattern towards HVAPs. There is an accelerated flow of export of high-value food commodities from developing to developed countries (Diaz- Bonilla and Recca, 2000) and thus, globalization has also created new opportunities for export of high- value products.

The geo-climatic situation of Meghalaya provides tremendous possibility for growing of different types of horticultural crops including fruits, vegetables, spices, plantation crops, medicinal and aromatic plants of high economic value. In the mountainous region, in general land utilization under high value agricultural commodities tends to yield the highest level of both social and economic benefits and thus an attempt has been made in this paper to understand high value agricultural commodity scenario and agribusiness opportunities in Meghalaya. This paper also tries to make a SWOT analysis of agribusiness sector in Meghalaya so as to find out strength, weakness, opportunities and threats related to agribusiness development in the state. Based on this institutional and policy requirements have been suggested to tap agribusiness opportunities and export potential of the state.

Methodology

The present study is based on secondary data. The time series data on area, production and yield of important high value agricultural commodities were obtained from Directorate of Horticulture, Government of Meghalaya, Shillong.

Statistical Tools Used for Analysis

Annual compound growth rate (ACGR) and coefficient of variation were computed to examine the trends and instability in area, production and yield over the years.

Annual Compound Growth Rate (ACGR)

The annual compound growth rate regarding the area, production and yield of above commodities were calculated by using the exponential function of the form

$$Y = a.b^t$$

Here, Y = Variable under study *i.e.* area, production and yield of high value agricultural commodities

a = Intercept

b = Regression coefficient of log y on t

t = Time variable (1, 2..,n) for each period.

The parameters a and b were estimated by the method of least square

By transforming the above equation to log linear form

$$\log y = \log a + t \log b$$

$$\text{or } Y = A + Bt$$

The compound growth rate r was given by

$$r = (b-1) \times 100$$

with b as antilog of b

Coefficient of Variation

To examine the stability with respect to area, production and yield, coefficient of variation were worked out with following formula.

$$\text{Coefficient of variation} = \frac{\sigma}{\overline{X}} \times 100$$

where, σ = Standard deviation

$\overline{X}$ = mean of sample data

$$\text{Standard deviation } (\sigma) = \sqrt{\frac{\sum X^2}{N}}$$

where, $X = (X_1 - X)$

N = No. of observations

Results and Discussion

Zonal Distribution of Horticultural Crops in Meghalaya

Meghalaya is located within the Eastern Himalayan Zone-II which is further sub- divided into 5 (five) sub-regions, taking into consideration topography, rainfall, temperature, soil type and cropping system of which Meghalaya comes within sub region-II. For the purpose of planning for development, research and extension, pending further detailed delineation of agro-climatic zones, the state can be divided into 3 (three) zones for working conditions. Broad outlines of the horticultural crops grown in each zone are as follows (Table 4.1).

Table 4.1: Zonal Distribution of Horticultural Crops in Meghalaya

Sl.No.	*Agro Climatic Zones*	*Horticultural Crops*
1	Tropical Zone (100- 300 msl)	Litchi, Papaya, Lemon, Orange, Pineapple, Banana Guava, Arecanut, Black pepper, Betel leaf, Coconut, Turmeric, Cinnamon, Chilli, Potato, Sweet Potato
2	Sub-tropical Zone (300- 1100 msl)	Litchi, Lemon, Orange, Guava, Grape fruit, Pineapple, Ginger, Turmeric, Cinnamon, Chilli, Potato, Sweet Potato, Vegetables
3	Temperate Zone (1100- 2000 msl)	Plum, Peach, Pear, Apricot, Chestnut, Potato, Spices, Off season vegetables

High Value Agricultural Commodities Scenario in Meghalaya

The state Meghalaya, due to its diversity in climate, topography, soil profile etc, offers scope for a wide range and variety of horticultural crops like fruits, vegetables, spices and a variety of colourful and attractive flowers, ornamental plants and orchids. The annual average and compound annual growth rate of some of the important high value agricultural commodities in Meghalaya during the period 1999- 2000 to 2009- 2010 are presented in the Table 4.2. The table reveals that ginger, citrus fruits, potato, areca nut and pineapple have highest annual average area under cultivation in comparison to other potential high value agricultural products in the region with an annual average production area of 8707.38 ha, 8510.77 ha, 183.92.23 ha, 11300.85 ha and 9698.08 ha respectively. However, as far as average annual production is concerned, potato showed highest annual production with 165646.54 MT followed by pineapple (89591.31 MT) and ginger (48995.15 MT). A perusal of the table 2 also shows that all the high value agricultural commodities (except potato) have positive compound annual growth rate as far as their area expansion is concerned *viz.* pineapple (1.1 per cent), citrus fruits (2.2 per cent), banana (2.4 per cent), papaya (2.1 per cent), ginger (2.2 per cent), turmeric (3.3 per cent), tea (19.2 per cent), areca nut (3.71 per cent) and cashewnut (1.9 per cent). With area, the production of the crops have also increased and most of the crops shown positive CAGR *viz.* pineapple (2.2 per cent), citrus fruits (1.1 per cent), banana (1.8 per cent), papaya (1.5 per cent), ginger (1.8 per cent), turmeric (3.5 per cent) and chilies (2.5 per cent), tea (20.6 per cent) and cashewnut (7.7 per cent). HVAP's viz pineapple (1.1 per cent), potato (0.5 per cent), turmeric (0.2 per cent), chili (2.0 per cent), tea (1.4 per cent), areca nut (1.1 per cent) and cashewnut (5.8 per cent) have shown positive growth rate for yield also. This is a very encouraging scenario for the state. This clearly implies that farmers in the region are shifting their focus towards cash crops. Mandarin orange is the most dominant economic crop of the state in citrus fruit category. Khasi mandarin can be considered as an important variety widely known in the north east region of India and outside having good acceptance among the consumers. In banana, varieties like jahaji, chenichampa, malbhog and indigenous varieties are commonly grown in the region. Recently tissue cultured banana saplings are introduced in the state and are also doing well. Pineapple (with varieties like giant kew and queen) is another commercial fruit crop found in the sub mountainous tract in the northern part of East and West Khasi hills as well as the northern and southern part of Garo hills. Potato is extensively grown in the East and West Khasi Hills District, Tapioca is found in Ri-Bhoi and Garo Hills District while sweet potato is cultivated throughout the state. Among the plantation crops, Areca nut, cashewnut and tea are important ones. Since time immemorial areca nut has been grown in Meghalaya as an important crop. As far as tea is concerned, at present, the under tea is being expanded in villages around tea nurseries. It has been found that tea area is increasing. Cashewnut is extensively grown in Garo hills. However the bulk of the produce is sold outside the state as raw products. The increase area under this plantation crop is due to the effort for reclamation of jhuming lands (Meghalaya agriculture profile, 2006).

Table 4.2: Annual Average and Compound Annual Growth Rate of some Important Crops in Meghalaya during the Period 1997–98 to 2009–10

Sl.No.	Category	Crops		Average	Equation	CAGR (per cent)	R^2	SD	CV (per cent)
1.	Fruit Crops	Pineapple	Area	9698.08	$y = 8965.e^{0.011x}$	1.1	0.797	475.63	4.90
			Production	89591.31	$y = 76109e^{0.022x}$	2.2	0.921	8346.05	0.93
			Yield	9222.08	$y = 8488.e^{0.011x}$	1.1	0.833	458.43	4.97
2.		Citrus fruits	Area	8510.77	$y = 7248.e^{0.022x}$	2.2	0.809	830.35	9.76
			Production	36054.23	$y = 33323e^{0.011x}$	1.1	0.500	2161.48	0.60
			Yield	4254.31	$y = 4594.e^{-0.01x}$	–1.0	0.498	268.68	6.32
3.		Banana	Area	5863.85	$y = 4904.e^{0.024x}$	2.4	0.887	605.13	10.32
			Production	68225.46	$y = 59870e^{0.018x}$	1.8	0.924	5155.29	0.76
			Yield	11669.69	$y = 12207e^{-0.00x}$	–0.00	0.437	444.57	3.81
4.		Papaya	Area	566.15	$y = 484.7e^{0.021x}$	2.1	0.761	54.02	9.54
			Production	4393.15	$y = 3927e^{0.015x}$	1.5	0.792	299.75	0.68
			Yield	7784.69	$y = 8101.e^{-0.00x}$	–0.0	0.204	378.57	4.86
5.	Tuber Crops	Potato	Area	18392.23	$y = 19797e^{-0.01x}$	–1.0	0.543	1088.34	5.92
			Production	165646.54	$y = 17085e^{-0.00x}$	–0.0	0.031	19717.10	1.19
			Yield	8991.77	$y = 8630e^{0.005x}$	0.5	0.063	739.89	8.23
6.		Tapioca	Area	4037.08	$y = 3928.e^{0.003x}$	0.3	0.413	95.03	2.35
			Production	21113.62	$y = 21004e^{0.000x}$	0.0	0.012	526.81	0.25
			Yield	5232.85	$y = 5347.e^{-0.00x}$	–0.0	0.631	79.44	1.52
7.	Spices Crops	Ginger	Area	8707.38	$y = 7386.e^{0.022x}$	2.2	0.779	847.55	9.73
			Production	48995.15	$y = 42892e^{0.018x}$	1.8	0.702	4289.33	0.88
			Yield	5642.15	$y = 5807e^{-0.00x}$	–0.0	0.082	332.25	5.89

Contd...

Table 4.2–Contd...

Sl.No.	Category	Crops		Average	Equation	CAGR (per cent)	R^2	SD	CV (per cent)
8.		Turmeric	Area	1667.38	$y = 1310.e^{0.033x}$	3.3	0.942	223.56	13.41
			Production	9306.77	$y = 7128.e^{0.035x}$	3.5	0.579	1851.40	1.99
			Yield	5567.15	$y = 5439.e^{0.002x}$	0.2	0.009	643.46	11.56
9.		Chillies	Area	1827.46	$y = 1769.e^{0.004x}$	0.4	0.684	39.14	2.14
			Production	1237.38	$y = 1029.e^{0.025x}$	2.5	0.899	130.77	1.06
			Yield	677.00	$y = 583.1e^{0.020x}$	2	0.884	58.99	8.71
10.	Plantation Crops	Tea	Area	934.92	$y = 192.5e^{0.192x}$	19.2	0.906	547.51	58.56
			Production	3167.77	$y = 551.7e^{0.206x}$	20.6	0.743	1992.72	6.29
			Yield	3296.46	$y = 2864.e^{0.014x}$	1.4	0.032	995.92	30.21
11.		Arecanut	Area	11300.85	$y = 9351.e^{0.026x}$	2.6	0.902	1210.96	10.72
			Production	14812.62	$y = 11254e^{0.037x}$	3.7	0.940	2277.56	1.54
			Yield	1305.38	$y = 1203.e^{0.011x}$	1.1	0.739	67.78	5.19
12.		Cashew nut	Area	6789.00	$y = 5918.e^{0.019x}$	1.9	0.786	602.73	8.88
			Production	8901.46	$y = 4931e^{0.077x}$	7.7	0.873	2988.64	3.36
			Yield	1290.31	$y = 833.5e^{0.058x}$	5.8	0.805	332.03	25.73

A: Area (Ha); P: Production (MT); Y: Yield in Kgs/hect.

Source: Computations are based on data available from Directorate of Horticulture, Government of Meghalaya, Shillong.

Agribusiness Opportunities in Meghalaya and Related Challenges

The north- eastern region of India with its endowment of vast untapped natural resources along with an attractive virtual tax free fiscal package offered by the government is being envisaged as a major growth area in the country, creating opportunity for diversified investments in agribusiness sector in the region (Chakravarthy *et al.*, nil).

As far as Meghalaya is concerned, there is huge potential particularly in horticulture, floriculture, agro and food processing, organic farming and plantation (most important high value agricultural commodities with having vast agricultural potential in the sate are presented in Table 4.2). However, it is unable to fully harness this potential due to a number of constraints such as inadequate infrastructural facilities, use of traditional and outdated methods of production, lack of market access and insufficient credit. Table 4.3 presents the swot analysis of agribusiness sector in the state.

The marketing system in Meghalaya can be classified into two phases. Pre-mission and post- mission. In the pre-mission phase, the horticulture production was of kitchen garden scale. These were mostly retained by farm household or locally consumed. All though crops like potato ginger, pineapple and pear got entry into the marketing channels. Though this pattern still exist, it can be safe to say that market situation is undergoing slow but sure revolution with commercial cultivation of high value agricultural products promoted by mini mission II. Unlike other states of north-east India, these products are usually being purchased by agents at the farm gate or auctioned at the adjacent market yard. The prevailing marketing system of the state has been depicted in Figure 4.1.

Potato, tejpata, broom stick, rosin, ginger are notified for transaction in APMC yards. But these commodities are also taken for trade to the markets in the neighboring districts of states like Assam and Tripura. Meghalaya also has a huge system of primary and retail markets which function outside the domain of State Agriculture Marketing Act. Trading in these markets is run by local authorities, tribal chiefs or private parties. As per the directory maintained by the State Government there are 112 such markets. Ginger, potato (off season), pineapple, cabbage (off season), green chilli, passion fruit, chow chow and strawberry have potential for being marketed outside the state. A substantial part of this produce goes to Bangaldesh via Dawki/ Cherrapunjee route. In the deficiency of market arrivals data in most of these market outlets which are doing transactions in an un-regulated manner, it is impossible to accurately enumerate the marketed surpluses.

The issue of development of agribusiness sector in the region needs to be approached by fully exploring the potential of the region and strategizing the method of harnessing the same. It should be based on a holistic strategy to create a favourable environment by making available the infrastructural facilities and linkages necessary for economic productivity, market linkages, enhanced processing facilities, positioning of the product in domestic and export market and sustainable agricultural development.

Table 4.3: SWOT Analysis of Agribusiness Sector in Meghalaya

Strengths	Opportunities
☆ Rich diversity of climate creating potential for growing wide variety of crops ☆ Rich biodiversity ☆ Available of good quality usable timber ☆ Proximity to nearby export markets ☆ Natural resources like soil and water are not degraded with excessive use of agricultural inputs ☆ Availability of unique and exotic products	☆ Creation of agri infrastructural facilities with greater private sector participation ☆ Introduction of agri technology ☆ Organic farming ☆ Provision of greater market and credit access ☆ Design and application of appropriate policy instruments for agricultural development ☆ Focus on agro industry ☆ Potential international market due to geographical location ☆ Contract farming ☆ Value addition ☆ Off season vegetable and quality flower production ☆ Cold/bulk storage ☆ Micro and mass production of planting material ☆ Quality certification ☆ Common facility centres
Weakness	**Threats**
☆ Lack of agri infrastructural facilities along the food value chain ☆ Traditional and inefficient agriculture supply chain ☆ Weak capital base ☆ Lack of market access- domestic and international ☆ Limited flow of men and material ☆ Need for technological interventions ☆ Prevalent practice of Jhum cultivation and subsistence farming ☆ Excessive dependence on erratic rainfall, unpredictable and variable monsoon	☆ Competition from the other eastern and north eastern states ☆ Political instability ☆ Insurgency

Institutional and Policy Requirements to Tap Agribusiness Opportunities and Export Potential

A number of policy decisions have been taken by both central and state governments to bust internal agriculture and agribusiness. These include relaxation in regulation governing markets, fiscal incentives for food processing industry, increased availability of credit to farmers at lower interest rate, funding of contract

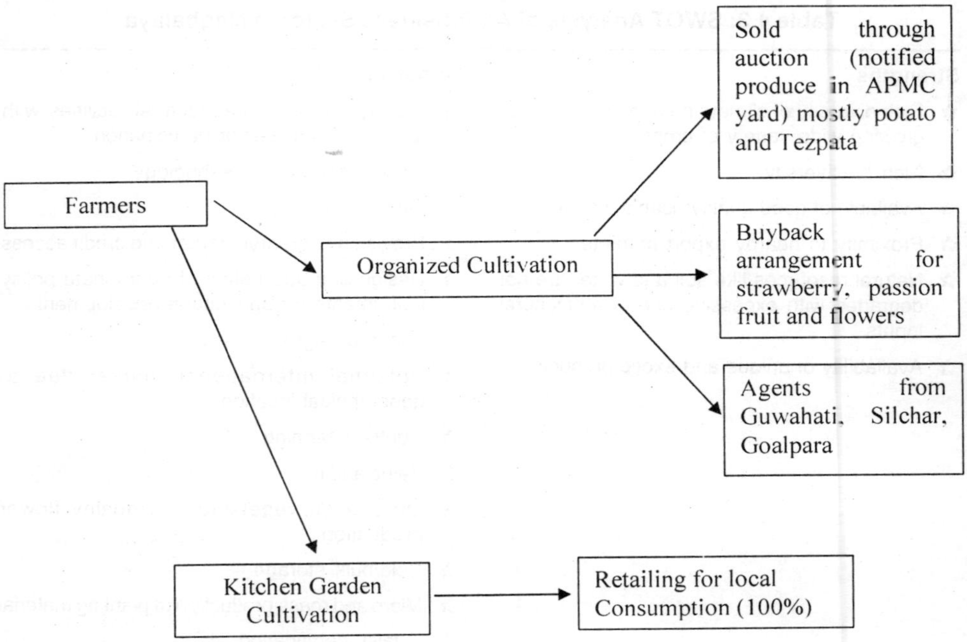

Figure 4.1: General Horticulture Supply Chain in Meghalaya and Garo Hills
***Source*: Adapted from marketing model presented in "An In-depth Market Study and Impact Assessment Report"; North Eastern Region**

farming schemes by institutional agencies, permission of FDI in single brand retailing etc. (Brithal *et al.*, 2007). However, since the existing supply chains are long and are dominated by a number of intermediaries like assembler, wholesaler, sub- wholesaler, commission agent and retailers. In case of fruits and vegetables, farmers receive one-third to one- half of the final price (Gandhi and Namboodiri, 2002), indicating high marketing cost and margins. Brithal *et al.* (2005) have estimated the marketing cost to be around 20 per cent of the sale price of vegetables. There is need to focus on institutional innovations to reduce the high marketing and transaction costs. Institutions such as cooperative, grower associations and contract farming are considered to reduce marketing and transaction costs and risks by providing 'market' to the farmers at their doorsteps (Eaton and Shepherd, 2001).

Since majority of the farmers in the state belong to marginal and small holder category, this section also discusses some of the institutional innovations attempted across the globe including India to integrate small holders into the market as recognizable players and can be with some modifications be implemented in the state also. This can also be considered as an attempt to give policy recommendations for the state.

One of those institutional innovations can be Enabling Rural Innovation (ERI) which provides a practical framework to link farmer participatory research and market research to empower farmer better manage their resources. It is based on participatory approach to build capacities of farmer' groups and rural communities in managerial areas to identify and evaluate market opportunities to develop profitable agro enterprises, identify production through experimentation. This was tested in pilot sites in Malawi, Uganda and Tanzania. It stimulates farmer to reduce risk, access new technologies and improve the production and competitiveness. Under a USAID funded grant in late 2003, CARE along with ministry of agriculture in Upper Egypt helped 100 farming community having progressive leader to organize and register voluntary into member-based, service-oriented Farmer Associations (FAs) as Non-Governmental Organisations (NGOs). CARE introduced the notion that no seed should be planted unless a farmer has a buyer for what he will grow. CARE also pioneered the idea that FAs should enter into forward contracts with buyers or exporters for high-value fruits and vegetables or medicinal and aromatic plants. The farmer came to realize that farming is a business and not merely a way of life. Now, there are more than 11,000 members of 105 FAs in 9 governorates of Upper Egypt that are involved in high- value production, especially for the EU market. In India also, there are number of examples of institutional innovations. Two institutional arrangements, namely, Anand pattern of Cooperatives and e- choupal have received wide attention for the benefits they have claim to have passed on to the small farmer.

In the era of LPG, *i.e.* Liberalization, Privatization and Globalization, it is very important to focus attention on the establishment and strengthening of regional cooperation across the globe to bust trade and business across the regions. Various forms of regional cooperation like EU, NAFTA and LAFTA confirms this need. Similarly, South Asian Association of Regional Co- Operation (SAARC) and Association of South East Asian Nations (ASEAN) are good examples in this region. As a validation of a conscious adoption of a "Look East" policy, India became a full dialogue partner of ASEAN and participated for the first time at the Post Ministerial Conference (PMC) of ASEAN in Jakarta in July 1996. The formation of BIMSTEC Group (Bangladesh, India, Myanmar, Singapore and Thailand- Economic Cooperation), which is likely joined by Bhutan and Nepal, is another step to forge regional cooperation which will benefit the North- East, to a large extent. The South Asian Free Trade Area Framework Agreement, 2004 and South Asian Preferential Trade Agreement (SAPTA) are also expected to bring the benefits of LPG.

However, much progress has not been achieved in reaping these benefits. Few suggestions which may help in increasing trade and cooperation among the member countries may be: establishment of transportation and communication network between these countries should be taken on priority basis, relaxation in foreign trade policy in favour of busting trade with member countries, bringing political stability in NER and peace to provide a favourable condition for investment, agribusiness development in the region and also organization of seminars, conferences and business summits may be helpful. However, before harnessing export potential of the state and region, there is a need of policies and strategies that would improve the farmer access to credit as well as education both through formal and non formal programs to increase the farmer participation decisions in growing and marketing of HVAPs. Here, there is a need to organize farmers into groups or farmer associations so as to enable them to take decision collectively and market collectively. In addition, improved farmers access to agricultural equipment is also needed. Also, due consideration should be given to the training and organization of HVAPs producer farmers to help them to get better income from the crop. All these along with the above suggestion of reaping the benefits of export potential will lead to agribusiness lead development of the state and the region.

Conclusion

The potential for growing various commercial crops such as pineapple, oranges, banana and ginger have been clearly demonstrated in the region. There is tremendous potential for cultivating horticultural crops in the hills, but the success depends upon various factors such as: development of rural infrastructure including marketing links, cold storages and processing facilities, improvement in extension services with better provision of good quality seeds, inputs and seedlings, expansion of horticulture in tribal areas to replace jhumming, formation of self- help groups comprising fruit, vegetable and flower growers, and providing training and micro- finance support, increase private sector participation. There is huge export potential also. Few suggestions which may help in increasing trade and cooperation among the neighboring countries may be: establishment of transportation and communication network between these countries on priority basis, relaxation in foreign trade policy in favour of busting trade with member countries, agribusiness development in the region and also organization of seminars, conferences and business summits may be helpful.

References

Brithal *et al.* (2007). Linking farmers to markets for high- value agricultural commodities, *Agricultural Economics Research Review,* Vol. 20(Conference Issue) 2007 pp 425- 439.

Brithal, P.S., Joshi, P. K. and Gulati, A. (2005). *Vertical Coordination in High- value Food Commodities: Implications for Smallholders,* MTID Discussion Paper 85. International Food Policy Research Institute, Washington D.C

CGIAR. (2004). Revised Summary Report on Developing CGIAR System Priorities for Research. Science Council working paper. Washington DC: CGIAR.

Chakravarthy, K., Dasgupta, S. and Singh, R. (nil). *North east region: Harnessing agribusiness potential.* Yes Bank Ltd. Mumbai, India.

Diaz- Bonilla, E., and Recca, L. (2000). Trade and agro- industrialization in developing countries. Agricultural Economics, 23(3): 219- 230

Department of Agriculture. (2006). Meghalaya agriculture profile. Government of Meghalaya. Shillong

Eton, C., and Shepherd, A. W. (2001). Contract farming: Partnerships for growth', *FAO Agricultural Services Bulletin 145,* Food and Agriculture Organisation, Rome, Italy.

Gandhi, V. and Namboodiri, N. V. (2002). Fruits and Vegetable Marketing and Its Efficiency in India: A Study in Wholesale Markets in Ahmedabad Area, Mimco, Indian Institute of Management, Ahmedabad.

Ravi, C., and Roy, D. (2006). Consumption patterns and food demand projections, a regional analysis. Draft Report submitted to IFPRI, New Delhi.

Chapter 5

Fish Production and Potential in North East India: Implications for Sustainable Management

Biswajit Debnath[1], Diganta Sharmah[2], Apu Das[3], Pankaj K. Mugaonkar[4] and M. Islam[5]

[1]SMS (Fisheries), KVK (ICAR), South Tripura,
[2]SMS (Plant Protection), KVK (ICAR), South Tripura
[3]M.F.Sc. Student, CIFE, Mumbai – 61
[4]Ph.D. Scholar, CIFE, Mumbai – 61
[5]KVK (ICAR), Tura, Meghalaya

ABSTRACT

Fisheries sector plays an important role in economic development in North East Region of India. The region has good potential in fisheries sector supplemented by huge water resources in the form of rivers, floodplains wetlands (*Beels*), lakes, reservoirs, ponds and mini-barrages and low-lying paddy fields. Total fish production from this region was estimated to be 296.72 t mt. during the year 2009–10 and experts estimated more potentiality of fish production. To realise the fish production potentiality, North East Region needs to overcome many constraints. Availability of quality fish seed, high rainfall, floods, fish diseases, critical inputs like fish feed, conflicts between different users of open water, river regulation etc. are major issues to be addressed for sustainable management.

Keywords: ***Fisheries in North East India, Fish production potential, Fisheries resources, Sustainable fisheries management.***

Introduction

Arunachal Pradesh, Assam, Manipur, Meghalaya, Mizoram, Nagaland, Sikkim and Tripura – are the states of north east India encompasses a geographic area of 2,62,179 km^2. The topography is wholly mountainous with varying altitudes from 300 m in the sub-tropical belt to 5500 m m.s.l. in the temperate alpine zone. Hills and mountains comprise over two-third of the region's territory (Barch, 2001). This undulating topography supplemented by high rainfalls has given rise to numerous forms of fisheries resources in the region. This region presents diverse topographical conditions as well as fisheries resources in the form of rivers, floodplains wetlands (*beels*), lakes, reservoirs, ponds and mini-barrages and low-lying paddy fields. Further, the peoples of North East India have high preference towards fish and hence, high demand for the same prevails in the region. Almost entire population of the region are fish eater. However, fish production and potentiality of different states of the region is quite diversified and sustainable management of the fisheries resources to achieve sufficient fish production was one the neglected areas till the recent past. Sugunan (1998), for the first time, summarised the fish production potential of fisheries resources of the region vis-à-vis their present production. Thus, it is imperative that fish production and potential from this region is increased by all possible means to satisfy its requirements as well to improve the socio-economic condition.

Fish Production Status of Different States of the Region

Presently the region is producing 296.72 t mt. (2009–10) fish with an annual average growth rate of 3.75 per cent for last decade (Table 5.1). Highest growth rate in fish production among these states was observed in Tripura (12.9 per cent), followed by Assam (5.01 per cent) and in case of Meghalaya, no regular trend was observed. The status and potentiality in fish production from different resources across the states were summarized by Sugunan, 2007 and state wise highlights of the same have been discussed in the following section.

Arunachal Pradesh

This hilly state with low temperature regime has pond productivity of 600 kg/ha/yr, which can be increased up to 2500 kg/ha/ yr by adopting appropriate species and other management measures. By bringing 1500 ha under pond culture, an estimated production of 3750 t can be realised. The state has 3 trout farms and 260 km of rivers (Tawang and West Kameng) under trout stocking and conservation. The *beels* of East Kameng, Lower Subansiri, East Siang, DibangValley, Lohit, Tirap and Changlang districts produce fish at the rate of 60 kg/ha. The Ranganadi reservoir (160 ha) can produce 16 t at a modest yield rate of 100 kg/ha. About 2800 ha of paddy fields are amenable for paddy-cum fish culture, of which only 575 ha are being utilized. The rivers of Arunachal Pradesh could produce 400 t at the yield rate of 200 kg/ha. The state has a production potential of 9416 t.

Assam

It is the most resourceful state (both in terms of area and potential) of the region, which can achieve a production level of 172110 t through planned development of its fisheries resources. *Beels* and swamps, which form the mainstay of Assam's fisheries, can produce 75000 t of fish through a combination of culture, culture-based and capture fisheries development approaches. Potential production from rivers, paddy-cum-fish culture and reservoirs are estimated at 9640 t (@ 2 t/ km), 14000 t and 171 t respectively. A total production of 181,311 t is attainable including the fish production from ponds.

Table 5.1: Fish Production of North East States during 2000 to 2010 (in thousand tonnes)

Year	*Arunachal Praesh*	*Assam*	*Manipur*	*Megh-alaya*	*Mizo-ram*	*Naga-land*	*Sikkim*	*Tripura*	*North East*
2000	2.4	159.77	15.51	4.68	2.89	5	0.14	14.02	204.41
2001	2.5	158.62	16.05	6.18	2.86	5.5	0.14	14.35	206.2
2002	2.6	161.45	16.45	4.97	3.15	5.2	0.14	14.7	208.66
2003	2.6	165.52	16.6	5.37	3.25	5.5	0.14	16.15	215.13
2004	2.65	181	17.6	5.15	3.38	5.56	0.14	17.98	233.46
2005	2.70	186.31	17.8	5.638	3.68	4.9	0.14	19.838	241.014
2006	2.75	188.01	18.22	4.12	3.75	5.5	0.15	23.87	246.37
2007	2.77	181.48	18.61	5.487	3.76	5.8	0.15	28.634	246.694
2008	2.83	190.32	18.8	4	3.76	5.8	0.175	36.245	261.93
2009	2.88	206.15	18.6	3.959	2.891	6.175	0.168	36.199	277.022
2010	2.65	218.82	19.2	4.21	3.042	6.36	0.168	42.268	296.72
CGR	1.35	5.01	2.13	—	1.12	1.92	2.31	12.9	3.75

Source: Indiastat, 2011.

Manipur

It is potentially the second most resourceful state in the region from fisheries perspective. The LoktakLake, *Pats* and natural water bodies, along with the proposed 40000 ha of reservoirs provide scope for substantial increase in fish yield in Manipur and can make it surplus state in fish production. Loktak Lake produced 1500 t if fish at a yield rate of 78 kg/ ha to achieve a production of 4395 t from 14, 650 ha of the lake area. The remaining 4500 ha can be converted into fishery estates for culture fisheries development. The state has terraced paddy fields in Ukhrul and Senapati districts, where people are known to grow fish in their paddy fields. In addition, about 3362 km rivers stretches can be brought under fishing to net a modest 672 t. thus, Manipur has tremendous potential to produce as much as 76327 t of fish annually, which can make it self-sufficient in fish production.

Meghalaya

With its limited aquatic resources, Meghalaya has the potential to produce 11623 t. if the state is to meet this ambitious target, most of the enhanced fish production should come from the ponds that have to increase in area from 500 ha to 2400 ha in the coming years and produce 6000 t of fish at a yield rate of 2500 kg/ha. Similarly,

5000 ha have to be brought under paddy cum fish culture to produce another 3000 t of fish. Conversion of marginal *beels* areas into fishery estates and efficient management of capture fisheries in the rest of the area can lead to an additional production of 713 t. yield rates from reservoirs and rivers had to be improved to 100 kg/ha and 200 kg/ km to achieve the targets of 843 t and 1120 t respectively from these resources. Thus, the state has fish production potential of 11623 t.

Mizoram

The modest fishery resources of Mizoram have the potential to produce 5771 t. with its hilly terrain giving little scope for digging ponds, the state has to create more mini-barrages to increase its culturable area. There is also scope for improving yield from rivers and bringing more areas under paddy cum fish culture and reservoir have potential to produce 4487 t, 340 t, 936 t and 8 t of fish respectively. Thus, there are scope exits for nearly doubling its production in coming years.

Nagaland

The state has to increase its pond area to 2000 ha in order to produce 5000 t of fish and improve the total fish yield from rivers (320 t) and *beels* (583 t) apart from bringing 5000 ha under paddy cum fish culture to produce 300 t. by achieving the potential fish production and requirement considerably.

Sikkim

The state has modest fisheries resources in the form of lakes, snow-fed rivers and streams, some of which have associated marshy areas. The total annual fish yield contributed by Sikkim's riverine fisheries is of order of 150 t. about 1500 - 2000 villagers living close to the river banks are engaged in part-time fishing. The main thrust area of fisheries development of the state at present is on conservation of existing fisheries resources of the state including ichthyo-faunistic resources. In order to maintain the yield at a sustainable level and to conserve some of the rare and endangered fish species of Sikkim Rivers, it is prioritised to implement conservation measures strictly.

Tripura

Tripura recorded steady growth in its fisheries sector and reportedly has the lowest deficit in the north east (Sugunan, 2003) and maintained a steady fish production levels for last decade. Fish yield rates from the ponds and reservoirs of Tripura are higher compared to that from other northeast states, which holds the key for its success. If the state can produce 40,500 t from ponds and mini-barrages, 1200 t from river (1 t /km) 600t from reservoir and 100 t from the *beels* and swamps it can achieve fish production to the order of 42400 t in near futurc, which will be quite satisfactory considering its geographical and water spread area.

Constraints and Implications for Sustainable Management to Realize Production Potential

The region has considerable potential for substantial improvement both in production and yield of fishes from its water bodies. There are many areas requiring management interventions to achieve higher fish yields. The strategy for realising this potential comprises efficient management of the natural water bodies such as rivers, *beels* and lakes; developing culture based fisheries in the *beels* and reservoirs; extending the proven aquaculture technologies to all the ponds; and bringing more areas under paddy-cum-fish culture. By following such a combined strategy, it will be possible to raise the region's fish production substantially, thereby narrowing the deficit down to modest levels from the current level.

Non-availability of quality fish seed in adequate numbers has always been a constraint in the region (Sugunan, 2003). Unspecified but considerable quantities of fish seed is brought into the region from west Bengal by road and through three fish seed markets set up in Barpeta road, Rangiya and Dhekiajuli in Assam. The problem is further confounded by low temperature regime of the high altitudes; where breeding and rearing are difficult (Sugunan, 2003). Even in Assam and Tripura where a large number of carp hatcheries have been operating, considerable regional variations in seed availability occurs. Adequate facilities are also not available for growing fish spawn into fingerlings stage to stock the *beels* and reservoirs. Availing quality fish seed is one of the important issues for consideration for sustainable management of the sector.

The North East Region receives very high rainfall during the southwest monsoon season (june to September), which coupled with its unique undulating topography and siltation of most its rivers and associated wetlands, wrecks havocs on the floodplain areas every year. Floods constitute one of the major constraints to aquaculture development on a commercial scale in the region especially in Assam and Tripura (which are more prone to floods) because it submerges the aquaculture ponds and pens constructed in marginal areas of *beels* and reservoirs. Though the problem created by floods can be tackled to some extent by raising the height of the pond dykes / pen screens, such a step will increase the capital investments necessary to construct an aquaculture pond / pen. A cheaper option is to erect a temporary net wall around the pond / pen on the dykes / pen screens in times of flood

As aquaculture and fisheries enhancements have progressed in the region, fish diseases in aquaculture system have emerged as a notable problem. Fortunately, the problem is not very acute except for occasional outbreak of Epizootic Ulcerative Syndrome (EUS). However, there is no fish disease laboratory in the entire north east to diagnose the etiology of fish disease. Further, non-availability of prophylactic and therapeutic agents for treatment of common fish disease also creates problem for fish farmers of the region.

Shortage of other critical inputs like fish feed as well as prevailing low temperature and acidic conditions also hinder aquaculture development. In addition, limited availability of plain lands, difficult terrains and aquatic pollutions also come in the way of fishery development in the northeast.

Construction of embankment along with the banks of rivers for flood control has resulted in negligible auto-stocking of riverine fish species and annual flushing of water in many *beels / pats*, lakes and swamps (Choudhury and Bhattacharjya, 2006). Further, since most riverine fish species use the open *beels* as spawning and nursery ground, river regulation also adversely affects the fish stocks of the parent rivers

(Bhattacharjya, 2004). The harmful effects of dams on the ecology and fisheries of rivers have been well known (Jhingran, 1991) especially on the fisheries of migratory species. However, even though river regulation has harmful effects on the fisheries on open water bodies, it is irreversible since flood control is given more importance than fisheries. Thus, construction and effective operation of sluice gates appears to be practicable option.

In most open water fisheries of the region, conflicts arise between lessees/ fishers and agriculturists/ local residents over the use of marginal land and water for cultivation, irrigation, jute retting, navigation and other uses (Choudhury and Bhattacharjya, 2006). In order to minimize multiple-users conflicts, an integrated basin development plan should be drawn for the resources through participation of all stakeholders *i.e.* through co-management. Ensuring equitable distribution of benefits accrued from the resource among the riparian communities will help reduce the resultant social problems (Bhattacharjya, 2004).

As a result of years of reckless over-exploitation, stocks of major fishes are depleted in most rivers and its associated water bodies of the region. This has resulted in reduced auto-stocking and fish yield in these water bodies. Regulation of fishing requires formulation of practicable conservation technique suiting local conditions (*e.g.*, minimum landing size is more practicable than mesh size regulation in case of multi-species fisheries). Empowerment of fisheries officials for enforcement, Strengthening fishers' cooperative societies and creation of mass awareness among the fishers and beel managers are different options (Choudhury and Bhattacharjya, 2006).

Non-availability of funds for fisheries development is a serious economic constraint for fisheries development in the region especially that in large open water bodies. A substantial part of the resources in the north east comprises capture or culture based fisheries, development and management of which requires higher initial costs, compared to pond aquaculture. Most of the North East States are facing financial crunch, which has affected fisheries developments. In addition, poor post-harvest infrastructure and exploited by middlemen results in low share of fishers/ fish farmers in consumer's rupee.

There are some issues specific to the northeast such as those related to land holding system, access to water bodies, etc. that come in the way of developmental plans. Prevalence of clan lands and absence of land ownership rights by individuals in the tribal/ hilly areas (except in flat areas of Dimapur, tribal reserves of Tripura and permanently cultivated flat lands of Karbi hills) causes serious problems in getting financial support from the financial institutions (Sugunan, 2003) because mortgaging of land as security for loan is particularly impossible under such a situation. A few more constraints having technical (including environmental) and socio-economic ramifications are: reduction in productive beel area because of inclusion under wildlife (mainly bird) sanctuaries; absence of clear policy on multiple uses of open waters; lack of alternative employment opportunities for fisheries resulting in increase in fishing pressure and so on.

Conclusion

Agricultural developments including fisheries has to play the pivotal role in the socio-economic development of the region as no corporate bodies of the country and abroad are reluctant to set up industries in northeast because of many facts. A congenial environment for fisheries development prevails in the region by the way of availability of rich fisheries resources, very high demand and price of fish. It is heartening to note that notwithstanding the constraints hindering desired fisheries development in the region, some commendable progress has been made by couple of states like Tripura and Assam in recent years. Research efforts to develop suitable technologies of aquaculture for the medium altitude areas (700 to 1400 m m.s.l.) are also initiated by the concerned fisheries research institutes of the IndianCouncil of Agricultural Research to properly utilise the available resources in such areas of the region. Contribution of Central Agricultural University, State Agricultural University, North East Hill University etc. are also mentionable. On the development side, the government department and financial institutions will have to play a facilitating role in setting up capital intensive fisheries enterprises like commercial feed plants, seed farms, net making plants, etc. The capital needs of the fisheries are also to be met by the government or institutional funding agencies (*e.g.* the World Bank), as development of capture and culture-based fisheries provides means of livelihood to the resource-poor fishers of the northeast. The open water fisheries being a community based development process, has a direct bearing on the rural populace of this socio-economically sensitive region of country. Fisheries development often comes into conflict with environmental conservation norms and, therefore, careful planning is required in achieving fish yield enhancement while accommodating the environmental concerns. This is particularly important in northeast, which has so much at stake in the form of pristine aquatic environments and precious endemic fish species to be preserved for the future. Thus, concerted efforts are needed on the part of all concerned to realise the rich fish production potential, thereby contributing to the overall socio-economic development of the region.

References

Bareh, H.M. (ed.) (2001). Encyclopaedia of Northeast India, Vol. I-VIII. Mittal Publications, New Delhi, India

Battachariya, B.K. (2002). An effective process approach and responsible fisheries management in the context of inland fisheries, In Sugunan, V.V., Jha, B.C. and Das, M.K. (eds.), Culture based fisheries for Inland fisheries development. Bull. No. 113, Cent.Inland Fish. Res. Inst., Barrackpore, India

Battachariya, B.K. (2004). Floodplain wetlands of Assam: management options and issues from the fisheries perspective. In Choudhury, M.C., Srivastava, N.P. and Manna, R.K. (eds), Participatory approach to management of inland fisheries resources of northeastern India, Bull. No. 135, CIFRI, Barrackpore, India

Battachariya, B.K., Choudhury, M. and Sugunan, V.V. (2003). Ichthyofaunistic resources of assam with a note on their sustainable utilization. In Mahanta, P.C. and Tyagı, L.K. (eds), participatory approach for fish biodiversity conservation in northeastern india. Workshop proc. NBFGR, Lucknow, India

Choudhury, M. and Bhattachariya, B.K. (2006). Fisheries resources of northeastern region of India: management issues and strategies. In: (Vass, K.K., Mitra, K.K., Suresh, V., Kathia, P.K. and Srivastava, N.P., eds). River fisheries in India: issues and current status. Inland Fisheries Society of Inland, Barrackpore, Kolkata, India

Indiastat (2011). State wise Inland Fish Production in India.Ministry ofAgriculture and Ministry of Statistics and Programme Implementation.Government of India. Available [online] at www.indiastat.com[accessed on March, 2011]

Jhingran, V.G. (1991). Fish and fisheries of India.3rd edition.Hindusthan publishing Corp., New Delhi, India

Kottelat, M. and Whitten, T. (1996). Freshwater biodiversity in asia with special reference to fish. World Bank Tech. paper No. 343. The World Bank Washington, D.C.

Sen, N. (2000).Occurance, distribution and status of diversified fish fauna of northeast India. In: Ponniah, A.G. and Sanakr, U.K. (eds), Fisha Bio-diversity of Northeast India. NATP Publ. 2, NBFGR, lucknow, India

Sinha, M. (1990). Reservoir fisheries – its present status and future potential in northeastern region. In: Jhingran, A.G. and Unnithan, V. K. (eds.). Reservoir fisheries in India. Proc. Nat. Workshop, 3-4 January, 1990. Asian Fisheries Society – Indian Branch, Mangalore, india

Shrivastava, N.P. and Bhattachariya, B.K., (2003). Status of fisheries of floodplain wetlands of Assam. In: Vinci, G.K., Jha, B.C., Bhaumik and Mitra, K. (eds.) Fisheries management of floodplain wetlands in India. Bull. No. 125, CIFRI, Barrackpore, India

Sugunan, V.V. (1998). Fisheries resources of northeast region and scope for development. Fishing Chimes, 18 (1)

Sugunan, V.V. (2003). Fisheries resources potential of Northeastern region. In: Mahanta, P.C. and Tyagi, L.K. (eds.). Participatory approach for fish biodiversity conservation in Northeast India, Workshop Proc., NBFGR, Lucknow, India

Sugunan, V.V. (2007). Development of fisheries in north-eastern India – potential and constraints. In: Compendium of lectures. HRD programmes on "Recent Advances andTechnologies for Fisheries development in North Eastern Regionfor KVK staff". CIFRI. Guwahati, India.

Chapter 6

Preparation of Paneer Using Buffalo and Soya Milk

C.S. Choubey[1] and B.K. Mishra[2]

[1]K.A.P.G. College, Allahabad, U.P.
[2]Department of RDAP, NEHU,
Tura Campus, Tura – 794 002, Meghalaya

ABSTRACT

Four different proportions of Murrah buffalo milk and soya milk in the ratio of 100:00, 85:15, 75:25 and 65:35, respectively were used to prepare paneer. The coagulating temperatures of 70°, 80° and 90° C were studied along with tartaric and lactic acids as coagulants. The yield of paneer (20.50 per cent), TS (57.47 per cent), TS recovery (57.23 per cent) and fat (18.44) were the highest (P/0.01) in paneer made from whole buffalo milk and all these parameters significantly (P/0.05) decreased with increasing levels of soymilk. The protein was lowest (30.79 per cent) in milk paneer and significantly (P/0.05) increased with increasing levels of soymilk. The yield of paneer, TS, fat and proteins values in paneer increased significantly (P0.05) with increasing temperatures of coagulation. The sensory evaluation scores of paneer decreased significantly (P/0.05) by increasing levels of soymilk.

Keywords: Soya milk, Paneer, Lactic acid, Tartaric acid, Citric acid, Yield.

Introduction

In augmenting market milk supplies and stabilizing the marketing of dairy products in India, soymilk supplies hold a great promise. In indigenous milk products of preparation of paneer is now gaining more popularity in the consumers which can be prepared with the help of soymilk leaving a substantial amount of cow or buffalo

---

* *Corresponding Author.*

milk for their consumption as fluid milk. Although, a number of workers (Babjee and Rathi, 1989; Grower *et al.*, 1989, Prakash and Iya, 1980; Chakraborty and Gangopadhyay, 1990 and Roy and Sing 1994 have tried to make paneer using soymilk, no specific recommendations have emerged that can be passed on to the people engaged in this business. Keeping in view an attempt was made to replace varying amount of buffalo milk with soymilk in preparing paneer, using traditional coagulants).

Material and Methods

Soymilk was prepared by the methods as suggested by Grower *et al.* (1983). The buffalo milk was replaced by soymilk having 2.8 per cent fat and 10.2 per cent total solids (TS). The buffalo milk on an average contained 7 per cent Fat and 17.1 per cent Total solids. The blends of soymilk and buffalo milk in the ratio of 50:50 showed incomplete coagulation and unsatisfactory body and texture of the finished products. It was therefore, decided to replace the whole buffalo milk with soymilk to the extent of 15, 25 and 35 per cent.

Four different admixtures of buffalo milk having 0, 15, 25, 35 per cent of soymilk were taken to prepare paneer and were designated as I, II, III and IV groups respectively. Each lot was prepared from 5 liters of mixed milk three coagulating temperatures 70, 80 and90*C was studied. The milk samples were heated to these temperatures for a period of two minutes. Lactic acid and citric having 1 per cent concentration was take in to precipitate the milk while preparing paneer. The parameters as mentioned above were replicated three times. Products prepared were subjected to sensory evaluation by panel of four on the basis of 9 points hedonic scale (with a score of 9 for liked extremely and 1 for disliked extremely). The products having score of 7 or more were considered acceptable. The chemical analysis of Paneer in respect to fat, protein and total solids were statistically analysed by the procedure described by Panse and Sukhatme (1978).

Results and Discussion

Effects of different admixture of buffalo milk and soymilk on the yield, chemical and organoleptic of paneer have been presented in Table 6.1. Average yield of paneer was 20.55 per cent when prepared with whole buffalo milk. Yield of paneer decreased significantly (P0.05) with increasing concentration of soya milk (vs. 15 per cent 25 per cent and 35 per cent). The minimum yield of 15 per cent was recorded when paneer was prepared with milk sample containing 35 per cent. The yield of paneer from the blended samples of soymilk and buffalo milk also different significantly (P0.01) amongst themselves. Our findings seem to be quite in agreement with the observation of Babje and Rathi (1989), who also noted a reduction in the yield of paneer with increase in proportion of soymilk in whole in whole buffalo milk. The yield of paneer from whole milk mainly increased due to relatively higher recovery of total solids from the whole milk. Proteins from soymilk, because of their different chemical nature might not have completely precipitated resulting in lower recovery in case of blended milk samples.

Table 6.1: Effect of Different Admixture of Buffalo Milk and Soymilk on Yield and Recovery of Total Solids in Paneer (per cent)

Group/ Milk : Soya Milk	*Coagulation Temperature (ºC)*	*Yield of Paneer by Tartaric Acid Coagulation*	*Yield of Paneer by Lactic Acid Coagulation*	*Total Solid Per cent (Tartaric Acid)*	*Total Solid Per cent (Lactic Acid)*
Group I/100:0	70	20.13a	20.00a	56.85a	56.81a
	80	20.83b	20.33b	57.21b	57.11b
	90	20.33c*	20.67c	57.77c	57.63c
Group II/85:15	70	17.33a	17.25a	54.23a	54.00a
	80	17.84b	13.31b	54.96b	54.63b
	90	18.83c*	18.25c	55.24c	54.73c
Group III/75:25	70	16.27a	16.07a	52.73a	52.63a
	80	17.07b	16.32b	53.23b	52.74b
	90	17.92c	16.65c	53.72c	53.04c
Group IV/65:35	70	14.72a	14.32a	49.73a	49.63a
	80	15.20b	14.82b	50.21b	50.01b
	90	15.91c	15.12c	51.07c	50.14c

Different letters a,b,c denotes significance at 5 per cent level.

* Significant at 1 per cent level.

It is clear from the data presented in the Table 6.1 that irrespective of temperature and coagulants, the average yield (20.54 per cent) and recovery of T.S (57:23) were significantly (P/0.01) higher in buffalo milk that its different combinations with soymilk. Significantly (P/0.05) higher yield of panner was recorded in group i when citric acid (21.33 per cent) and lactic acid (20.67 per cent) coagulants were used at 90* C. The difference in the yield between these two coagulants was statistically very high (P/0.01). There seems to be a positive direct relation between temperature of coagulation and the yield of paneer. Present findings are at par with the results of Singh and Kanawjia (1988), and Pal and Yadav (1991). As the temperature increased the yield and recovery of paneer also increased significantly (P/0.05) in all the groups. Sachdeva and Singh (1988) found that the average gain in the form of increased solids recovery of higher temperature was lost due to the lower yield and poor quality which is not found in our case. Probably the addition of soymilk in whole buffalo milk compensated the loss of yield at higher temperature. When other parameters were kept constant the yield and recovery both were significantly (P/0.05) higher when paneer was prepared with citric acid than the lactic acid. These results are corroborated with the findings of Chkraborti and Gangopadhyay (1990).

The fat content of paneer significantly (P/0.05) decreased with increasing levels of soymilk in the buffalo milk showing the peak in group I (18.44 per cent) and the lowest (13.69) in group IV (Table 6.2). Bhattacharya (1971) in his study did not notice a proportional increase in fat per cent was increased fat content in milk. The fat per cent was increased (P/_0.05) at increasing levels of temperature with citric acid than the paneer prepared with lactic acid. It seems the citric acid performed better towards the recovery of fat in soypaneer than lactic acid. The best performance of citric acid was also reported by Singh and Ray (1977) and Vishwashwaraih and Anant Krishnan (1985). All the treatment of Soymilk resembles with PFA requirement. The fat content in group 4 was reported to be only 13.69 per cent. The protein content of paneer was significantly (P/0.05) reduced with the decreasing concentration of soymilk in the milk (Table 6.2). The protein content was significantly (P/0.01) lower in pure buffalo milk (30.79 per cent) than the highest value (35.99 per cent) obtained with 35 per cent addition of soymilk irrespective of temperature and coagulants. The effects of higher temperature showed significantly (P/0.05) highest value as compared to paneer prepared on lower temperature using both the coagulants. Among the coagulants, citric acid showed the best (P/0.05) performance for the maximum recovery of protein irrespective of temperature and groups. Direct positive correlations between protein content and increasing levels of soymilk in the whole milk have also been reported by Babje and Rathi (1989). The higher protein content in soymilk may lead to higher protein contents in finished products.

Table 6.2: Effect of Different Admixtures of Buffalo Milk and Soymilk on Chemical Composition of Paneer (per cent)

Group	*Temperature of Coagulation*	*Protein Content*		*Total Solid Content*		*Fat Content*	
		Tartaric	*Lactic*	*Tartaric*	*Lactic*	*Tartaric*	*Lactic*
I	70	30.87	30.05	55.05	55.38	18.15	18.08
	80	31.29	30.15	55.15	55.11	18.85	18.25
	90	31.87	30.62	56.36	55.79	18.95	18.38
II	70	32.27	32.15	51.36	50.76	16.75	16.65
	80	32.32	32.32	52.26	51.36	17.08	16.83
	90	32.78	32.78	52.78	52.65	17.25	17.01
III	70	34.19	33.96	49.33	49.92	14.98	14.75
	80	34.98	34.20	49.45	49.37	15.10	14.85
	90	35.22	34.72	49.88	49.94	15.23	15.08
IV	70	35.85	35.65	46.28	46.22	13.71	13.71
	80	36.34	35.75	46.87	46.81	13.81	13.35
	90	36.47	35.92	47.34	47.13	14.08	13.93

Different letters denotes significance at 5 per cent level.

* Significant at 1 per cent level.

The total solids in paneer significantly (P/0.05) decreased with increasing levels of soymilk (Table 6.2). The maximum TS (46.77 per cent) was recorded on addition of 35 per cent soymilk and maximum (55.47 per cent) without addition of soymilk in buffalo milk. The difference in the values was highly significant (P/0.01). The individual performance of high temperature in all the groups was significantly (P/0.05) higher when paneer was prepared with citric acid than lactic acid. Grower *et al.*

(1989) found that addition of cow milk to the extent of 10-30 per cent to soy slurry increased the yield of paneer with higher TS and lower protein content.

Flavour score (according to hedonic scale) of paneer significantly (P/0.05) decreased with increasing level of soymilk (Table 6.3) showing the highest value in group I (3.96) and lowest in group IV (3.47). It was further noted that flavour significantly (P/0.05) increased with increased with levels of temperature. The improvement of flavour at 90° C could be attributed to the heat induced change occurring in milk, resulting in number of flavourful compound, particularly the sulphur compound (-SH group and hydrogen sulphide) which are responsible for cooked pleasing flavour. In this case flavour score is significantly (P/0.05). Higher when lactic acid was used.

Table 6.3: Effect of Different Admixture of Buffalo and Soymilk on Organoleptic Scores of Paneer

Group	*Temperature of Coagulation*	*Colour and Appearance*		*Flavour*		*Body and Texture*	
		Tartaric	*Lactic*	*Tartaric*	*Lactic*	*Tartaric*	*Lactic*
I	70	1.5	1.5	3.70	3.75	3.72	3.70
	80	1.49	1.5	3.85	3.95	3.80	3.75
	90	1.5	1.5	3.95	4.00	3.72	3.70
II	70	1.45	1.45	3.71	3.73	3.69	3.68
	80	1.43	1.42	3.72	3.74	3.71	3.70
	90	1.42	1.45	3.73	3.75	3.69	3.69
III	70	1.21	1.22	3.55	3.60	3.55	3.53
	80	1.21	1.22	3.60	3.65	3.61	3.60
	90	1.23	1.22	3.75	3.75	3.55	3.55
IV	70	1.12	1.11	3.00	3.00	3.50	3.49
	80	1.11	1.12	3.40	3.40	3.51	3.50
	90	1.13	1.14	3.44	3.48	3.50	3.49

Different letters denotes significance at 5 per cent level.

* Significant at 1 per cent level.

The overall interaction effects of temperature and coagulants on body and texture were significantly (P/0.05) higher in group I (3.73 + 0.003) and decreased significantly score was also higher (P/0.05) in paneer prepared with citric acid than lactic acid.

The score of colour and appearance was significantly (P/0.05) higher in group I (1.49+ 0.005) than other groups. Irrespective of temperature and coagulant, the score was significantly higher (P/0.05) in group II (1.431+0.003) than the group III and IV. The overall sensory scores was the highest in paneer prepared with lactic acid and decreased significantly (P/0.05) on decreased levels of blended soymilk in pure milk. Singh and Ray (1997), Nasim *et al.* (1986) and Chakraborty and Gangopadhyay (1990) have also observed the best sensory score with lactic acid than citric acid coagulants. Irrespective of temperature and coagulants, the total score was significantly (P/0.05 higher in group I (9.08) than the other group. The total score significantly (P/0.05) decreased with increasing levels of soymilk in the milk. Similar result was also reported by Babje and Rathi (1989)

References

A.O.A.C. (1985). Associatation of official agricultural chemist, 14 the Edi, Washington.

Babje, J.S. and Rathi, S.D. (1989). Effect of blending soymilk with buffalo milk on quality of paneer. J.Food, Sci. and Tech. (Mysore: 1992), 29 (2)119-120

Bhattacharya, D.S. (1971). Studies on the method on production and selflife of paneer, J. Food Sci. and Tech. 8 (5)225.228.

Chakraborty, S.R. and Gangopadhyay, S.R. (1990). Innovation of Technology for preparation of Rassogolla analogue for Soymilk, J. Food Sci. and Technology 27 (4): 242-243.

Grower, V., Singh S. and B.A. (1983). Studies on extraction shell life of soybeen curd. J. Food Sci. and Tech. 20 (6): 298.

Grower, M., Tyagi, S.M. and Bajwa, V. (1989). Studies on soy paneer J. Food Sci. and Tech. 26 (4): 194-197.

Nasim, M., Mittal, B.K. and Tyagi, S.M. (1986). Development of process for preparation of soy paneer. J. Food Sci. and Tech.21: 286-290.

Pal, M.A. and Yadav, P.L. (1991). Physico-chemical and sensory Characteristics of low fat paneer form hogh heated milk Indian Dairy Sci. 44 (7:437).

Panse, V.G. and Sukhatme, P.V. (1978). Statistical method for Agricultural worker, 3rd Ed, I.C.A.R, New Delhi

Prakash, M.S. and Tya, K.K. (1986). Preparation of Paneer form blended soymilk and mixed milk, Indian Dairy Man. 3 (6): 433-436.

Singh, S. and Kannawjia, S.K. (1988). Development of manufacturing technique for Paneer form cow milk. J. Dairy Sci. 41 (3): 322.

Chapter 7

Angora Rabbit Farming: A Suitable Venture for High Altitude Areas of Himalayan

Sunil Kumar[1] *and Pushpanjali*[2]

[1]*Assistant Professor (Animal Science),*
College of Forestry and Hill Agriculture,
G.B. Pant University of Agriculture & Technology, Hill Campus,
Ranichauri, Tehri Garhwal – 249 199, Uttarakhand
[2]*Guest Faculty, SRT Campus, Badshaithaul,*
HNB Garhwal Central University, Srinagar, Uttarakhand

ABSTRACT

Livestock production is a major component of farm economies in developing countries, contributing not only milk and meat but also hides, fur, fibers, fertilizer, fuel and fertilizer. In addition, livestock, whether large or small, are part of the social and cultural reality of several million small farmers, for whom husbandry represents an element of economic stability and sustainability. Both human and livestock populations have grown considerably in the last 30 years, but the rates in developed and developing countries are not comparable. Whereas the global human population has risen by 75 per cent since 1960, in the developing countries the rate of increase was 97 per cent and in the industrialized countries 28 per cent. All species of livestock populations increased, but monogastrics (pigs and poultry) much more than ruminants. Till the year 2010, the world population is 7.2 billion, moving past nine billion by 2025. This increase will be felt mainly in the developing countries, where the corollary will be sizeable growth of the peripheries of urban conurbations, increased pressure on available land and major changes in the composition of animal populations. There will also be substantial impact on available natural resources and on the future demand for livestock products. This will have a profound effect on the choice of feed resources and livestock systems. Angora farming can play a very vital role in various socio-economic developments by providing employment and supplementary income to all the sections of people of hilly areas. Angora rabbits are reared for different purposes such as meat, wool and experimental animal for medical research. In Himachal Pradesh, Garhwal Kumaon hills of Uttarakhand and Nilgiri hills of Tamil Nadu in recent years a large numbers of farm families have diversified their agriculture by introducing Angora rabbits for wool production. Angora rabbit is a viable substitute of sheep due to more income, labour saving and eco friendly nature in hilly areas. Angora farming can be very profitable and rewarding venture provides appropriate management practices are followed.

Keywords: Angora rabbits, Management, Wool, Wool products.

The angora rabbits are a variety of domestic rabbit bred for its long, soft hair. The Angora is one of the oldest types of domestic rabbit, originating in Ankara, Turkey, along with the Angora cat and Angora goat. The rabbits were popular pets with French royalty in the mid 1700s, and spread to other parts of Europe by the end of the century. They first appeared in the United States in the early 1900s. They are bred largely for their long Angora wool, which may be removed by shearing, combing, or plucking (gently pulling loose wool). Angoras are bred mainly for their wool because it is silky and soft. They have a humorous appearance, as they oddly resemble a fur ball. Most are calm and docile but should be handled carefully.

Best known for being prolific, rabbits are also herbivores which efficiently convert fodder to food. The whole point of meat production is to convert plant proteins of little or no use to people as food into high-value animal protein. In efficient production systems, rabbits can turn 20 per cent of the proteins they eat into edible meat. Comparable figures for other species are 22 to 23 per cent for broiler chickens, 16 to 18 per cent for pigs and 8 to 12 per cent for beef. Rabbits can also easily convert the available proteins in cellulose-rich plants, whereas it is not economical to feed these to chickens and turkeys - the only animals with higher energy and protein efficiency. The traditional grain and soy cakes fed to these domestic poultry put them in direct competition with humans for food. For countries with no cereal surpluses, rabbit meat production is thus especially interesting.

Angora rabbit farming can act as a supplementary source of employment and income for subsistent, agricultural labours, educated unemployed youth related civil and army personnel and school going children of hilly areas. It needs small land than needed for other livestock enterprises. In the hilly areas where most of the farmers are marginal and small can choose Angora rabbit farming as alternative method for generating good income with their limited land and water use. Rabbit manure being rich in Nitrogen, Potash and Phosphorus are much better than the manure obtained from other animals and highly suitable for vegetable and fruit crop. Angora hair is about 7 times warmer than Marino sheep's wool and is usually blended with other fiber for a fabric that allows the skin to breath without becoming clammy.

Angora rabbit lacks functionally sweat glands, which cause thermo-regulation problem during hot and humid climates. The optimum temperature humidity and altitude range for successful rearing of Angora range between 15° to 20°C, 60 to 80 per cent and 1500-2000 meter msl, respectively.

Table 7.1: Commercial Angora Rabbit Breeds

Breed	*Size*	*Use*	*Mature Weight (kg)*
German Angora	Medium	Wool	3.50-4.00
Russian Angora	Medium	Wool	3.50-4.00
British Angora	Medium	Wool	3.00-3.50
Hybrid Angora	Medium	Wool	3.00-3.50
French Angora	Medium	Wool	3.00-4.50
Swish Fox	Small	Wool	2.00-2.50
Netherlands Dwarf	Small	Wool	0.9-1.25
English Angora	Small	Wool	2.00-3.00

Table 7.2: Performance of Various Angora Breed and its Crosses at Ranichauri, Tehri Garhwal Conditions

Characters	*GA*	*RA*	*BA*	*GRA*	*GBA*	*GBR*
Litter Size at birth	5.4	6.1	5.3	5.9	5.4	5.0
Litter size at weaning (35days)	4.6	5.4	5.0	5.3	4.0	4.3
Weight at Weaning (gm)	500	550	590	610	615	660
Weight at birth (gm)	230	295	260	270	250	240
Kit survival (per cent)	60-80	70-90	60-90	70-95	70-90	75-90
Annual wool yield (gm)	750	390	390	410	370	415
Staple length (cm)	4.5	6.5	6.0	6.0	6.0	6.1
Medullation (per cent)	2.5	2.8	3.00	2.8	2.6	2.6

GA: German Angora; RA: Russian Angora; BA: British Angora; GRA: German Angora x Russian Angora; GBA: German Angora x British Angora; GBR: German Angora x British Angora x Russian Angora.

The above observations have been recorded at high altitude (1600-1800 meters MSL) during 2002-2004 of the Garhwal region. During the study period the temperature and Relative Humidity were recorded as 20-28 °c and 60-90 per cent respectively.

Housing Management

The rabbitory should be an enclosed building that has proper ventilation, lighting, heating and cooling systems. Heating and ventilation are crucial because rabbits do not tolerate temperature extremes. It is required for maintaining the herd on a year-round schedule of 12 hours each of light and dark to keep the rabbits breeding throughout the year.

Many different type of cages are being used, however, all metal cages help to prevent unsanitary conditions that can lead to health problems. The cages should be made of 1 –by-2 inch mesh for the side and top and ½ - by-1-inch mesh for the floor. Hanging the cages from the ceiling in single layers makes management easier for the producer. Mature buck and doe should have individual cages at least 30x30 inches

Figure 7.1: German Angora

Figure 7.2: British Angora

Figure 7.3: Russian Angora

Figure 7.4: German Angora X Soviet Chinchilla

wide and 20 inches high. Junior doe may be kept in small groups in one pen. Each cage should have a feed hopper and a watering system attached to the out side of the cage.

Nest boxes should provide enough space for each doe and her litter, but should be small enough to keep the litter close together. Nest boxes can be made of non treated wood, wire mesh or sheet metal. The box should be except for a small opening on top for the doe to enter.

Maintaining a sanitary operation will help prevent diseases. The best waste management systems should have porous pits under the cages with layers of gravel and drainage tile. Earth and concrete floor are acceptable, but require more frequent cleaning. House should have concrete walkways between the cages and accumulated manure should be removed at least twice a year. Cages and nest boxes should be cleaned and sanitized after each use and the hair should be burned off the cages. New additions to the herd and sick animals should be kept in separate cages isolated from the rest of the herd.

Breeding

Medium- weight breeds (3.00 to 4.50 kg) are able to start breeding at 6 to 7 months of age, with males maturing one month later than females. Because outward signs of heat are not always evident in mature doe, hence it requires following a strict breeding schedule. One buck can service about 10 does, but no more than two to three a week. Place the female in the buck's cage for breeding. Never bring the buck to the doe's cage, because she will fight to protect her territory. Mating should occur immediately and the doe should then be returned to her cage.

The average gestation period lasts 29 to 32 days. 28 days after breeding, place a nest box in the doe's hutch. The average commercial litter consists of 6 to 9 kits. 48 hours after birth observe and count the kits, and remove dead kits. Remove the nest box 15 to 21 days after birth. The young should be weaned in about 30 days this doe will produce five litters in a year.

Nutritional Requirements

Angora rabbit are fully housed animals and as such need to be fed a balanced diet to produce 650 to 1200 gm per year of wool. This represents approximately 0.30 of the rabbit's body weight, the highest keratin production per kg live weight found in any fiber-producing animal. In sheep, goat or camel, this ratio is generally less than 0.10. Protein utilization (ratio of protein retained in fiber to protein used for fiber growth) in Angora rabbits is estimated to be 0.43, double the efficiency of an estimated 0.20 to 0.25 for Merino sheep to produce.

Not only the Angora has a high rate of wool production, they are also shorn at 3 monthly intervals by scissors of plucking. Wool production rate varies over the 3 month period between wool harvests and thus feed allowances are adjusted to meet both changes in fiber growth and energy requirement for body heat losses incurred immediately after shearing. Complete well- balanced pellet diets have been developed and the feed offered to the rabbit varies according to the time after shearing to maintain live weight and maximum wool production.

The nutrient requirements for the reproducing doe are higher than non-reproducing does. Dietary protein concentrations need to be increased by 10 g/kg and digestible energy content increased to 11.7 MJ/kg. Dietary concentrations of lysine, Methionine plus Cystine and Arginine are also suggested to be increased by 1g/kg of the values shown in table 4. The diet is fed as a pellet on a daily basis with restricted rates of feeding to prevent obesity and nutritional problems with the rabbits.

☆ First month after shearing 1,200 g per animal per week
☆ Second month after shearing 1,100 g per animal per week
☆ Third month after shearing 1,000 g per animal per week

The weekly ration must be distributed equally over six days of the week with a one- day fast. Angora rabbits are manual fed so that the rabbits are individually checked for any animal health problems, while they are being fed. The one-day fast is to prevent the development of hairballs in the stomach (trichobezoard). The one-day fast facilities the rabbits in voiding of ingested hair in hard faces.

The feed requirements for the different stage of growth and reproduction are shown in Table 7.3.

Table 7.3: Live Weight, Weight Gains and Daily Intake for Rabbits during Growth and Reproduction

Physiology Status	*Live Weight (kg)*	*Weight Gain (g/day)*	*Intake (g/day)*
Weaning rabbits 3 months and older	0.5 to 2	25 to 30	60 to 135
Growing rabbits	2.5 to 3.5	10 to 15	150 to 170
Pregnant does	3.5 to 4.5	–	>170
Lactating does	3.5 to 4.5	–	>220

Rabbit require fresh, clean, water every day. Automatic watering systems offer a continuous water supply, while reducing waste and contamination. A doe and her litter need 5 litter of water a day in warm weather.

Studies on the quantities needed have been virtually confined to arginine, lysine and the sulphur amino acids (methionine and cystine). Expressed as a percentage of the ration, the lysine requirements for growing rabbits are 0.6 and, for sulphur amino acids, 0.7 per cent. The lysine intake of breeding does should be considerably higher under intensive milk production to feed nine to 12 young. The arginine intake should be at least 0.8 per cent and a little more for growing rabbits. The toxicity thresholds of lysine and arginine are well above the recommended intake levels. For the sulphur amino acids, however, there is a slender margin between the amount the rabbit needs and an excess dose that would diminish its performance.

The recommended amounts of other essential amino acids have been estimated simply on the basis of regular satisfactory diets. Where these essential amino acids are supplied by protein in the diet, 15 to 16 per cent crude proteins should be enough for fattening rabbits. Rabbits will always eat more of a balanced feed containing essential amino acids than the same feed without amino acids. Amino acid balance can easily be achieved with plant protein alone as in almost all balanced European feeds. Proteins of animal origin can be used by rabbits but are absolutely unnecessary: all that counts is the amino acid intake, not the substratum.

Table 7.4: Recommended Chemical Composition of Feeds for Intensively Reared Rabbits of Different Categories

Components of Feed, Assumed to Contain 89 per cent Dry Matter	*Young Rabbit (4 to 12 weeks)*	*Lactating Doe*	*Peri-weaning*	*Mixed (Maternity + Fattening)*
Crude proteins (*per cent*)	16	18	15	17
Digestible proteins (*per cent*)	11.5	13.3	10.8	12.4
Amino acids				
Methionine + cystine (*per cent*)	0.60	0.60	0.55	0.60
Lysine (*per cent*)	0.70	0.90	0.65	0.70
Arginine (*per cent*)	0.90	0.80	0.80	0.90
Threonine (*per cent*)	0.55	0.70	0.55	0.60
Tryptophane (*per cent*)	0.13	0.20	0.12	0.13
Histidine (*5*)	0.35	0.43	0.35	0.40
Isoleucine (*per cent*)	0.60	0.70	0.67	0.65
Phenylalanine + tyrosine (*per cent*)	1.20	1.40	1.10	1.25
Valine (*per cent*)	0.70	0.85	0.68	0.80
Leucine (*per cent*)	1.05	1.25	1.00	1.20
Energy and bulk				
Digestible energy (*kcal/kg*)	2500	2650	2400	2550
Metabolizable energy (*kcal/kg*)	2380	2520	2280	2420
Fats (*per cent*)	3-5	4-5	3	3-4
Crude fiber (*per cent*)				
Indigestible crude fibre (*per cent*)	12	10	14	12
ADF (*per cent*)	18	14	20	18
Ratio digestible proteins/digestible energy (*g/1000 kcal*)	45	51	46	48
Minerals				
Calcium (*per cent*)	0.40	1.20	1.00	1.10
Phosphorus (*per cent*)	0.30	0.50	0.50	0.60
Potassium (*per cent*)	0.60	0.90	0.60	0.90
Sodium (*per cent*)	0.30	0.30	0.30	0.30
Chlorine (*per cent*)	0.30	0.30	0.30	0.30
Magnesium (*per cent*)	0.25	0.25	0.25	0.25
Vitamins				
Vitamin A (*IU/kg*)	6000	10000	10000	10000
Vitamin D (*IU/kg*)	1000	1000	1000	1000
Vitamin E (*ppm*)	50	50	50	50
Vitamin K (*ppm*)	0	2	2	2

Contd...

Table 7.4–Contd...

Components of Feed, Assumed to Contain 89 per cent Dry Matter	*Young Rabbit (4 to 12 weeks)*	*Lactating Doe*	*Peri-weaning*	*Mixed (Maternity + Fattening)*
Vitamin C (*ppm*)	0	0	0	0
Vitamin B_1 (*ppm*)	2	–	2	2
Vitamin B_2 (*ppm*)	6	–	6	4
Vitamin B_6 (*ppm*)	2	–	2	2
Vitamin B_{12} (*ppm*)	0.01	0	0.01	0.01
Folic acid (*ppm*)	5	–	5	5
Pantothenic acid (*ppm*)	20	–	20	20
Niacin (*ppm*)	50	–	50	50
Biotin (*ppm*)	0.2	–	0.2	0.2

Source*:* Lebas, 1989.

Shearing, Grading and Fleece Handling

Angora rabbits are first shorn at 40-50 days of age and then at approximately 82-90 days intervals. Angora rabbits are sheared using either clippers or scissors. In cold weather a strip can be left along the back line until the rest of the body re-grows wool and then the backline wool is removed. Alternatively a box is provided in the cages to protect the rabbit from cold until there is sufficient hair growth to insulate the rabbit. Shearing can also be timed to reduce heat stress on the animal during the peak of the hot summer. Shearing time is estimated to be between 15 and 30 minutes per rabbit depending upon the skills of the shearer.

The recognized grades of Angora wool in India are given as under:

- ☆ *Grade I or "A"* Pure white, completely clean, unmated, free from any vegetable matter and fiber length 6 cm and above.
- ☆ *Grade II or "B"* Pure white, free from any foreign matter, non- felted and length between 3 and 6 cm.
- ☆ *Grade III or "C"* White, clean, untangled, free from vegetable matter and; length less than 3 cm.

On a well run and organized Angora farm about 70-80 per cent wool should be of "A" grade, 20-25 per cent, "B" and "C" grade and rest 3-5 per cent as dirty and matted wool. The fleece is packed into plastic bags to prevent felting during transport.

Health Care Management

The most important factors for maintaining a healthy rabbit herd are cleanliness, good ventilation, close observation and rain. Rabbits are susceptible to several diseases that reduce production to unprofitable levels. The respiratory disease caused by *Pasturella multicida* is responsible for decreased productivity and a high mortality rate in does. *Pasturella* – free animals should be purchased from authentic sources.

To help in prevent disease problems, do not permit visitors inside the rabbitry. They may introduce diseases and cause additional stress to the animals. Isolate any sick or injured rabbit immediately. Disinfect the isolation cage and the rabbit's regular cage to avoid spreading the disease. For a good health program, keep accurate health record of each animal. Provide each rabbit with a tattoo identification card with health and breeding information to its hutch.

Management practices such as good sanitation, hygienic conditions and proper feeding and housing management are very important of Angora rabbit farming. Under proper management conditions rabbit farmers can achieve more profit from the rabbits, so important management should be followed by the rabbit farmers obtaining more profit from the units.

Some important diseases which are very common in rabbit are discuss briefly as under:

Pasteurella

Symptoms

The first signs of the disease are sneezing and discharge from the nose and or eyes. Not every sneeze is indicative of the presents of Pasteurella. A rabbit may sneeze when it gets water up it's nose while drinking, or it may have an allergy to something in the area. The Pasteurella germ can be carried on caretaker's clothing and person. It is of extreme importance that you change your clothes and wash thoroughly before going near any other rabbits. Isolate any sick rabbit immediately and care for the herd first and the isolated animal last. Disinfect it's cage and any other equipment the rabbit came in contact with.

Treatment

Rabbits can be treated with a number of antibiotics but to this point none have been successful in bringing about a cure. The best treatment for Pasteurella is prevention. For now, strict sanitation, good ventilation and culling will go a long way in helping to prevent the spread of this dreaded disease.

Prevention

Ventilation is important in snuffles control since both humidity and ammonia are involved in the spread and growth of this condition. Ammonia is present in rabbit urine, having it build up in the atmosphere has a bad effect on both humans and rabbits. If you can smell it while walking through your rabbitry, think how it is affecting your rabbits. By removing the urine and feces from the rabbitry you are decreasing the amount of ammonia in the surrounding area.

Coccidiosis

Very common disease found in rabbitry. There are two types of Coccidiosis Intestinal and Live Form. The disease is caused by a protozoan parasite which attacks the bile duct or the intestinal tract. There are ten different species of the genus Eimeria which may infect the intestine. Generally these are not of much concern, unless a case of enteritis makes an apperance. This can lead to enterotoxemia and or Mucoid enteropathy, although enterotoxemia has been reported in rabbits which are free of Coccidiosis. Liver Coccidiosis is a problem as it causes white spots on the liver which renders the rabbit uneatable.

Symptoms

Include lack of appetite, rough coat, loss of weight, poor weight gain, potbelly, listlessness and diarrhea. All of which often lead to death.

Treatment

The best treatment is the use of sulfur drugs in the drinking water as the only available water for 5 days on, 5 days off then 5 days on again.

Prevention

Brush the cling-ons from the cage floor within 24 hours. This will aid greatly in preventing the spread of Coccidiosis. Do not allow fecal contamination of the feed or water.

Ear Mites

Are caused by the parasite Psoroptes Cuniculi which is a mite that likes to make it's home in the rabbit's ear canal. They irritate the rabbit to the point that the scratching can lead to infection.

Symptoms

Shaking of the head, scratching the ears, and a brown waxy crusty substance will appear after awhile if not attended to.

Treatment

Place a few drops of any type of oil in the ears and massage the base of the ear gently to work the oil in. The oil will drown the mites. Treat one to three times a day for three days. Repeat in ten days to get hatchlings..

Fur-Mites

Will look like flakes of dandruff. It is caused by: Chyletiella Parasitivorax and Listrophorus Gibbus. They occur most commomly on the back of the neck. An over-population may cause Dermatitis.

Symptoms

An unthrifty coat, loss of fur around the face, neck and back.

Treatment

A good Cat flea powder, Listerine mouthwash rubbed into the infected area will make short order of the Mites. You can also use a droplet the size of a small green-pea of Ivermectine/Zimectrine Horse Paste Wormer. Repeat treatment in ten days. Do NOT use treated animal as a meat source for at least 30 days after last treatment.

Sore-Hocks

Sore-Hocks is a genetic problem. The condition is easily noticed by the open wounds on the footpads of the afflicted rabbit. Part of the cause is poorly furred footpads and long often uncut toenails.

Symptoms

Noticed by the open sores on the feet and which can rapidly lead to secondary infections in rabbits. Often the rabbit sits in the corner and for the most part is inactive.

Treatment

You should breed toward well furred footpads. Use of Preparation H will help heal the sores. Place a board or piece of drywall large enough for the rabbit to sit on. Do not allow it to get dirty or it will become a source of infection. Change or clean it often.

Woolblock or Hairball

It is caused by the rabbit when they ingesting too much fur during the cleaning process.

Symptoms

Rabbit fails to eat and becomes listless. The stomach becomes taut, firm to the touch as the condition worsens. A quick response is essential if you are to save the rabbit.

Treatment

Regular grooming of normal furred rabbit and shearing of the wool on all fiber animals will go far in helping to prevent this condition. You can use Cat hairball remedies. You can also use the enzyme papain or bromelain which is found in Pineapple and Papaya. Use the fresh fruit as the caning process kills the enzyme. You can also use Adolph's meat tenderizer [same enzyme] mixed with enough water to feed with a needless syringe. Give one to three times a day until rabbit returns to normal.

Prevention

You should provide grass hay to the rabbit's diet on a regular basis at least two to three times in a week. The long fiber keeps the hindgut working properly. This is a sure sign that the hay is doing it's job. Brush the cling-ons from the wire within twenty-four hours to aid in the prevention of coccidiosis.

Pneumonia

We observe a bluish color to the lips, tongue and ears due to lack of oxygen.

Treatment

There is little chance of survival for the rabbit stricken with pneumonia. Many broad spectrum drugs have been tried to no avail. Prevent the possibility of heat stress. The use of fans to create air movement through the rabbitory helps. Try to house your rabbits in an area where they get some shade in the warmer months, or create a shaded area to protect from the suns heat creating rays. Treat your rabbits immediately at the first signs of heat stress. Lungs the into air more get to effort an in back thrown be will heard breathing, difficulty exhibit rabbit The [Pasteurella}. Snuffles of stage final is it often Most situation. Stressful type any by weakened system defence when encountered usually in the system.

Heat Stress

Stress can greatly affect your rabbit's health. It is something you should always be aware of. Conditions which can cause stress are; a long day away from familiar surroundings, radical change in temperature, fright and loud noises and the birthing process. If you know your rabbit will be going through one of these ordeals or one has just occurred unexpectedly you can use one of these remedies: You can mix one part Gatorade to four parts water and give as drinking water. This will add electrolytes or you can add one or two teaspoons of Tang to a gallon of drinking water.

Heat stress occurs during extremes of hot weather especially if there is not enough air circulation through the rabbitry. If you notice your rabbit getting wet around the face and or breathing hard then dip the rabbit in a bucket of tepid water. In emergency extreme measures must be taken if there is to be any chance at all of saving the rabbit. Don't be afraid to spray cool water on a rabbit or in an emergency dip the rabbit into a bucket of tepid water not ice water. Keeping the rabbits nose and mouth above water. During warmer months you can freeze soda bottles or milk jugs half full of water. Place these in the cage with your rabbit[s] when temperatures exceed 80 to 85degrees. They will lie against them to cool themselves.

Diarrhea

Take immediate action at the first sign of diarrhea. This could be a life and death situation for any rabbit especially the very young. STOP ALL treats as they are the most likely cause of the problem. Feed plenty of long-fiber grass-hay and oats will work well. You may even want to stop the pelleted feed when the stools return to normal and slowly add the pelleted feed back to the diet while continuing to give grass hay on a regular basis. Two to three times a week if not more often.

Pregnancy Toxemia

Pregnancy Toxemia is a general problem with rabbits and some are more prone to it than others. Some of the more common reasons are overweight Does, hairball blockage and stress, brought on by high heat and humidity.The symptoms most often appear a few days before the Doe kindles or right after. The problem is associated with a combination of high nutritional demandand conditions that limit intake. The Dam can not meet it's energy demands as she is unable to matabolize properly. As a result of this inability to matabolize the fat reserves accumulate in the liver and cause the problem. If not treated the rabbit goes into a coma and will die.

Because of the inadaquate energy supply, the blood glucose level is depleted. A sub-Q injection of Glucose or Dextrose is an effective treatment. Give a small breed two 3cc sub-q injections under the skin at the neck area between and behind the ears., [one in the AM and one in the PM] It will take three or four injections for the larger breeds. Sub-Q injections may actually pull the skin away from the injection site. You must treat and they heal very nicely. As a follow up, feed with a curved tip feeding syringe either Karo syrup or molasses. 1cc every four hours until they start eating pelleted feed again. While they remain off pellets, mix 1-2 teaspoons of baby oatmeal or high protein cereal, 1cc syrup, 1tblsp nutrical and feed with curved tip syringe.

Some tips for keeping your rabbits healthy

Build a rabbit cage which can be easily cleaned and will keep your rabbit safe from the elements of weather. Build a roomy and comfortable hutch or cage. Your rabbit must remain dry and free of drafts. Place the hutch in an area where it will avoid dampness and excessive heat.

1. Keep all your cages and equipment clean. This will help you to avoid most infectious diseases and injuries.
2. Learn to recognize rabbit diseases and treat at the first sign of illness should it become necessary.
3. Isolate all new arrivals for two to four weeks. Water and feed them after the rest of the herd has been cared for.
4. Beware of lending or borrowing rabbits for breeding. This is a good way to bring disease to your herd.

Production of Hair for Textiles

Angora rabbits, however, are produced solely for the hair. The only way the producer can be sure of quality hair is to apply a very specific methodology quite different from that used in meat-rabbit production. Angora, the hair of Angora rabbits, is one of the five keratin textile fibers of animal origin of significant economic value. Wool from sheep is of course by far the main fiber, at over 1.3 million tones per year (thoroughly washed). The four others: mohair, angora, cashmere and alpaca, each at outputs of 5 000 to 30 000 tones, exhibit original qualities of fineness, lustre and feel for the production of high value added luxury items. Angora is often considered one of the "noble" fibers.

Angora Wool and its Uses

Angora wool differs from sheep wool in several respects. Angora is a modulated fiber, which makes it light and soft. Angora is a hetero type fleece and contains some coarse hair called bristle. The bristle content varies with breed and is usually considered desirable, as they prevent felting on the rabbit and imparts the characteristic fluffiness to yarn. The down fiber is very fine, with an average fiber diameter of about 11 to 13 microns; however, most of the commercial breeds have some guard hairs that increase the average fiber diameter of the fleece. Angora fiber has a low density of about 1.15 to 1.18 gm/cm^3 compared to 1.33 gm/cm^3 for wool and 1.50 gm/cm^3 for cotton. This gives the Angora garments a feeling of being very light but warm. Angora wool can be used in several ways such as : Knitted clothes usually with a moderate fluffing effect, such as pullovers, woolen hats, socks, gloves, scarves, sweaters, kid clothes, jacket sport garments and coat, etc. Knitted product is the common fabric type produced, as the bristle content of most Angora fleeces makes it unsuitable for the production of next to skin wear or high quality suiting material.

Table 7.5: Wool Quality Attributes of Angora Rabbits

Angora Wool Type	*Average Fiber Length (cm)*	*Range (cm)*	*Average Fiber Diameter (μ)*	*Range (μ)*	*Average Guard Hair Per cent*	*Range*
German	6.50	3.53–8.15	11.50	9.45–13.00	2.15	1.56–6.57
Russian	5.76	3.66–5.79	16.28	13.27–19.42	20.85	13.35–28.00
British	7.30	4.73–8.23	12.66	11.36–13.97	3.92	2.00–7.66
British × Russian	7.10	4.92–9.20	13.59	12.10–16.32	5.25	1.66–9.00

Textile Properties of Angora Wool

In the matter of textiles, "angora" without any other qualification refers solely to the hair produced by Angora rabbits. Its International Organization for Standardization (ISO) symbol is WA: W for wool, reserved for noble textile hair, as opposed to H used for ordinary hair. The letter A is for the Angora rabbit and distinguishes it from the mohair produced by the Angora goat, M. The symbol for mohair is thus WM. The short hair of the ordinary rabbit is designated HK (K = Kaninchen which is "rabbit" in German).

Length

Angora hair is unusually long owing to the prolongation of the active phase of the hair follicle cycle: the hair grows for approximately 14 weeks, whereas that of the rabbit with ordinary (short) hair grows at the same rate but for only five weeks. This is due to the presence of a recessive gene in Angora rabbits.

Apart from this great length, there is no other modification either in the hair's structure or in the composition of the coat, which contains the three classic types of rabbit hair.

Guide hairs: the longest (10 to 11 cm) and the roughest; they cover and guide the coat;

Guard hairs ("barbes"): shorter than guide hairs (8 cm); their rough points lie on the coat and hermetically seal it (covering hair); four to each guide hair;

Down: shortest hair (6 cm); rounded point, hardly visible, very fine body (14 m). Very numerous, 60 to a guide hair, they constitute the thermic isolation undercoat. The length of angora hair accounts for its textile value, because it permits cohesion in the thread.

Friction Coefficient

The rabbit's hair has a characteristically low friction coefficient owing to the very slight relief of the cuticle scales. This results in a particular softness to the touch, but also an exceptional capacity for slipping. This is why the length of angora is important; the hair is twisted and stays in the thread. The use of ordinary rabbit hair to replace angora produces threads of bad quality which spread everywhere: this is a fraudulent process which reflects badly on the Angora industry.

Because of its softness angora hair is used for the manufacture of insulating underclothes (keratin). Ten per cent angora in a mixture of wool, cotton and synthetic fibres makes an extremely soft fabric, very easy on the skin.

The kemp points and the covering hairs, which are more rigid, rise from the fabric, giving it a fluffy appearance which is much prized. Whole angora hairs obtained by depilation are the most suited for this purpose.

Other Characteristics of Angora Wool

Although the Angora rabbit exists in all colors, only the albino strain is produced now. Its coat is entirely white, which is an advantage for dyeing. Colored Angora rabbits are raised in India for the manufacture (by breeders themselves) of undyed artisanal fabric with muted color motifs. The hairs are all modulated (hollow), which makes them lighter than wool (density 1.1 against 1.3) and increases their insulating properties. They have all the properties of keratin, notably insulation, water absorption and good dyeing quality.

The Angora rabbit's coat is 98.5 per cent pure as cutaneous secretions (restricted to those of the sebaceous glands) are very slight and the animal grooms itself frequently (a sheep's fleece is only 50 per cent pure because of the presence of suint). Angora wool goes straight to the card without previous washing: it is imperative that the producer keep constant control over the cleanliness of the animals.

Commercial Qualities

There are several grades of hair, identified by length, type of animal and cleanliness. First-quality hair which represents 70 per cent of the coat must be over 6 cm in length (down) and clean. This grade was worth 950 FF a kg in 1984, but only 300 FF in 1981 to 1982. Since 1988, the price has ranged from 300 to 380 FF.

Second-quality hair is clean but too short (down less than 6 cm) or too woolly. It is grown on the belly and extremities and is worth about 20 per cent less than the first-quality wool. The hair of the young Angora rabbit is shorter and softer. It is the product of the first and sometimes the second collection. The clean but felted hairs collected on the necks of females or breeding animals are worth only 15 per cent of the value of first-quality hair.

Dirty hair of any length is virtually worthless. At best, it is worth less than shorn hair from ordinary rabbit breeds. Its value would be no more than 5 or 6 per cent of the first quality. Clean hair is therefore absolutely essential in angora wool production.

Prospects for Angora Wool Production

A point to be considered very carefully is that Angora rabbit production is labor-intensive and also requires great expertise. The slightest mistake can mean the loss of productive adults: the animals have to be over a year old to return a profit. Hair collection is always a delicate operation and careless sorting irredeemably downgrades the product. Above all, not all climates are suitable: excessive heat and intense light (albinos) are very bad elements. In cold countries, or in countries with cold winters, the solution is to use buildings that shelter the animals against the rigours of the winter. Recently denuded animals require special care, however. The feed requirements of Angora rabbits are important: a poor, deficient diet will always mean qualitatively and quantitatively poor hair production.

Last and probably most important, the price of angora wool fluctuates: first, according to fashion, with a cycle of three to five years, but also and more abruptly, in classical supply and demand terms, when world production is structurally either excessive or insufficient compared with average utilization of the fibre. The price of angora (sheared wool) suddenly doubled between 1976 and 1978 (from US$13 to $28 per kilogram) because world production, estimated at 900 tonnes in 1977, was clearly insufficient. The price remained at this high level for about ten years, following the dollar; up to US$45 to $50/kg, and then in 1988, when world production had increased by a factor of ten to 9 000 tonnes, the market collapsed and the price had fallen to US$20/kg by the summer of 1991. There was a recent reversal of this slump in Chile, Argentina, Hungary and France (and China to a lesser extent), bringing the price up to US$30 in 1992. The volumes traded, and hence angora utilization, continue to rise: the production figure is likely to reach 10 000 tonnes per year again.

As for France, the only developed country to have maintained an angora output of original quality (the guard hair), the situation is one of unprecedented crisis. Production costs no longer permit the sale of French angora wool at less than US$75/kg and the gap between that and the world price appears immense to foreign buyers (the difference between world prices and the price for French angora conventionally being 40 to 50 per cent). Quality French angora hair has remained virtually unexported since 1988, therefore, and is very difficult to market internally, either in the unprocessed form for manufacture or in manufactured form (*e.g.* sweaters).

Clearly this is a highly speculative production and should be approached with great caution. The utilization of the noble textile fiber, angora, continues to grow despite competition from other natural fibers and particularly from synthetic fibers. This is partly due to the new sectors that have opened, particularly for fabric, in combination with cashmere and silk. The price slump from 1987 to 1991 did indeed follow ten very favorable years, after decades of good angora prices. Better times could return again.

The Latest Fashion Using Angora rabbit Products

When it comes to rabbit products, there are the ordinary, like meat, and then there are the extraordinary, like the ultra luxurious rabbit products made from angora rabbit wool. In this article, we look at what type of products that are made from angora rabbits are making big news.

Knitted Garments

One of the most common, and popular uses for angora rabbit fur is in knitted garments. Here, the rabbits are shaved or plucked of their fur, which is then spun into yarn, either by hand or by machine, and then knitted into sweaters, scarves, caps and more.

The most exclusive (and expensive) of these garments are those that have been entirely hand made, while there are cheaper options that are made from machine processed angora, sometimes of a lower grade.

Felted Products

Another way that angora rabbit fur is used to make rabbit products is to felt it. This process uses water, and the fur is wet and then pressed to make a soft and simple fabric, that can be used to make all kinds of garments. This is a particularly useful technique to use on wool that has been poorly cared for while on the rabbit, or that is shorter and of lower quality, as it can be done with almost any wool off the rabbit.

Woven Products

Another way that angora is used in the manufacture of clothing and textiles is as woven products. Here again, the wool is plucked or shorn, and then spun into yarn. The yarn is then woven into fabric, which can be used to manufacture coats, suits and other items of clothing.

Blends

The cheapest type of angora wool products is those that are blends. Here, angora is mixed with other fibers (either synthetic or natural) before being processed as mentioned above. Blends of angora may be knitted, felted or woven, and can be used to create all kinds of garments, from thermal underwear to soft sweaters.

Commercial Appeal of Angora

When it comes to the commercial side of rabbit farming, breeding and raising angora rabbits has some very specific appeal. Instead of the relatively low maintenance meat rabbit, angoras require specific and labor intensive care. However, the wool that they produce can sell for as much as $10 to $16 per ounce, depending on quality.

As far as rabbit products go, angora is prized as a luxury item, and it is exclusive and expensive. Designer goods manufactured from angora can sell for hundreds of dollars in stores, and the cost of the finished rabbit products offsets the increased labor costs of production.

Garments made of 100 per cent angora wool are considered too warm and the fibers are too fine to provide density. It is therefore usually blended with other fibers, such as wool (*right*) to improve its processing performance, elasticity and wear ability. French angora products usually contain up to 20 per cent sheep's wool.

Angora rabbit fiber is a specialty animal fiber. It is softer, finer and warmer than sheep's wool. Angora rabbit fiber is processed generally in long staple system, but spinning in short staple system becomes more interesting day by day. Because of low fiber cohesion the spinning of 100 per cent Angora rabbit fibers is difficult. So angora fiber is commonly used in blends with other fibers.

The aim of this work is to investigate possibilities of angora rabbit fiber utilization in short staple spinning system, to determine optimum spinning conditions for Angora rabbit fiber (with suitable staple length for short staple spinning) and cotton blended yarns and to choose suitable spinning oil with optimum concentration. For this purpose, angora/cotton blended yarns with different blend ratios and 100 per cent cotton yarns produced in OE rotor and ring spinning systems. Yarn evenness, yarn strength and yarn hairiness are measured. The evaluation of results indicates that blend ratio and spinning method have significant affect on yarn properties

Constraints

The following constraints faced by rabbit owners. The identified constraints were lack of high yielding Germplasm, lack of balanced feed and its high cost, non availability of medicine/vaccine, lack of veterinary health care facilities, sudden out break of disease, high kits/adult mortality, government policy and credit facilities etc. The critical issue in Angora wool production is the cost of feeding which has to be reduced so that rabbit farming can pick up in right earnest. There is need to vitalize the Angora rabbit farming and making it a profitable enterprise and protect the interest of producers against price fluctuation

The future prospects of Angora rabbit farming for socio-economic enhancement of small and marginal farmers of hilly areas. In hilly areas there are many possibilities on the part of resourceful farmers to develop over 18-20 subsidiary industries based on Angora rabbit farming like wool processing, spinning, dying, weaning finishing, cage fabrication, feed formulation, skin processing, garment preparing, wool collection and storage etc. Introduction of high yielding good quality wool breed and providing basic health care facilities can bring a significant improvement in sustainable Angora Rabbit farming in rural areas of Uttarakhand.

References

Angora Rabbits http://en.wikipedia.org/wiki/Angora rabbit.

Annual Report 2010-11, Ministry of Textiles, Government of India.

Lebas, F., Coudert, P., de Rochambeau, H. and Thébault R.G. (1997). The Rabbit - Husbandry, Health and Production, FAO Animal Production and Health Series No. 21, FAO, Rome.

King, J.O.L. (1974). The effects of pelleting rations with or without an antibiotic on the growth rate of rabbits. *Vet. Rec.*, 94: 586-588.

Lebas, F. (1969). *Où en sont les essais d'alimentation des lapins aux granulés de luzerne déshydratée.* Technical data from the Bureau de la nutrition animale et de l'élevage. Marseille. Doc. IT 436 D.

Vision 2025. 2007. Central Institute for Research on Cotton Technology (Indian Council of Agricultural Research) Adenwala Road, Matunga, Mumbai-400 019.

Vision 2030. 2011. Central Institute for Research on Cotton Technology (Indian Council of Agricultural Research) Adenwala Road, Matunga, Mumbai-400 019.

Chapter 8
Economics of Ginger Cultivation in Dimapur District of Nagaland

L.T. Sangtam[1], D.C. Kalita[2] and R. Nakhro[1]

[1]Department of Agricultural Economics, Nagaland University, SASRD, Medziphema Campus, Nagaland
[2]Reader, Department of RDAP, NEHU, Tura Campus

ABSTRACT

The present study was conducted in Dimapur district covering two blocks, Niuland and Medziphema. The study was comprised of 60 sample farmers with ginger cultivation. Multi stage random sampling was adopted for the selection of sample farmers. The randomly selected farmers were stratified into 3 size group *viz.*, Group I (2 – 4.00 ha), Group II (4.01 – 5.00 ha) and Group III (5.01 – 6.00 ha) respectively based on the area under land holding by using cumulative root frequency rule. Cost of cultivation was estimated based on various cost concepts like Cost-A, Cost-B and Cost-C along with per quintal net return, per hectare gross and net income following tabular analysis. The results revealed that the per hectare total cost of ginger production in the average farm situation was estimated to be Rs. 92794.03. The share of labour was highest (19.27 per cent) in the total variable cost. The average gross income per hectare from ginger production was Rs. 55261.98. Ginger cultivation was profitable in the area and more profitable particularly for larger size groups in the area.

Keywords: Ginger, Economics, Fixed cost, Variable cost.

Introduction

Ginger (*Zingiberi Officinale* Rosc.) is an important and one of the principal spice crop all over India and the world. Ginger is mainly used for seasoning food to give an agreeable flavour and aroma. It is also used for the manufacture of ginger oil, oleoresin, essence and tincture. Fresh ginger is an ingredient of whole mixed pickling and is used in the preparation of various pickles, soup sauces etc. A number of alcoholic beverages are prepared from ginger such as ginger brandy, ginger wine, ginger beer and gingerale. According to the Ayurvedic medical system, ginger is valued for its manifold medicinal and spicy properties and is useful in gastritis, dyspepsia, flatulence and in cold and cough as an expectorant. It has properties to check cancer and other maladies.

India is the largest producer, consumer and exporter of ginger in the world. India's share in the world production is about 50 per cent. During 2002 – 03, the estimated production of ginger is 307307 tonnes from an area of 85,930 hectares with productivity of 3577 kg/hectares (Anonymous, 2004). There has been an increased in production of ginger over the years. Being the largest producer of ginger, India has capabilities to cater to the requirements of the consuming countries in terms of quantity and quality matching to international standards. There is a tremendous scope to increase the yield per unit area and thereby the total production.

In north eastern region, the area, production and productivity of ginger also increased from 9.7 thousand hectare, 39.3 thousand tonnes, 4052 kilograms per hectare to 33.24 thousand hectare, 191.04 thousand tonnes, 5800 kilograms per hectare, during 1984 – 85 to 2002 (Anonymous, 2002). The compound growth rate of area, production and productivity of ginger during 1984 to 2002 was 11.91, 26.72 and 6.77 per cent respectively, which was statistically significant at 5 per cent probability level. This reveals the growing importance of ginger cultivation in North Eastern Region. In Nagaland, the area and production of ginger are 1130 hectare and 13818 respectively (Anonymous, 2004). The production of ginger in Nagaland increased significantly especially during the last few decades. Hence, an attempt is made in this paper to study the economics of ginger in Dimapur district of Nagaland.

Methodology

The present study was conducted in Dimapur district covering two blocks, Niuland and Medziphema. The study was comprised of 60 sample farmers with ginger cultivation. Multi stage random sampling was adopted for the selection of sample farmers. The randomly selected farmers were stratified into 3 size group *viz.*, Group I (2 – 4.00 ha), Group II (4.01 – 5.00 ha) and Group III (5.01 – 6.00 ha) respectively based on the area under land holding by using cumulative root frequency rule (Gupta, and Kapoor, 2001).

Cost of cultivation was estimated based on various cost concepts like Cost-A, Cost-B and Cost-C along with per quintal net return, per hectare gross and net income following tabular analysis (Sharma *et al., 1989*). The cost and income concepts used in the present study are discussed below:

Cost Concept

Cost A_1: Cost A_1 includes hired human labour, value of seed of ginger, marketing transportation cost and depreciation on equipments and interest on working capital.

Cost B_1: Cost A_1 + interest value on fixed capital asset.

Cost C_1: Cost B_1 + imputed value of family labour

Cost C_2: It includes 10 per cent of the total Cost (C_1) on account of managerial function performed by the ginger farmers.

Variable Cost

Variable cost used in the present study includes the cost of labour, seed value and implements.

Fixed Cost

Fixed cost includes the depreciation cost of land, implements and interest on fixed capital.

The family labour income was calculated by deducting cost B_1 from gross income.

Net Income

Net income was calculated by deducting Cost C_1 from gross income.

Return from Management

It was calculated by deducting Cost C_2 from gross income.

Benefit Cost Ratio

Benefit cost ratio based on variable cost

$$BCR = \frac{\text{Gross income}}{\text{Variable cost}}$$

$$\text{Benefit cost ratio on total cost (BCR)} = \frac{\text{Gross income}}{\text{Total cost}}$$

Results and Discussion

Per Hectare Cost and Return of Ginger Production Across Size Groups

The per hectare cost of ginger production across size groups is shown in Table 8.1. From the Table it is cleared that the total cost of production of ginger was Rs. 92794.03 in the average farm situation which was found to highest in group III farm (Rs. 36364.91). Out of the total cost, the share of variable cost was highest (36.26 per cent) in average as well as in each size groups of farm. Out of the total variable cost, the share of human labour was highest in all the size groups and it was rental value of land in case of fixed cost.

Cost A_1

The average cost A_1 per hectare was estimated at Rs. 26094.85 which was Rs. 28658.38, Rs. 26719.19 and Rs. 22906.99 in group I, II and III respectively. It was observed from the Table that among the various items of Cost A_1 in ginger production, value of hired labour was highest at Rs. 10257.64 per hectare followed by transportation at Rs. 439.90 per hectare. The Cost A_1 per hectare decreased with the increase in the size of the growers. This might be due to more human labour hired by the lower size groups of farmers.

Table 8.1: Item Wise Break Up of per Hectare Cost of Ginger Production Across Various Size Groups (in Rs.)

Sl.No.	*Particulars*	*Group I*	*Group II*	*Group III*	*Average*
(A)	Variable cost				
1.	Human Labour				
	(i) Family labour	8653.25 (6.32)	8666.67 (8.24)	5566.67 (15.31)	7628.86 (8.22)
	(ii) Hired labour	11657.89 (8.52)	10115.02 (9.61)	9000 (24.75)	10257.64 (11.05)
2.	Marketing cost	1965.94 (1.44)	1607.98 (1.53)	1500 (4.12)	1691.31 (1.82)
3.	Transportation	4698.14 (3.43)	4650.23 (4.42)	3833.33 (10.54)	4393.90 (4.74)
4.	Seed	5977.55 (4.37)	6204.22 (5.90)	5233.33 (14.39)	5805.03 (6.26)
5.	Interest on working capital	4283.86 (3.13)	4061.74 (3.86)	3267.33 (8.98)	3870.98 (4.17)
6.	Total variable cost	37236.63 (27.22)	35305.86 (33.56)	28400.66 (78.10)	33647.72 (36.26)
(B)	Fixed cost				
1.	Rental value of land	88050 (64.36)	61775 (58.72)	6975 (19.18)	5266.67 (56.33)
2.	Depreciation on implements	75 (0.05)	80 (0.08)	73 (0.20)	76 (0.08)
3.	Interest on fixed capital excluding land	11456.25 (8.37)	8041.15 (7.64)	916.25 (2.52)	6804.55 (7.33)
4.	Total fixed cost	99581.25 (72.78)	69896.15 (66.44)	7964.25 (21.90)	59147.22 (63.74)
(C)	Total cost (A + B)	136817.88 (100.00)	105202.01 (100.00)	36364.91 (100.00)	92794.03 (100.00)

Figures in the parentheses indicate percentage to the total cost.

Cost B_1

Cost B_1 was computed by adding the interest on value of own capital asset to Cost A_1. The average Cost B_1 per hectare was Rs. 32899.40 which increased from Rs. 23823.24 in group III to Rs. 40114.63 per hectare in group I. Like the Cost A_1, the Cost B_1 per hectare also decreased with the increase in the size of growers. This was mainly because of higher rental value of land in lower size groups.

Cost C_1

Cost C_1 was worked by adding the imputed value of family labour to Cost B_1. The Cost C_1 per hectare was Rs. 48767.88, Rs. 43427.01 and Rs. 44261.44 in group I, II and III respectively. The average Cost C_1 per hectare was found to be Rs. 45485.44.

Cost C_2

Cost C_2 was computed by taking 10 per cent of the total cost on account of managerial function performed by ginger growers. The average cost C_2 was found to be Rs. 4810.68 per hectare. The Cost C_2 per hectare was Rs. 5737.15, Rs. 5204.07 and Rs. 3490.83 for group I, II and III respectively. Like the other costs, the Cost C_2 per hectare decreased with the increase in the size of the growers. This might be due the fact of higher rental value of land and variable cost incurred in lower size groups of farm.

Gross Income

The average cost per hectare was worked out to be Rs. 55261.98. The gross income per hectare was Rs. 61281.25, Rs. 58004.69 and Rs. 46500 in group I, II and III respectively. It was observed that gross income per hectare decreased with the increase in size of growers.

Family Labour Income

The family labour income was calculated by deducting Cost B_1 from gross income. The average family labour income per hectare was estimated to be Rs. 22362.58. It was observed that per hectare family labour income was highest in group II (Rs. 23244.35) and lowest in group I (Rs. 21166.32).

Net Income

On an average the net income per hectare of ginger was Rs. 9776.54. The net income per hectare was highest in group II (Rs. 14577.68) and lowest in group III (Rs. 2238.56). This was mainly due to higher cost incurred by sample farmers in lower size groups.

Return from Management

The average return to management per hectare was found to be Rs. 50451.30. It was highest in group I (Rs. 55544.10) and lowest in group III (Rs. 43009.17).This might be due to higher cost incurred by the sample farmers in lower size groups. Further, the rental value of land was less in group III farm due to having less area under ginger cultivation which ultimately affected the return from management.

Benefit Cost Ratio

The average benefit cost ratio based on variable cost was found to be 1.58 while the average benefit cost ratio based on total cost was 0.76. The benefit cost ratio over variable was highest in group II and III (1.64) and lowest in group I (1.46). The benefit cost ratio based on total cost was highest in group III (1.29) and lowest in group I (0.45).

The cost of ginger production was also computed by working out the variable cost, fixed cost and total cost incurred in one crop which is given in Table 8.2.

Variable Cost

The average variable cost in ginger production accounted for 36.26 per cent of the total cost. The variable cost incurred per hectare was Rs. 37236.63, Rs. 35305.86

and Rs. 28400.66 in group I, II and III respectively. The average variable cost incurred per hectare was found to be Rs. 33647.72. It was observed that the variable cost per hectare increased with the increase in the size of growers.

Table 8.2: Per Hectare Cost and Return of Ginger Production Across Size Group (in Rs.)

Sl.No.	*Particulars*	*Group I*	*Group II*	*Group III*	*Average*
1.	Cost A_1	28658.38	26719.19	22906.99	26094.85
2.	Cost B_1	40114.63	34760.34	23823.24	32899.40
3.	Cost C_1	48767.88	43427.01	44261.44	45485.44
4.	Cost C_2	5737.15	5204.07	3490.83	4810.68
5.	Gross return	61281.25	58004.69	46500	55261.98
6.	Family labour income	21166.62	23244.35	22676.76	22362.58
7.	Net income	12513.37	14577.68	2238.56	9776.54
8.	Return from management	55544.10	52860.62	43009.17	50451.30
9.	BCR based on variable cost	1.46	1.64	1.64	1.58
10.	BCR based on total cost	0.45	0.55	1.29	0.76

Fixed Cost

The average cost in ginger cultivation accounted 63.74 per cent of the total cost. The fixed cost incurred per hectare was Rs. 99581.25, Rs. 69896.15 and Rs. 7964.25 in group I, II and III respectively. The results thus showed that more the area more was the fixed cost. Like the variable cost, the fixed cost per hectare increased with the size of growers.

Total Cost

The average cost per hectare was estimated at Rs. 92794.03.Per hectare total cost were Rs. 136817.88, Rs. 105205.01 and Rs. 36364.91 in group I, II and III respectively. Like the variable and fixed cost, the total cost per hectare increased with the increase in the size.

Conclusions

From the study the following conclusions can be drawn:

1. The per hectare total cost of ginger production in the average farm situation was estimated to be Rs. 92794 03. The share of labour was highest (19.27 per cent) in the total variable cost.
2. The average gross income per hectare from ginger production was Rs. 55261.98. The per hectare gross income from ginger production showed a decreasing trend.
3. The net income per hectare of ginger production was Rs. 9776.54. The net income per hectare was highest (Rs. 14577.68) in group II farm and lowest (Rs. 2238.56) in group III farm.

4. The average per hectare return from management was Rs. 50451.30. It was highest (Rs. 55544.10) in group I farm and lowest (Rs. 43009.17) in group III farm.
5. The average benefit cost ratio from ginger production on the basis of variable cost was 1.58 while the average benefit cost ratio based on total cost was 0.76. The benefit cost ratio over variable was highest (1.64) in group II farm and III farm and lowest (1.46) in group I farm.
6. Ginger cultivation was profitable in the area and more profitable particularly for larger size groups.

Policy Implications

The above results depicted that ginger cultivation was economically profitable in the area and therefore, appropriate strategy should be undertaken by the concerned department for better marketing of ginger along with value addition so that the farmer would get better price of the product. The existence of intermediaries should be eliminated in marketing process.

References

Anonymous (2002). Basic Statistics of North Eastern Region. North Eastern Council, Ministry of Home Affairs, Shillong, pp. 89.

Anonymous (2004). Statistical Handbook of Nagaland. Govt. of Nagaland.

Gupta, S.C. and Kapoor, V.K. (2001). Fundamentals of Applied statistics. Sultan Chand and Sons, pp-35.

Sharma, Y.P., Kumar, A., Yadav, D.S., Sasikumar, B., Tripathi, A.K., Anandraj, M., Sanwal, S.K. and Parthasarthy, V.A. (2006). Production and Marketing of Ginger. ICAR Research Complex for NEH region, Umiam. 1 – 27.

Chapter 9

An Economic Analysis of Jute Production in Barpeta District of Assam

G.K. Sarma[1] *and D.C. Kalita*[3]

[1]*Farm Manager, Krishi Vigyan Kendra, Darrang, Assam*
[2]*Reader, Department of RDAP, NEHU, Tura Campus, Meghalaya*

ABSTRACT

The present study was conducted in Barpeta district of Assam with the objectives to determine the production variability between the size groups of farms and to investigate the constraints in jute production. Random sampling method was followed for selection of farmers and total 105 farmers were selected from 15 villages from three development blocks. The negative and the non-significant coefficient of family labour showed in the result revealed that there was excessive use of family labour in the farm work than the requirement. The same case was happened in case of hired labour but the factor was significant at 10 per cent probability level. This indicated that the sample farmers were operating in production zone III of the production function. The factors like use of chemical fertilizers and jute area have positive coefficients and highly significant at 1 per cent probability level. This means that farmers are not using the chemical fertilizers as per requirement of the crop. Thus, more production can be obtained with the increase in application of chemical fertilizers and area under jute. The factors manure and plant protection were significant at 5 per cent probability level and the coefficients were positive. This means that there is possibility to increase the production of jute fibre with the increase use of manures and plant protection chemicals as per recommendation rate. The coefficient l (lambda) was significant at 5 per cent probability level with a coefficient greater than 1 which indicated that the production variability was mainly due to the technical inefficiency rather than random disturbance. The significant difference between γ and zero indicated that there was variation of farmers' yield and the maximum potential yield. The estimated value of γ was 0.6022563 and this explained that approximately 60 per cent of total variability was due to the technical inefficiency and this technical efficiency may be due to biological factors or the socio-economic factors. The major constraints of jute production in the study area are non-growing of HYV, use of plant protection chemical, insufficient use of manures and chemical fertilizers, labour scarcity at time, market price of jute fibre, non-production jute seed, improper extension services.

Keyweords: Jute, Economic analysis, Productivity, Efficiency.

Introduction

Jute is the second most important crop next to cotton both in terms of area and production in India as well as in the world. India ranks second next to Bangladesh in terms of production in the world. Till 1947, the production of jute was monopoly in agriculture sector, but after division of India in to Pakistan and India, India retains most of the jute mills and major jute growing areas were went to Pakistan. In India major jute growing states are West Bengal, Assam, Bihar, Orissa, Uttar Pradesh, Tripura and Meghalaya. The climate of these states suits for luxurious growth of the crop. In Assam jute is grown mostly in Lower Brahmaputra Valley zone, North Bank Plain zone and Central zones. The area and productivity of jute from 2006-07 to 2008-09 in the zone has increased by 7 and 15 per cent respectively. Thus, it is clear that the area and productivity of jute crop is gradually increasing which is a good sign for the state of Assam.

Jute is a labour intensive crop. It also generates paid employment to the extent of 11.1 million man days (Burger and Smith, 1985) and provides substances to 0.25 million industrial workers and about two million people in the secondary and tertiary sectors of the country. In case of Assam, more than 0.20 million farm families are directly involved in cultivation of jute out of which 80 per cent are marginal and small.

Keeping view the above facts, the present study is an attempt to examine the production variability between the size groups of farms and the constraints in jute production.

Methodology

The study area was conducted in Barpeta district of Assam considering three community development blocks namely, Barpeta, Bhawanipur and Chenga community development block where jute crop is mostly grown. The lists of villages under each community development block were collected from the concerned Agriculture Development Officers. Random sampling method was followed for selection of villages. Out of 379 revenue villages under the three community development blocks, a total of 15 villages were selected randomly. Thus, each community development block covered 5 villages and one village was selected from one Agriculture Development Officer's circle. The list of households with respect to their operational holdings was prepared in consultation with the Agriculture Development Officer. Seven numbers of households were selected from each village. The households were selected randomly. A Total 35 numbers of households from each development block were selected. Finally from three community development blocks, a total 105 households were selected for the present study. The selected farmers were then stratified on the basis of their operational holdings into three groups *viz.*, small (Below 2 hectares), medium (2-4 hectares) and large (4 hectares and above). The data were collected with the help of pre-tested schedules keeping in view the specific objectives of the study by Personal interview method of sampling. The data pertain to the year 2009-10. Both tabular and functional analyses were carried out in the present study. The present status of production technology adopted by the farmers was derived by computing arithmetic average of the resource utilized for the production process. The Stochastic production Frontier was used to evaluate the productivity difference between the size groups of farms. Linear regression method was followed to investigate the factors affecting in yield variability in jute

Conceptual Model and Estimation Procedure

The difference in yield between the size group of farms may arise from the three sources like

1. Difference in technology used
2. The farmers have excess to the same level of technology but all are not equally efficient using the available technology.
3. Due to the effect of random factors which are beyond the control of the farmers

Assuming the Cobb-Douglass technology a Stochastic Production Frontier, which allows to examine the technical inefficiency of farmers, can be written as:

$$\text{Log}Y= \alpha + \Sigma \beta \text{ Log}x + v-u \qquad \text{.................(1)}$$

where,

Y = Yield of jute fibre

x_j, j = 1, 2, 3, ……,n; stands for inputs used in jute production

β_j, j = 1, 2, 3, ……,n; are coefficient associated with input x_j

v = Error term

u = Random term having one sided distribution, and represent technical inefficiency

v = Intercept

Estimation of u

The estimates of 'u' in equation (1) can be obtained by the method given by Jondrow *et al.* (1989). The showed that the mean of 'u' given (u+v) is:

$$E(u/v) = \mu - \frac{f(e\lambda/\delta)}{1-F(e\lambda/\delta)} - \frac{e\lambda}{\delta} \qquad \text{.................(2)}$$

where, e = u + v

$$\lambda = \frac{\delta^2}{\delta v^2}$$

$$\delta^2 = \delta u^2 + \delta v^2$$

$$\mu = \frac{\delta u^2}{\delta u^2.\delta\delta^2}$$

δu^2 = variance of individual variability (u)

δv^2 = variance of random variability (v)

λ>1 indicates that the variation in production is due to the technical inefficiency of the farmers; λ<1 indicates that the variation in production is due to the random variability

Results and Discussion

Size Group of Farms and Land Holding Pattern

Out of the total 105 numbers of selected farm families 47 were small farmers, 39 were medium and 19 were large farm families which accounted 44.76 per cent, 37.14 per cent and 18.10 per cent, respectively (Table 9.1). The average land holding in case of small farmers was 1.14 hectare and that of medium and large farmers were 2.67 hectares and 5.23 hectares, respectively.

Table 9.1: Distribution of Sample Population and Average Land Holding Pattern According in Different Size Group of Farms

Particulars	*Small Farmers (< 2 ha)*	*Medium Farmers (2–4 ha)*	*Large Farmers (> 4 ha)*	*All Farms*
Sample size	47 (44.76)	39 (37.14)	19 (18.10)	105 (100.00)
Average operational size (ha)				
Upland	0.67 (50.00)	1.19 (44.57)	1.77 (37.03)	3.63 (38.99)
Medium land	0.28 (20.89)	0.68 (25.47)	1.57 (32.85)	2.53 (27.18)
Low land	0.46 (34.33)	0.80 (29.96)	1.89 (39.43)	3.15 (33.83)
Total	1.41 (100.00)	2.67 (100.00)	5.23 (100.00)	9.31 (100.00)

Figures in the parentheses indicate the per cent to the total.

Productivity Difference Between the Size Groups of Farms

The productivity difference between the size groups of farms is presented in Table 9.2. The table clearly shows that the productivity of all categories of farmers was less than the potential yield and the productivity of farms increased with increase in farm size. It was mainly due to the better cultural practice, use of more HYV seeds, use of higher dose of manure and chemical fertilizers in comparison to the small farmers.

Table 9.2: Productivity Difference of Jute Fibre Between Size Groups of Farms

Particulars	*Small Farmers (< 2 ha)*	*Medium Farmers (2–4 ha)*	*Large Farmers (> 4 ha)*	*All Farms*
Productivity (qt per hectare)	16.05	18.88	21.69	19.10

Technical Efficiency

The variation between the productions occurs because of variation in technical inefficiency among the farms or due to the effect of random factors which are beyond the control of the farmers. The Table 9.3 shows the estimated parameters of maximum yield function for the sample farmers. The negative and the non-significant coefficient of family labour indicated that there was excessive use of family labcur in the farm work than the requirement. It might be due to the fact that there was no alternative employment opportunity of the family labour during jute growing period. The same case was happened in case of hired labour but the factor was significant at 10 per cent probability level. This indicated that the sample farmers were operating in production zone III of the production function. Thus, it is advisable to reduce the use of hired family labour in jute cultivation.

Table 9.3: Estimated Parameters of Maximum Yield Function for Sample Farmers

Variables	Coefficients	
	Maximum Likelihood	Ordinary Least Square
Intercept	4.17347 (0.0968999)	3.91943 (0.0769762)
Log (family labour)	– 0.00302046 (0.0108769)	– 0.00313734 (0.0113499)
Log (hired labour)	– 0.00929248*** (0.00537876)	– 0.0105011 (0.00531172)
Log (manure)	0.0174444** (0.00737013)	0.0171546 (0.00694885)
Log (chemical fertilizer)	0.0296932* (0.00624516)	0.0323165 (0.00677346)
Log (jute area)	1.12491* (0.0595193)	1.13916 (0.0572619)
Log (plant protection measures)	0.0146113** (0.00650635)	0.0164642 (0.00593636)
λ (lambda)	1.51424** (0.758775)	
δ (sigma)	0.401500* (0.0592397)	
Log likelihood function =	– 21.444	
γ =	0.6022563	
δ^2 =	0.1612021	
Variance of individual variability (δu^2) =	0.0970865	
Variance of random variability (δv^2) =	0.0641156	

Figures in parentheses indicate asymptotic standard error of estimates.

* Significant at 1 per cent probability level.

** Significant at 5 per cent probability level.

*** Significant at 10 per cent probability leve.l

The factor manure was significant at 5 per cent probability level and the coefficient was positive. This means that there is possibility to increase the production of jute with the increase of manure. Chemical fertilizer is a very important factor in crop production. The analysis revealed that the chemical fertilizer had positive coefficient and highly significant at 1 per cent probability level. Thus, it is clear that farmers are not using the chemical fertilizers as per requirement of the crop. This indicated that the sample farmers were operating in production zone I of the production function and more production can be obtained with the more use of chemical fertilizers.

The jute area showed a positive coefficient and it was highly significant at 1 per cent probability level like chemical fertilizer. Thus, more production can be obtained with the increase in area under jute. The factor plant protection had also a greater influence on jute production in the study area as indicated by its significant (at 5 per cent probability level) and positive coefficient. This clearly indicated that farmers did not apply plant protection chemicals against insect-pests and diseases as per requirement. It is due to the fact that farmers are ignorant about the use of these chemicals against the insect-pests and diseases. Thus the study revealed that proper use of the plant protection chemicals might increase the production of the crop.

The coefficient λ (lambda) was significant at 5 per cent probability level with a coefficient greater than 1 which indicated that the production variability was mainly due to the technical inefficiency rather than random disturbance..

From the analyzed Table 3, it was observed that the estimate of γ was large (0.6022563) and it was significantly larger than zero. The significant difference between γ and zero indicated that there was variation of farmers' yield and the maximum potential yield. Such a case is generally observed when there is variation in adoption of best cultivation practices of crop production; that is individual variability in adoption of best practices for cultivation was the major cause of yield variability in jute among the size group of farms. The estimated value of δ^2 (0.1612021), γ (0.6022563), variance of individual variability (δu^2) (= 0.0970865) and variance of random variability (δv^2) (= 0.0641156) showed that variance of individual variability exceeds the variance of random variability. The estimated value of γ was 0.6022563 which explained that approximately 60 per cent of total variability was due to the technical inefficiency and this technical efficiency may be due to biological factors (like soil characteristics, moisture stress etc.) and socio-economic factors (like education, tenure, jute farming experience, labour, flood etc.)

Constraints in Jute Production

There are so many constraints faced by the farmers in jute cultivation. Out of them the major constraints are discussed below Table 9.4.

Non-Growing of HYV

From Table 9.4, it was observed that out of 105 sample farmers 66 farmers (62.86 per cent) used HYV for jute cultivation and the rest used the local varieties for which the farmers could not harvest the crop up to the maximum potential yield.

Use of Plant Protection Chemical

Due to climatic condition, so many insect-pests and diseases appear in the jute crop. Due to the ignorance and lack of awareness of farmers about the positive effect of plant protection chemicals farmers hardly use these chemicals. It was observed from the table that only 33.33 per cent farmers used the plant protection chemicals. The economic condition is the main bottle neck to get rid of this problem.

Table 9.4: Constraints of Jute Farming Faced by the Sample Farmers

Particulars	*Small Farmers (< 2 ha)*	*Medium Farmers (2–4 ha)*	*Large Farmers (> 4 ha)*	*All Farms*
No. of farm families	47 (44.76)	39 (37.14)	19 (18.10)	105 (100.00)
Varieties used				
Only HYV	16 (34.04)	19 (48.72)	12 (63.16)	47 (44.76)
Both HYV and local variety	7 (14.89)	8 (20.51)	4 (21.05)	19 (18.10)
Only local variety	24 (51.06)	12 (30.77)	3 (15.79)	39 (37.14)
Use of plant protection chemicals (No. of households)	12 (25.53)	11 (28.21)	12 (63.16)	35 (33.33)
Manure used (No. of households)	9 (19.15)	6 (15.38)	5 (26.32)	20 (19.05)
Chemical fertilizers used (No. of households)	32 (68.09)	34 (87.18)	17 (89.47)	83 (79.05)
Availability of seeds				
Owned	6 (12.77)	12 (30.77)	11 (57.89)	29 (27.62)
Purchased from local market	41 (87.23)	27 (69.23)	8 (42.11)	76 (72.38)
Technical assistance	5 (10.64)	6 (15.38)	5 (26.32)	16 (15.24)

Figures in parentheses indicate percentage to the total.

* Figures in parentheses indicate the area under cultivation (in hectare).

Insufficient Use of Manures and Chemical Fertilizers

Due to the replacement of bullock plough by the tractors/power tillers, farmers are now a day can not collect and use cow dung for application in the crop field. Our study reveals that only 20 numbers of farmers (19.05 per cent) farmers used cow dung/compost/FYM in the crop field which is a very very bad sign in agriculture. Similarly, out of the sample farmers, 83 (79.05 per cent) farmers used chemical fertilizers in the crop field but not as per requirement of the crop.

Labour Scarcity at Time

The crop jute is a labour intensive and time consuming crop. Moreover, the growing period of autumn paddy (*ahu* paddy) also coincide with the growing period of jute crop. Thus, the farmers faced labour scarcity especially during harvesting and retting period. Since, harvesting and retting is a very hard work, the labourers do not prefer to do the job. Since most of the farmers are under small(44.76 per cent) and medium (37.14 per cent) in the study area, so some times it become a major problem to invest more money for intercultural and post harvest activities of jute crop.

Market Price of Jute Fibre

As per the secondary as well as the primary data, it is clear that the price of jute fibre fluctuates at a very high range for which the farmers cannot obtain an assured return. Some times they sell the fibre products at an unremunerative price by which they could not meet the cost of cultivation. To get rid of this problem the Government has to fix the procurement price of jute fibre so that farmers can meet the cost of cultivation with some profit.

Non-Production Jute Seed

Like rice, pulse crops, farmers usually do not produce seed of jute crop. Our study reveals that out of the total sample farmers only 27.62 per cent farmers produced seeds of jute and the remaining 72.38 per cent purchased seeds from the local market. As the NSC seeds are not available in the local market and due to a higher price of NSC seeds the farmers are unable to purchase the seeds.

Improper Extension Services

Jute cultivation being a labour intensive and costlier practice, management of input is very important factor and for this proper extension service is must. In the study area only a few farmers got technical assistance from the extension personnel.

Conclusion

From the above study it is clear that the average yield gained by the farmers was too low than the potential yield and the productivity of farms can be increased with increase in farm size. There are so many factors that are responsible for non achievement of potential yield. The farmers used the inputs like manures and chemical fertilizers, area under jute cultivation and plant protection chemicals at a very small quantity for which farmers could not obtain the maximum potential yield. The major constraints of jute production in the study area are non-growing of HYV, use of plant protection chemical, insufficient use of manures and chemical fertilizers, labour scarcity at due time, market price of jute fibre, non-production jute seed, improper extension services.

References

Ahmed, A.U. and Bhowmick, B.C. (1991). Behaviour of price, productivity and acreage response of some important crops in Assam. Agricultural Situation in India, 46(4): 225-229.

Basak, K. and Dutta, K.K. (1984). Some factors Influencing Jute Acreage in West Bengal: Role of Jute Price and the Credit Market. Indian Journal of Agricultural Economics, 39 (4): 649-660.

Goswami, P.C. and Bora, C.K. (1974). Cultivation of Jute vis-à-vis Autumn Paddy in Assam: A Case Study in Nagaon District, Assam, Indian Journal of Assam, 34th Session of the Annual Conference Issue, 29: 174.

Kalirajan, K. (1981). An Econometric Analysis of Yield variability in Paddy Production. Canadian Journal of Agricultural Economics, 29 (3): 283-294.

Rathore, M.S. (1984). Contribution of Factors to the Productivity difference between small and Large Farms, Indian Journal of Agricultural Economics, 39(1): 70-76.

Chapter 10
Hill Farming in Udalguri District of Assam: An Economic Study

G.K. Sarma[1], P.S. Baruah [2] and D.C. Kalita[3]

[1]*Farm Manager, Krishi Vigyan Kendra (KVK), Darrang*
[2]*Assistant Professor, B.H. College, Howly*
[3]*Associate Professor, Department of RDAP, NEHU, Tura Campus*

ABSTRACT

The district Udalguri is located in the northern part of the state of Assam with a hilly and foot hill boundary in the northern part. Numbers of communities are the inhabitant in the study area. A study was conducted in the Udalguri district of Assam to know the demographic pattern of the people, to analyze their economic condition and to identify the major constrains in the field of agriculture and allied sectors. Both tabular and statistical analyses were done to analyze the data. Mainly people from Boro, Nepali, Assamese, Saotal (*Adibashi*), Bhojpuri, Rabha communities are predominant in the study area. The main source of income of the farmers are agricultural farming. The rice based farming, animal husbandry and sericulture have been found as the main source of income of the farming community. Farmers faced a number of costraints like their poor economic condition, lack of awareness, ignorance on technical know how and so on.

Keywords: Hill farming, Economics, Land use pattern, Farm families.

Introduction

Udalguri district is located in the northern part of the state of Assam sharing the international boarder with Bhutan and Arunachal Pradesh in the North. The district has been divided into three agro-ecological situations based on physiography, hydrology, soil, climate and cropping pattern. They are (*i*) Plain Zone, (*ii*) Hill and hillock Zone and (*iii*) Foot Hill Zone. The district comprises of 12.00 per cent hill and hillock areas and the foot hill area enshrines to 24.85 per cent. The climate of the district is hot summer followed by monsoon of heavy rainfall and relatively cold winter with scanty rainfall. The average rainfall is about 3027 mm per annum. The mean maximum and minimum temperature varies from 33-38° C and 8-10° C, respectively. The soils are predominantly sandy loam to clay loam in texture and acidic to neutral reaction

Major field crops grown in the district are paddy, jute, mesta, black gram, green gram, sesame, wheat, toria, niger, linseed, lentil, pea, buckwheat in the district. Under horticultural crops, both *kharif* and *rabi* vegetables, like cabbage, cauliflower, knolkhol, potato, radish, brinjal, tomato, cucurbits, lady's finger etc. Major fruit crops that are grown in the district are banana, citrus, pineapple, papaya, jackfruit, orange, mango etc. Chilli, black pepper, turmeric, ginger are the major spice crops in the district. Plantation crops like coconut, areca nut, tea, rubber are mostly grown. Besides these, the fish rearing, rearing of poultry birds, cows, pigs are very common amongst the farmers. Rearing of *eri, muga*, mulberry silk is a very popular practice among the scheduled cast and scheduled tribe communities.

Table 10.1: Land Use Pattern and Land Holding in the District

Sl.No.	*Land Use Pattern*	*Area*
1	Total geographical area	1,85,215 ha
2	Total Cultivable area	36.16 per cent
3	Total Cultivable Waste land	0.8 per cent
4	Current Fallow land	0.8 per cent
5	Area under forest	44.36 per cent
6	Area under pasture	6.08 per cent
7	Area under non-agricultural use	8.10 per cent
8	Area under miscellaneous plantations	3.70 per cent
9	Net area sown	91,681 ha (20.33 per cent of total geographical area)

The average size of land holding in the district has been calculated as 0.87 ha (6.53 *bigha).*

Considering all possible aspects, the study attempted on the few objectives to know the demographic pattern of the people living in the hill and foot hill areas of the district, to analyze their socio-economic condition and to identify the major constrains in field of agriculture and allied sectors.

Methodology

The study was conducted in Udalguri district of Assam. Out of three agro-ecological situations in the district, two namely, (*i*) Hill and hillock Zone and (*ii*) Foot Hill Zone were purposively selected. Five villages from each agro-ecological situation were selected at random. Fifteen households were selected randomly from the 10 selected villages that comprised the ultimate sample respondents of 150 numbers. Based on size of land holding, the selected samples were further stratified into three categories such as small farmers (land having less than 2.0 ha), medium farmers (land having 2.0–4.0 ha) and large farmers (land having more than 4.0 ha) groups. The data collected pertaining to the year 2010–11. Both tabular and statistical analyses like mean, standard deviation were done to analyze the data.

Results and Discussion

Population According to Communities in Different Size Group of Farms

The district is having inhabitants like Assamese, Boro Kachari, Saotal (*Adibashi*), Marwari, Bengali Hindu and Muslims, Nepali, Garo, Bhojpuri, Rabha and other tribes. Out of these, Assamese, Boro, Saotal (*Adibashi*), Nepali, Bhojpuri. Rabha have been found predominant in the study area (Table 10.2). Boro community has been found as the highest percentage of (53.33 per cent) of inhabitants followed by Nepali (16.00 per cent), Bhojpuri (10.00 per cent), Saotal (Adibashi) (8.67 per cent) and the Assamese (8.00 per cent). Only 4.00 per cent Rabha communities has been calculated.

Table 10.2: Distribution of Population According to Communities in Different Size Group of Farms

Communities	Small Farmers	Medium Farmers	Large Farmers	Total
Boro	44 (53.66)	33 (55.00)	3 (37.50)	80 (53.33)
Nepali	12 (14.63)	10 (16.67))	2 (25.00)	24 (16.00)
Assamese	8 (9.76)	4 (6.67)	0 (0.00)	12 (8.00)
Bhojpuri	8 (9.76)	6 (10.00)	1 (12.50)	15 (10.00)
Saotal (Adibashi)	7 (8.54)	6 (10.00)	1 (12.50)	13 (8.67)
Rabha	3 (3.66)	1 (1.67)	1 (12.50)	6 (4.00)
Total	**82**	**60**	**8**	**150**

* Figures in the parentheses indicate percentage to the total.

Population Pattern and Family Size According to Size Group of Farms

Altogether 150 (one hundred and fifty) numbers of farm families have been covered for the study. Out of which 82 farm families are under the category of small farmer which has been accounted as 54.66 per cent of the total farm families. 60 farmers (40.00 per cent) are medium and 8 (5.37 per cent) farm families are large farmers (Table 10.3). The distribution of population has revealed that the per cent of male and female population in all groups of farmers have been found almost equal. The highest percentage (55.01 per cent) of people has been found in small farmers group followed by medium size of farmers (39.75 per cent). Only 5.24 per cent of large farmers have been found in the study area. The average family size in all categories of farmers has been calculated as 5.82. Individual categories of farmers also showed almost similar average family size. This average family size is similar to that of the state average of Assam

Table 10.3: Population Pattern and Family Size According to Size Group of Farms

Size Groups	No. of Farm Families	Population (no.)			Average Family Size
		Male	Female	Total	
Small farmer	82 (54.66)	257 (54.68)	226 (55.39)	483 (55.01)	5.89
Medium farmer	60 (40.00)	189 (40.21)	160 (39.22)	349 (39.75)	5.81
Large farmer	8 (5.37)	24 (5.11)	22 (5.39)	46 (5.24)	5.75
Total	**150 (100.00)**	**470 (100.00)**	**408 (100.00)**	**878 (100.00)**	**Av. 5.82**

* Figures in the parentheses indicate percentage to the total.

Age Wise Distribution of Farm Families

The age groups of the farm families have been categorized into four different categories such as less than 15 years, 15 years to less than 30 years, 30 years to less than 60 years and above 60 years. From Table 10.4, it has been observed that out of the total population, 20.16 per cent populations are under the category of less than 15 years which are considered as the children and they are dependent. Similar percentage of population has also been observed in individual farmers' groups. Highest percentage (36.56 per cent) of populations has been found under the category of 30 years to less than 60 years and it has been followed by the category 15 years to less than 30 years (36.10 per cent). These two age groups have been considered as the most active and people belonging to theses categories can perform hard work. Except in large farmer group under the category of 15 years to less than 30 years (26.09 per cent), all categories of farmers are having almost equal percentage of population under these two age groups. Only 7.18 per cent populations have been found under the category 60 years and above and the population under this category has also been considered as dependent to some extent.

Distribution of Farm Families According to Educational Standard

Table 10.5 presents the distribution of farm families according to their educational standard. The table depicts that out of the total populations, 9.68 per cent has been found as illiterate and illiteracy has been increased with the increased in size group of farmers. Highest illiteracy has been observed in the large farmers group (30.43 per cent) and the highest literacy has been observed in primary level of education (33.26 per cent) followed by class X standard (28.47 per cent) and class XII standard (23.58 per cent), respectively. Only 5.01 per cent populations amongst the selected farmers has been accounted who have passed the degree standard. Highest percentage of

Table 10.4: Age Wise Distribution of Farm Families

Age Groups	*Small Farmer*			*Medium Farmer*			*Large Farmer*			*Total*		
	Male	*Female*	*Total*	*Male*	*Female*	*Total*	*Male*	*Female*	*Total*	*Male*	*Female*	*Total*
Less than 15 years	51 (20.73)	43 (18.14)	**94** **(19.46)**	38 (21.23)	35 (20.59)	**73** **(20.92)**	5 (21.74)	5 (21.74)	**10** **(21.74)**	94 (20.98)	83 (19.30)	**177** **(20.16)**
15 to less than 30 years	86 (34.96)	92 (38.82)	**178** **(36.85)**	64 (35.75)	63 (37.06)	**127** **(36.39)**	7 (30.35)	5 (21.74)	**12** **(26.09)**	157 (35.04)	160 (37.21)	**317** **(36.10)**
30 to less than 60 years	95 (38.62)	87 (36.71)	**182** **(37.68)**	63 (35.20)	59 (34.71)	**122** **(34.96)**	8 (34.78)	9 (39.19)	**17** **(36.96)**	166 (37.05)	155 (36.05)	**321** **(36.56)**
60 years and above	14 (5.69)	15 (6.33)	**29** **(6.00)**	14 (7.82)	13 (7.65)	**27** **(7.74)**	3 (13.04)	4 (17.39)	**7** **(15.22)**	31 (6.92)	32 (7.44)	**63** **(7.18)**
Total	**246** **(100.00)**	**237** **(100.00)**	**483** **(100.00)**	**179** **(100.00)**	**170** **(100.00)**	**349** **(100.00)**	**23** **(100.00)**	**23** **(100.00)**	**46** **(100.00)**	**448** **(100.00)**	**430** **(100.00)**	**878** **(100.00)**

* Figures in the parentheses indicate percentage to the total.

Table 10.5: Distribution of Farm families according to Educational Standard

Educational Level	*Small Farmer*			*Medium Farmer*			*Large Farmer*			*Total*		
	Male	*Female*	*Total*	*Male*	*Female*	*Total*	*Male*	*Female*	*Total*	*Male*	*Female*	*Total*
Illiterate	12 (4.88)	29 (12.24)	**41 (8.49)**	9 (5.03)	21 (12.35)	**30 (8.60)**	6 (26.09)	8 (34.78)	**14 (30.43)**	27 (6.03)	58 (13.49)	**85 (9.68)**
Primary level	81 (32.93)	98 (41.35)	**179 (37.06)**	36 (20.11)	65 (38.24)	**101 (28.94)**	6 (26.09)	6 (26.09)	**12 (26.09)**	123 (27.06)	169 (39.30)	**292 (33.26)**
X standard	63 (25.61)	76 (32.07)	**139 (28.79)**	45 (25.14)	58 (34.11)	**103 (29.51)**	4 (17.39)	4 (17.39)	**8 (17.39)**	112 (25.00)	138 (32.09)	**250 (28.47)**
XII standard	69 (28.05)	34 (14.35)	**103 (21.33)**	70 (39.11)	26 (15.29)	**96 (27.51)**	5 (21.74)	3 (13.04)	**8 (17.39)**	144 (32.14)	63 (14.65)	**207 (23.58)**
Degree and above standard	21 (8.54)	0 (0.00)	**21 (4.35)**	19 (10.61)	0 (0.00)	**19 (5.44)**	2 (8.70)	2 (8.70)	**4 (8.70)**	42 (9.38)	2 (0.47)	**4 4 (5.01)**
Total	**246 (100.00)**	**237 (100.00)**	**483 (100.00)**	**179 (100.00)**	**170 (100.00)**	**349 (100.00)**	**23 (100.00)**	**23 (100.00)**	**46 (100.00)**	**448 (100.00)**	**430 (100.00)**	**8 7 8 (100.00)**

* Figures in the parentheses indicate percentage to the total.

Table 10.6: Nature of Employment of Farm Families as Major Occupation

Employment Nature	*Small Farmer*			*Medium Farmer*			*Large Farmer*			*Total*		
	Male	*Female*	*Total*	*Male*	*Female*	*Total*	*Male*	*Female*	*Total*	*Male*	*Female*	*Total*
Farming	136 (55.28)	168 (70.89)	**304 (62.94)**	116 (64.80)	146 (85.88)	**237 (67.91)**	13 (56.21)	17 (73.91)	**30 (65.22)**	265 (59.15)	331 (76.98)	**571 (65.03)**
Govt. Service	3 (1.21)	0 (0.00)	**3 (0.62)**	5 (2.79)	0 (0.00)	**5 (1.43)**	5 (21.74)	1 (4.35)	**6 (13.04)**	13 (2.90)	1 (0.23)	**14 (1.59)**
Pvt. Sector Service	6 (2.44)	0 (0.00)	**6 (1.24)**	11 (6.15)	0 (0.00)	**11 (3.15)**	3 (13.04)	0 (0.00)	**3 (6.52)**	20 (4.46)	0 (0.00)	**20 (2.28)**
Business/Contractor	6 (2.44)	0 (0.00)	**6 (1.24)**	10 (5.59)	1 (0.59)	**11 (3.15)**	5 (21.74)	0 (0.00)	**5 (10.87)**	21 (4.69)	1 (0.23)	**22 (2.51)**
Labourer	25 (10.16)	27 (11.39)	**52 (10.77)**	7 (3.91)	0 (0.00)	**7 (2.01)**	0 (0.00)	0 (0.00)	**0 (0.00)**	32 (7.14)	27 (6.28)	**59 (6.72)**
Total population	**246 (100.00)**	**237 (100.00)**	**483 (100.00)**	**179 (100.00)**	**170 (100.00)**	**349 (100.00)**	**23 (100.00)**	**23 (100.00)**	**46 (100.00)**	**448 (100.00)**	**430 (100.00)**	**878 (100.00)**

* Figures in the parentheses indicate percentage to the total.

degree passed farm family members have been observed in the large farmer category (8.70 per cent) followed by medium farmers (5.44 per cent) and small farmers (4.35 per cent).

Nature of Employment of Farm Families as Major Occupation

As a source of income, the farmers have taken various activities like farming, Govt. services, private sector service, business/contractor and labourers as their major occupation (Table 10.6). Out of the selected farm families, 65.03 per cent populations are being employed under farming activities. Highest percentage (67.91 per cent) of population has been engaged in medium farmer group followed by large farmer group (65.22 per cent) and small farmers group (62.94 per cent). It has been observed that the percentage of engagement of female is more than the male in all categories of farmers. Only a few percentages of populations are found to be engaged in Govt. services (1.59 per cent), private sector services (2.28 per cent) and business/contractor (2.51 per cent). There are 6.72 per cent populations who are engaged as labourers in agriculture, industry or other sectors.

Types of Farming Performed by Farm Families

Major farming systems followed by the farm families in the study area have been presented in Table 10.7. Five major farming systems have been observed in the study area. Out of these, field crops + horticulture + plantation crops + animal husbandry + sericulture has been mostly followed (42.00 per cent) by the selected farm families followed by field crops + horticulture + plantation crops + animal husbandry (24.67 per cent), field crops + horticulture + animal husbandry + sericulture (20.00 per cent). The farming system field crops + animal husbandry has been least followed by the farmers (5.33 per cent).

Table 10.7: Major Farming System Followed by the Farm Families

Sl.No.	*Farming System Followed*	*No. of Farm Families*
1.	Field crops + Animal Husbandry	8 (5.33)
2.	Field crops + Horticulture + Animal Husbandry + Sericulture	30 (20.00)
3.	Field crops + Horticulture + Plantation Crops + Animal Husbandry	37 (24.67)
4.	Field crops + Horticulture + Plantation Crops + Sericulture	12 (8.00)
5.	Field crops + Horticulture + Plantation Crops + Animal Husbandry + Sericulture	63 (42.00)
	Total	**150 (150.00)**

* Figures in the parentheses indicate percentage to the total.

Types of Farming Practiced by Farm Families

Based on the economic condition, the farmers have been practicing different types of farming. Different types of farming have been practiced by the farmers has been presented in Table 10.8. The farming namely, rice and other cereals has been practiced by all farmers because rice is the staple food for the people of the district as well as the state of Assam. On an average, 90.00 per cent farmers have grown animals and poultry birds followed by sericulture (83.33 per cent) and fruit crops (78.67 per cent). From the last few years farmers have been started to grow tea and rubber in the foot hill areas and hilly regions. Till now it has not been found in a large scale. Out of the three different farmers categories, it has been observed that except sericulture, the farmers belonging to the large farmers group have covered more percentage of farming..

Table 10.8: Types of Farming Practiced by Farm Families

Types of Farming	*Small Farmers*	*Medium Farmers*	*Large Farmers*	*Total*
Rice and other cereals	82 (100.00)	60 (100.00)	8 (100.00)	150 (100.00)
Pulses	11 (13.41)	17 (28.33)	3 (37.50)	31 (20.67)
Vegetables	25 (30.49)	19 31.67)	4 (50.00)	48 (32.00)
Fruit crops	68 (82.93)	42 (70.00)	8 (100.00)	118 (78.67)
Sericulture	69 (84.15)	51 (85.00)	5 (62.50)	125 (83.33)
Animal Husbandry	74 (90.24)	53 (88.33)	8 100.00)	135 (90.00)
Fishery	32 (39.02)	37 (61.67)	6 (75.00)	75 (50.00)
Tea	27 (32.93)	20 (33.33)	5 (62.50)	52 (34.67)
Rubber	9 (10.98)	7 (11.67)	3 (37.50)	19 (12.67)
Total	**82 (100.00)**	**60 (100.00)**	**8 (100.00)**	**150 (100.00)**

* Figures in the parentheses indicate percentage to the total.

Engagement of Family Labour in Different Farming Components According to Size Group of Farms (Man day/year)

The family labours are mainly responsible for looking after and management of day to day activities of a farm. Table 10.9 describes the detail of engagement of family labour in different farming components according to size group of farms (Man day/year). It has been observed that the highest percentage of family labour (18.43 per cent) has been engaged in rice and other cereal crop farming followed by animal husbandry (15.18 per cent) and tea plantation (14.14 per cent). Lowest percentage (5.63 per cent) has been engaged in pulse crop farming. In case of total man days engagement, it is clear from the table that except a few cases, the engagement of total number of female man day of family labour is more in all types of farming than the male man day.

Table 10.9: Engagement of Family Labour in Different Farming Components According to Size Group of Farms (Man day/year)

Types of Farming	*Small*			*Medium*			*Large*			*Total*		
	Male	*Female*	*Total*	*Male*	*Female*	*Total*	*Male*	*Female*	*Total*	*Male*	*Female*	*Total*
Rice and other cereals	156 (24.61)	195 (19.66)	**351 (21.59)**	102 (16.89)	126 (14.57)	**228 (15.52)**	11 (14.65)	16 (13.44)	**27 (13.99)**	269 (19.46)	337 (17.68)	**606 (18.43)**
Pulses	57 (8.99)	61 (6.15)	**118 (7.26)**	51 (8.44)	72 (8.32)	**123 (8.37)**	6 (8.11)	8 (6.72)	**14 (7.25)**	114 (8.24)	71 (3.73)	**185 (5.63)**
Vegetables	102 (16.09)	116 (11.69)	**218 (13.41)**	75 (12.42)	86 (9.94)	**161 (10.96)**	7 (9.46)	13 (10.92)	**20 (10.36)**	184 (13.31)	215 (11.28)	**399 (12.14)**
Fruit crops	74 (11.67)	57 (5.75)	**131 (8.06)**	68 (11.26)	49 (5.66)	**117 (7.96)**	3 (4.05)	5 (4.20)	**8 (4.15)**	145 (10.49)	111 (5.82)	**256 (7.79)**
Sericulture	0 (0.00)	154 (15.52)	**154 (9.47)**	0 (0.00)	206 (23.82)	**206 (14.02)**	0 (0.00)	29 (24.37)	**29 (15.03)**	0 (0.00)	389 (20.41)	**389 (11.83)**
Animal Husbandry	116 (18.30)	171 (17.24)	**287 (17.65)**	89 (14.74)	101 (11.68)	**190 (12.93)**	9 (12.16)	13 (10.92)	**22 (11.40)**	214 (15.48)	285 (14.95)	**499 (15.18)**
Fishery	29 (4.57)	46 (4.64)	**75 (4.61)**	48 (7.95)	51 (5.90)	**99 (6.74)**	16 (21.62)	18 (15.13)	**34 (17.62)**	93 (6.73)	115 (6.03)	**208 (6.33)**
Tea	102 (16.09)	133 (13.41)	**235 (14.45)**	100 (16.56)	109 (12.60)	**209 (14.23)**	12 (16.22)	9 (7.56)	**21 (10.88)**	214 (15.48)	251 (13.17)	**465 (14.14)**
Rubber	68 (10.73)	59 (5.95)	**127 (7.81)**	71 (11.74)	65 (7.51)	**136 (9.26)**	10 (13.51)	8 (6.72)	**18 (9.33)**	149 (10.78)	132 (6.93)	**281 (8.55)**
Total Man days employed	**634 (100.00)**	**992 (100.00)**	**1626 (100.00)**	**604 (100.00)**	**865 (100.00)**	**1469 (100.00)**	**74 (100.00)**	**119 (100.00)**	**193 (100.00)**	**1382 (100.00)**	**1906 (100.00)**	**3288 (100.00)**

* Figures in the parentheses indicate percentage to the total.

Income Received and Expenditure Incurred at Different Farming Components (per annum)

The detail of income received and expenditure incurred at different farming components (per annum) has been presented in Table 10.10. Farmers have been receiving their livelihood income from different farming components namely, rice and other cereals like wheat, maize etc.; pulses like green gram, black gram, lentil, pea etc.; different winter and summer vegetables; fruit crops like oranges, litchi, jackfruit etc.; sericulture; live stocks; fishery and plantation crops like tea and rubber. For convenient, different strata of income groups has been made to discuss the findings like income (*i*) less than Rs. 5000.00, (*ii*) Rs. 5,000.00 to 8,000.00, (*iii*) Rs. 8,000.00 to 10,000.00, (*iv*) Rs. 10,000.00 to 15,000.00 and (*v*) above Rs. 15,000.00 per annum.

Income Received and Expenditure Incurred from Major Field Crops

From Table 10.9, it has been observed that in case of rice and other cereals, more than 50 per cent small farmers income have been confined to the income groups Rs. 8,000.00 to 10,000.00 (28.05 per cent) and Rs. 10,000.00 to 15,000.00 (36.59 per cent) per annum. Only 12.20 per cent farmers have been calculated in the income group above Rs. 15,000.00 per annum. While, in case of medium and large farmers groups 23.33 per cent and 62.50 per cent farmers have been accounted in the income group above Rs. 15,000.00 per annum, respectively. No large farmers have been accounted in the income groups less than Rs. 5000.00 and Rs. 5,000.00 to 8,000.00. It has also been clear from the table that both the mean net income received and mean expenditure incurred (excluding family labour) by the farmers have been increased with increase in size groups with an average mean of net income of the three groups Rs. 20,336.00 per annum with an average standard deviation Rs. 4,488.00. The standard deviation has been found decreased with increase in farmers' size group which indicated that there is comparatively more stable in production process in large size group of farmer.

Pulses are not the major field crops in the district. There is a very small area under pulse crop in the study area. Farmers also do not have interest to grow the pulse crops extensively like rice or other cereals. Only a few farmers have been found to grow the pulse crops with a net annual income of Rs. 2,247.00 per annum. It has also been observed from the table that the small and medium groups of farmers prefer to grow the pulse crops than the large farmers. Here, the standard deviation has been found decreased with increase in farmers' size group with an average mean standard deviation Rs.791.00.

Income Received and Expenditure Incurred from Major Horticultural Crops

Vegetable is a very popular crop in the study area. Some farmers grow vegetable commercially as well as domestic consumption. It has been observed from the table that, large farmers groups have been confined to two income groups Rs. 10,000.00 to 15,000.00 and Rs. 15,000.00 and above while, in case of small and medium farmers groups, more than 60 per cent farmers have been getting income at the income groups Rs. 10,000.00 to 15,000.00 and above Rs. 15,000.00. In this case, the mean net income received has been increased with the increase in farmers group with an average

Table 10.10: Income Received and Expenditure incurred at Different Farming Components (per annum)

Income Groups (Per annum)	*No. of*			
	Small Farmers	*Medium Farmers*	*Large Farmers*	*Total*
	I. Major Field Crops			
A. Rice and other cereals				
Less than Rs. 5,000.00	7 (8.54)	4 (6.67)	0 (0.00)	11 (7.33)
Rs. 5,000.00 to 8,000.00	12 (14.63)	7 (11.67)	0 (0.00)	19 (12.67)
Rs. 8,000.00 to 10,000.00	23 (28.05)	18 (30.00)	1 (12.50)	42 (28.00)
Rs. 10,000.00 to 15,000.00	30 (36.59)	17 (28.33)	2 (25.00)	49 (32.67)
Above Rs. 15,000.00	10 (12.20)	14 (23.33)	5 (62.50)	29 (16.67)
Total	**82 (100.00)**	**60 (100.00)**	**8 (100.00)**	**150 (100.00)**
Mean Gross Income received (Rs.)	**21,030.00**	**27,230.00**	**42,270.00**	**30,246.00**
Mean Expenditure incurred (excluding family labour) (Rs.)	**7,490.00**	**9, 280.00**	**12,760.00**	**9,910.00**
Mean Net income received (Rs.)	**13,540.00**	**17,950.00**	**29,520.00**	**20,336.00**
Standard Deviation on Mean Gross Income received	**4,810.00**	**4,570.00**	**3,085.00**	**4,488.00**
B. Pulses				
Less than Rs. 5,000.00	9 (81.82)	14 (82.35)	3 (100.00)	26 (83.87)
Rs. 5,000.00 to 8,000.00	2 (18.18)	3 (17.65)	0	5 (16.13)
Rs. 8,000.00 to 10,000.00	0 (0.00)	0 (0.00)	0 (0.00)	0 (0.00)
Rs. 10,000.00 to 15,000.00	0 (0.00)	0 (0.00)	0 (0.00)	0 (0.00)
Above Rs. 15,000.00	0 (0.00)	0 (0.00)	0 (0.00)	0 (0.00)
Total	**11 (100.00)**	**17 (100.00)**	**3 (100.00)**	**31 (100.00)**
Mean Gross Income received (Rs.)	**5,110.00**	**5,230.00**	**4,450.00**	**4,930.00**
Mean Expenditure incurred (excluding family labour) (Rs.)	**2,390.00**	**2,780.00**	**2,880.00**	**2,683.00**
Mean Net income received (Rs.)	**2,720.00**	**2,450.00**	**1,570.00**	**2,247.00**
Standard Deviation on Mean Gross Income received (Rs.)	**940.00**	**750.00**	**545.00**	**791.00**

Contd...

Table 10.10–Contd...

Income Groups (Per annum)	*No. of*			
	Small Farmers	*Medium Farmers*	*Large Farmers*	*Total*
II. Major Horticultural Crops				
C. Vegetables (Winter and Summer)				
Less than Rs. 5,000.00	2 (8.00)	1 (5.26)	0 (0.00)	3 (6.25)
Rs. 5,000.00 to 8,000.00	4 (16.00)	3 (15.79)	0 (0.00)	7 (14.58)
Rs. 8,000.00 to 10,000.00	2 (8.00)	3 (15.79)	0 (0.00)	5 (10.42)
Rs. 10,000.00 to 15,000.00	11 (44.00)	6 (31.58)	1 (25.00)	18 (37.50)
Above Rs. 15,000.00	6 (24.00)	6 (31.58)	3 75.00)	15 (31.25)
Total	**25 (100.00)**	**19 (100.00)**	**4 (100.00]**	**48 (100.00)**
Mean Gross Income received (Rs.)	**25,570.00**	**33,190.00**	**42,980.00**	**33,913.00**
Mean Expenditure incurred (excluding family labour) (Rs.)	**8,720.00**	**10,890.00**	**12,680.00**	**10,763.00**
Mean Net income received (Rs.)	**16,850.00**	**22,300.00**	**30,300.00**	**23,150.00**
Standard Deviation on Mean Gross Income received (Rs.)	**4,355.00**	**3,980.00**	**5,165.00**	**4,500.00**
D. Fruit crops				
Less than Rs. 5,000.00	34 (50.00)	19 (45.24)	5 (62.50)	58 (49.15)
Rs. 5,000.00 to 8,000.00	22 (32.35)	16 (38.10)	2 (25.00)	40 (33.90)
Rs. 8,000.00 to 10,000.00	7 (10.29)	6 (14.29)	1 (12.50)	14 (11.86)
Rs. 10,000.00 to 15,000.00	5 (7.35)	1 (2.38)	0 (0.00)	6 (5.08)
Above Rs. 15,000.00	0 (0.00)	0 (0.00)	0 (0.00)	0 (0.00)
Total	**68 (100.00)**	**42 (100.00)**	**8 (100.00)**	**118 (100.00)**
Mean Gross Income received (Rs.)	**5693**	**7,325.00**	**7,740.00**	**6918**
Mean Expenditure incurred (excluding family labour) (Rs.)	**890**	**1,175.00**	**1,290.00**	**1,118.00**
Mean Net income received (Rs.)	**4,800.00**	**6,150.00**	**6,450.00**	**5,800.00**
Standard Deviation on Mean Gross Income received (Rs.)	**1,120.00**	**1,275.00**	**1,490.00**	**1,278.00**

Contd...

Table 10.10–Contd...

Income Groups (Per annum)	No. of			
	Small Farmers	Medium Farmers	Large Farmers	Total
III. Sericulture				
E. Sericulture				
Less than Rs. 5,000.00	32 (46.38)	27 (52.94)	4 (80.00)	63 (50.40)
Rs. 5,000.00 to 8,000.00	25 (36.23)	20 (39.22)	1 (20.00)	46 (36.80)
Rs. 8,000.00 to 10,000.00	11 (15.94)	4 (7.84)	0 (0.00)	15 (12.50)
Rs. 10,000.00 to 15,000.00	1 (1.45)	0 (0.00)	0 (0.00)	1 (0.80)
Above Rs. 15,000.00	0 (0.00)	0 (0.00)	0 (0.00)	0 (0.00)
Total	**69 (100.00)**	**51 (100.00)**	**5 (100.00)**	**125 (100.00)**
Mean Gross Income received (Rs.)	**5,660.00**	**7,260.00**	**7,540.00**	**7,503.00**
Mean Expenditure incurred (excluding family labour) (Rs.)	**450**	**900**	**1050**	**800**
Mean Net income received (Rs.)	**5,210.00**	**6,360.00**	**6,490.00**	**6,020.00**
Standard Deviation on Mean Gross Income received (Rs.)	**1,130.00**	**1,330.00**	**1,010.00**	**1,015.00**
IV. Livestock				
F. Animal Husbandry				
Less than Rs. 5,000.00	0 (0.00)	0 (0.00)	0 (0.00)	0 (0.00)
Rs. 5,000.00 to 8,000.00	0 (0.00)	0 (0.00)	0 (0.00)	0 (0.00)
Rs. 8,000.00 to 10,000.00	20 (27.03)	10 (18.88)	0 (0.00)	30 (22.22)
Rs. 10,000.00 to 15,000.00	32 (43.24)	26 (49.06)	5 (62.50)	63 (46.67)
Above Rs. 15,000.00	22 (29.73)	17 (32.08)	3 (37.50)	42 (31.11)
Total	**74 (100.00)**	**53 (100.00)**	**8 (100.00)**	**135 (100.00)**
Mean Gross Income received (Rs.)	**27,290.00**	**30,150.00**	**32680**	**30,330.00**
Mean Expenditure incurred (excluding family labour) (Rs.)	**9,990.00**	**11,200.00**	**11,900.00**	**11,030.00**
Mean Net income received (Rs.)	**17,300.00**	**18,950.00**	**21,650.00**	**19,300.00**
Standard Deviation on Mean Gross Income received (Rs.)	**3,210.00**	**2,990.00**	**2,790.00**	**2,996.00**

Contd...

Table 10.10–Contd...

Income Groups (Per annum)	No. of			
	Small Farmers	Medium Farmers	Large Farmers	Total
V. Fishery				
G. Fishery				
Less than Rs. 5,000.00	15 (46.88)	11 (29.73)	0 (0.00)	26 (34.67)
Rs. 5,000.00 to 8,000.00	6 (18.75)	8 (21.62)	1 (16.67)	15 (20.00)
Rs. 8,000.00 to 10,000.00	3 (9.38)	5 (13.51)	1 (16.67)	9 (12.00)
Rs. 10,000.00 to 15,000.00	6 (18.75)	6 (16.22)	1 (16.67)	13 (17.33)
Above Rs. 15,000.00	2 (6.25)	7 (18.92)	3 (50.00)	12 (16.00)
Total	**32 (100.00)**	**37 (100.00)**	**6 (100.00)**	**75 (100.00)**
Mean Gross Income received (Rs.)	**5,805.00**	**8,510.00**	**10,540.00**	**8,285.00**
Mean Expenditure incurred (excluding family labour) (Rs.)	**895**	**1,210.00**	**1640**	**1,248.00**
Mean Net income received (Rs.)	**4,910.00**	**7,300.00**	**8,900.00**	**7,037.00**
Standard Deviation on Mean Gross Income received (Rs.)	**1,145.00**	**1,376.00**	**1,561.00**	**1,303.00**
VI. Major Plantation Crops				
H. Tea				
Less than Rs. 5,000.00	13 (48.15)	14 (70.00)	2 (40.00)	29 (55.77)
Rs. 5,000.00 to 8,000.00	9 (33.33)	5 (25.00)	1 (20.00)	15 (28.85)
Rs. 8,000.00 to 10,000.00	5 (18.52)	1 (5.00)	2 (40.00)	8 (15.38)
Rs. 10,000.00 to 15,000.00	0 (0.00)	0 (0.00)	0 (0.00)	0 (0.00)
Above Rs. 15,000.00	0 (0.00)	0 (0.00)	0 (0.00)	0 (0.00)
Total	**27 (100.00)**	**20 (100.00)**	**5 (100.00)**	**52 (100.00)**
Mean Gross Income received (Rs.)	**7,900.00**	**9,220.00**	**11,980.00**	**9599**
Mean Expenditure incurred (excluding family labour) (Rs.)	**1,600.00**	**1,970.00**	**2,180.00**	**1,916.00**
Mean Net income received (Rs.)	**6,300.00**	**7,250.00**	**9,800.00**	**7,783.00**
Standard Deviation on Mean Gross Income received (Rs.)	**1,140.00**	**1,290.00**	**1,360.00**	**1,263.00**

Contd...

Table 10.10–Contd...

Income Groups (Per annum)	No. of Small Farmers	Medium Farmers	Large Farmers	Total
I. Rubber				
Less than Rs. 5,000.00	7 (77.78)	6 (85.71)	3 (100.00)	16 (84.21)
Rs. 5,000.00 to 8,000.00	2 (22.22)	1 (14.29)	0 (0.00)	3 (15.79)
Rs. 8,000.00 to 10,000.00	0 (0.00)	0 (0.00)	0 (0.00)	0 (0.00)
Rs. 10,000.00 to 15,000.00	0 (0.00)	0 (0.00)	0 (0.00)	0 (0.00)
Above Rs. 15,000.00	0 (0.00)	0 (0.00)	0 (0.00)	0 (0.00)
Total	**9 (100.00)**	**7 (100.00)**	**3 (100.00)**	**19 (100.00)**
Mean Gross Income received (Rs.)	**4,360.00**	**5,780.00**	**4,750.00**	**4,963.00**
Mean Expenditure incurred (excluding family labour) (Rs.)	**680**	**890**	**1,050.00**	**873**
Mean Net income received (Rs.)	**3,680.00**	**4,890.00**	**3,700.00**	**4,090.00**
Standard Deviation on Mean Gross Income received (Rs.)	**890**	**910**	**780**	**860**
Grand Total of Mean Net Income (Rs.) {I (A+B) + II (C+D) + III (E) +IV (F) + V (G) +VI (H+I)}	**86,130**	**109,100**	**148,380**	**114,535**
Per month Mean Net Income (Rs.)	**7,177.00**	**9,091.00**	**12,365.00**	**9,529.00**

* Figures in the parentheses indicate percentage to the total.

mean net income Rs. 23,150.00 per annum. Highest average of standard deviation on mean gross income received has been calculated in the large farmer group and the lowest has been calculated in medium size group of farms.

In case of fruit crops, highest percentage of large farmers have been calculated in the income group less than Rs.5000.00 and lowest has been calculated in case of medium size group of farmers. The main reason behind this may be due to that the farmers are not growing the fruit crops scientifically and commercially. Most of the fruit crops are grown in the backyard of the homestead and only a few number of fruit crops are grown. In this case, the income received per month has been increased with the increase in size group of farms. The average of standard deviation on mean gross income received has been found as Rs. 1,278.00 and the same has been increased with the increased size of farmers group.

Income Received and Expenditure Incurred in Sericulture

Sericulture is a very important component of farming in the study area. Almost all the farm families of *boro* community rears *eri*. On account of unscientific rearing, reeling and weaving system, the growers are not receiving their income up to the desired level. Out of the total selected farmers, 50.40 per cent have been receiving income less than Rs. 5,000.00 per annum. Only one farmer from small farmer group receiving the income more than Rs. 10,000.00 per annum and no farmer has been observed in the income group of Rs. 15,000.00 and above with a mean net income of Rs. 6,020.00 per annum. Here, the standard deviation on mean gross income has been found lowest in large farmer group followed by small and medium farmers group with an average of Rs. 1,015.00.

Income Received and Expenditure Incurred in Livestock

Livestock is also a very important income generating component of the farmers in the study area. The farmers of Nepali community have taken rearing of cows as a livelihood profession. Similarly, the people of *boro* community rear pigs and poultry birds. It has been observed from the Table 10.9 that on an average no farmer earned less than Rs. 8,000.00 per annum from this enterprise. Out of the total farmers, on an average 46.67 per cent farmers are under the income group of Rs. 10,000.00 to Rs.

15, 000.00 and 31.11 per cent are under the income group of above Rs. 15,000.00. The mean net income received by the farmers increased with the increase in the farmers' size group. On the other hand, the standard deviation of mean gross income received has been decreased with the increased size group of farms with an average standard deviation of mean gross income received of Rs. 2,996.00.

Income Received and Expenditure Incurred in Fishery

Except a few scientific fisheries, most of the farmers rear fishes in traditional method. Even they do not have technical know how about the scientific culture of fish. Out of the total farmers selected for the study 34.67 per cent farmers earned less than Rs. 5000.00 per annum and 20.00 per cent earned Rs. 5,000.00 to Rs 8,000.00 per annum. The mean net income received by all the farmers has been calculated as Rs. 7,037.00 per annum and it has been increased with the increase of farmers' size group. Due to unscientific manner of fish farming, the standard deviation on mean gross income received has been found to be increased with the increase in size group of farms.

Income Received and Expenditure Incurred in Major Plantation Crops

The major plantation crops that are found popular among the farmers of the study area have been identified as tea and rubber. The plantations of these two crops have been started by the farmers recently in the fallow lands like very high lands, foot hills and hillocks. The farmers have not yet received a good income from these plantations due to lack of proper marketing facilities and tea manufacturing industries in the locality. From the table, it has been observed that 55.77 per cent farmers have been receiving less than Rs. 5,000.00 per annum from tea enterprise and no farmers have been receiving more than Rs. 10,000.00 per annum. The average standard deviation on mean gross income received has been accounted as Rs. 1,263.00 with the mean net income received of Rs. 7,783.00 per annum.

In case of rubber farming, no farmer group has received more than Rs. 8,000.00 per annum. The medium size group of farmers has been received highest mean net income followed by large and small size group of farms with an average of Rs. 4,090.00 per annum.

From the detail analysis it has been observed that the per month net mean income has been increased with the increase of farmers size group with an average per month net mean income of Rs. 9,529.00.

Constraints in Adopting the Agricultural Practices

The major constraints that have come into notice during the study period has been discussed below.

1. *Poor economic condition of the farmers*: More than 50 per cent farmers are small and marginal. Moreover they are depending upon agriculture. There are less than 2 per cent farm families whose family members works in Government sector. Hence, they can not meet up the need of their requirements.
2. *No irrigation facilities*: Since the study area covers hilly, foot hill and high land region, the ground water table is located at the very deep layer for which it is not possible to lift water by the shallow tube well. Only some natural water steams are observed.
3. *High incidence of insect-pests and diseases*: The farmers not yet utilized the agro chemicals to control the insect-pests and disease in their crops. Even the don't have knowledge on scientific cultivation practices like crop rotation, fallowing of land and so on.
4. *Non-availability of quality seeds/fingerlings/breeds*: The agricultural technology has not yet reached to the area. All seeds of crops, animals, poultry and fingerlings are used by the farmers which are locally available.
5. *Poor marketing/storage infrastructure*: The farmers are facing a great problem in marketing and storing of agricultural products after harvest. They have to sale their products in the local market at a low price.
6. *Non-availability of reeling and weaving machine*: There is a vast scope of establishment of *eri* industry in the study area. They are rearing, reeling and weaving the sericulture traditionally for which they are not earning their income up to the desired level. Even they are not getting the market to sale their products to get a high price.

Conclusion

From the discussion made in above study it is clear that the farming method in the hilly and foot hill areas is more or less similar to that of the plain areas in the district. *Jhum* cultivation is not practiced in the area. The major farming systems in the study are has been observed as rice and other cereal crops, sericulture and animal husbandry and the farming have been practiced in a traditional method. To improve the economy of the study area, the farmer must be aware in adoption of improved agricultural technology. To overcome all the problems the farmers must be very much aware about the new technologies and they should get all sorts of support from the Government.

References

Bhowmick, B.C., Sarma, A.K. and Talukdar, K.K. (1999). Farming System in Assam – A publishehed Report on NABARD Chair Scheme Unit, Department of Agricultural Economics, Assam Agricultural University, Jorhat.

Comprehensive District Agriculture Plan, Udalguri. Published by the District Agriculture Planning Init, Udalguri, 2010.

Panomtarinichigul, M. (2004). *Research on Sustainable Hill Farming in Northern Thailand.*

Saikia, A. (1985). Status of female agricultural labourers in Assam, Kurukshetra. 30(6): 11-13.

Sharma, P., Ogale, N. and Singh, S. (2010). *Women in Hill Agriculture.* Concept Publishing Company Pvt. Ltd. New Delhi.

Chapter 11

Resource Conservation Technologies for Sustainable Development in Hill Agriculture

Md Mokidul Islam[1], L.K. Nath[2], D.P. Patel[3], Anup Das[3], G.C. Munda[4], Tanmay Samajdar[5] and S.V. Ngachan[6]*

[1]Subject Matter Specialist (Agronomy), [2]Programme Assistant, KVK, Tura
[3]Senior Scientist, [4]Principal Scientist, [5]Programme Coordinator, KVK Tura
[6]Director, ICAR Research Complex for NEH Region, Umiam (Meghalaya)

ABSTRACT

The resource conservation technologies like system of rice intensification (SRI) and integrated crop management (ICM) in the context of farmer's realities in lowland rice ecosystem is necessary for the development of sustainable hill agriculture in South Garo Hills district of Meghalaya. The technology of SRI and ICM in rice cultivation could conserve resources like seeds, water, nutrients, soil health and environment, reduce cost, higher yield etc compared to conventional rice culture (CRC). The main reason for non-adoption of SRI and ICM technologies in rice cultivation was related to more efforts, faith towards traditional weed control methods, lack of knowledge and water management. It is important to understand the performance of any new technology in the context of farmer's realities, which are always more complicated and more diversified than on any research station. The net return of Rs. 36,788, Rs. 31,830 and Rs. 2,319/ha was realized in rice cultivation under SRI, ICM and CRC, respectively. The higher benefit cost ratio on variable cost was fetched by SRI farmers (2.78) followed by ICM farmers (2.51) and conventional farmers (1.14). Hence, the resource conservation technologies (SRI and ICM) along with integrated weed management (IWM) could replace the conventional method for sustainable development in valleys and foot hills rice farmers in terms of higher productivity and income, which may lead to

* *Corresponding Author:* E-mail: mislam01d@yahoo.co.in

improvement of livelihood of the farmers at Sibbari cluster of South Garo Hills district in Meghalaya.

*Keywords: **Resource conservation, Sustainability, System of rice intensification, Integrated crop management.***

Introduction

Rice (*Oryza sativa* L) is the major cereal crop and staple food for the people of the people of Eastern and North Eastern India. The North Eastern Region of India comprising of Assam, Arunachal Pradesh, Meghalaya, Manipur, Mizoram, Nagaland, Sikkim and Tripura having about 3.5 million hectare area under rice, with a production of about 5.38 million tonnes is still deficit of about 24 per cent rice grain. Gasuapara Community Development Block of South Garo Hills district of Meghalaya, India cultivate rice normally in an area of about 1177 hectare with the productivity of only 986 kg/ha which is lower than the district productivity of 1053 kg/ha (Table 11.1). The productivity of rice is very low due to traditional crop establishment methods, use of long duration local varieties, imbalanced fertilizer/manure application, pesticides as well as lack of appropriate weed management. Majority of the area is under hilly terrain with foot hills and valley lands where mostly *sali* (5.5 months) rice is grown in traditional methods. In traditional age old rice cultivation methods, farmers are using seedlings of 35-40 days old with 3-4 seedlings/hill which lead to increase insect pest infestation, diseases incidence and low yield. Since, rice is grown only with rain water and stream diversion water, it is necessary to develop and introduce the resource conservation technology for improvement of *sali* rice productivity and income. Rice being the staple food is grown in all possible locations and conditions. In addition to the water shortage, non availability of agricultural inputs like, seeds, fertiliser, labour, ploughing implements etc that increases the cost of cultivation. Even some farmers are giving up *sali* rice cultivation owing to its lower productivity and higher cost of cultivation. The poor farmers losing interest in rice cultivation as factor productivity is declining (Das *et al.*, 2010) and its profitability is in question with the rise in input costs. In this context, new resource conservation technologies like system of rice intensification (SRI) and integrated crop management (ICM) appears to have potential that saves inputs, protects the environment and could improve productivity and soil health (Satyanarayana *et al.*, 2006, Balasubramanian *et al.*, 2007). The System of Rice Intensification (SRI), developed in Madagascar over a 20-year period and synthesized in the early 1980s offers opportunities to researchers and farmers to expand their understanding of potentials already existing in the rice genome. Experience with SRI methods suggests that average rice yields could be about double the present world average without replacement of cultivar or the use of purchased inputs. Of late, certain modifications in SRI has been suggested by IRRI scientists in Philippines named Integrated Crop Management (ICM) to suit the local needs involving integrated use of best management practices to increase the productivity and income (Rajendran *et al.*, 2005).

Table 11.1: Block Wise Area Production and Productivity *of* *Sali* Rice in South Garo Hills, Meghalaya

CD Block	*Area (ha)*	*Production (tonne)*	*Productivity (kg/ha)*
Chokpot	800	1156	1445
Rongara	598	368	616
Gasuapara	1177	986	1161
Baghmara	763	867	1136
Total	**3338**	**3515**	**1053**

Source: District Agriculture Office, South Garo Hills (2009-10).

Hence, ICAR Research Complex for NEH Region, Umiam in collaboration with Krishi Vigyan Kendra (KVK), Tura has undertaken interventions on crop establishment methods *viz.*, System of Rice Intensification (SRI) and Integrated Crop management (ICM) along with Integrated Weed Management (IWM) on *sali* rice during 2008-10 to increase the rice productivity and income of the farmers of South Garo Hills district of Meghalaya. With the encouraging result from demonstrations, the present study was undertaken to assess the socio-economic impact of resource conservation technologies (SRI and ICM) along with integrated weed management(IWM) for improvement of productivity and income for sustainable development in valleys and foot hills of rice farmers.

The Resource Conservation Technologies

The field demonstrations on SRI and ICM in rice paddies were carried out during consecutive three rainy seasons of 2008-09 (introduction), 2009-10 and 2010-11 in Sibbari cluster of South Garo Hills, Meghalaya, India consisting of 9 villages from Gasuapara Community Development Block (Table 11.2). The socio-economic impacts of SRI, ICM vis-a-vis conventional rice cultivation practices in the context of farmer's realities were evaluated during 2010-11. The data were collected by using well structured questionnaire survey in three different groups of farmers' *viz.*, SRI farmers, ICM farmers and conventional rice farmers (CRC). A representative sample population of 134 rice farmers including 41 SRI and 41 ICM farmers and 52 conventional rice farmers were selected by following the criteria of stratified multistage sampling procedure. The information was collected on socio-economic profile of farmers, general information about SRI, ICM methodology and its advantages, disadvantages, extension process, ranking of advantages, comparison of SRI and ICM with CRC in terms of kinds, costs of production, weed management practices, input utilization, economic viability etc. The introduction and popularization of SRI and ICM methodology of rice cultivation in Sibbari cluster of South Garo Hills, Meghalaya was led by ICAR, Umiam and Krishi Vigyan Kendra (KVK), Tura during 2008-09 as a base year (training, demonstration in Institute field, community field, and 2 progressive farmers field, full cost borne by the Institute) and actual demonstration (Only cost of seed, fertilizer, manure, paddy weeder etc. borne by the Institute) was started during 2009-10 for the first time with 10 farmers only. However, the number of interested farmers was increased to 41 during 2010 by seeing the technology in 9 different villages under the Gasuapara Community Development Block. The methodology for resource conservation technology of rice demonstrated which involves 10 days old seedlings at 1 seedling/hill was transplanted in square geometry with 25 x 25 cm spacing under System of Rice Intensification (SRI) and 20 days old seedlings at 2 seedlings/hill in 20 x 20 cm spacing under Integrated Crop management (ICM) method. However, in conventional rice cultivation (CRC) method, 30 days old seedlings at 3 seedlings/hill with a spacing of 20 x 15 cm were demonstrated. Due to the success in demonstration in the Sibbari cluster which began to extended in more and more farmers and adopt these crop establishment methods for higher productivity, income and improve livelihood. There are 301 household in 9 different villages of Sibbari cluster under rice cultivation and 10 of them were SRI and ICM farmers in 2009 and 41 in 2010 from none in 2008. Since 2009 was the first year of SRI and ICM adoption for most farmers in the cluster of villages, the impact on livelihoods cannot be very evident yet. SRI is gaining popularity in many countries including India due to its advantages over conventional method *viz.* water and seed saving, high yield and less dependence on chemicals (Uphoff 2003, Rao *et al.*, 2007). A more in depth socio-economic assessment should be undertaken in a few years. However, the merits of SRI and ICM such as saving seeds, labour, water, and high yield etc has already been seen from the demonstration and the questionnaire survey that provided the data for this analysis.

Table 11.2: List of Villages Along with the Altitude and Longitude of Sibbari Cluster, Gasuapara Block, South Garo Hills, Meghalaya

Sl.No.	*Name of the Village*	*Longitude*	*Latitude*	*Altitude (m)*	*Household*
1.	Jadugre	25 ° 09′ 55′′ N	90 °26′ 00′′ E	20	36
2.	Mandanggre	25° 10′ 08′′ N	90 °28′39′′ E	27	43
3.	Defuliapara	25° 10′ 13′′ N	90 °29′ 00′′ E	28	54
4.	Onaijora	25 °10′ 14′′ N	90° 29′ 25′′ E	24	16
5.	Dimagre	25°10′ 26′′N	90° 29′26′′ E	31	54
6.	Dimachigitchak	25° 10′ 31′′ N	90 °29′30′′ E	32	36
7.	Chambil tolejan.	25°11′ 27′′ N	90°29′ 59′′ E	19	31
8.	Kongkona	25° 10′26′′ N	90° 30′04′′ E	17	36
9.	Batabari	25° 10′ 32′′ N	90° 30′29′′ E	15	40

Source: NAIP Base line survey report (2007-08), KVK (ICAR), Tura.

Adoption and Dissemination of SRI and ICM Technology

SRI and ICM technology was rather new to most of the farmers in Sibbari cluster villages. Among the 134 interviewees, 24.39 per cent (41 SRI and 41 ICM farmers) heard of it for the first time only in 2009, while 75.61 per cent first learned of it in 2010. Most of the 10 household/farmers each who used SRI and ICM in the first year were the member of the *Apex Body of NAIP*, Component-III of Sibbari cluster villages or relatives of the village committee. These numbers increased to 41 each in SRI and ICM technology during 2010. Out of 301 household/rice farmers in the Sibbari cluster,

62.13 per cent were medium wealth and 28.57 per cent were poor farmers. According to a wealth ranking of rice farmers by the village committee, poor farmers accounted 28.83 per cent of SRI users, 24.39 per cent ICM users compared with being 34.33 per cent of the total rice farmers in the cluster villages, while the proportion of poor SRI and ICM farmers was a little bit lower than average for the cluster villages, the difference was not really significant. SRI and ICM adopters were roughly proportional to the distribution of farmers (households) according to wealth ranking in the cluster villages.

The average size of paddy field for the households in the cluster villages was 1.18 ha before the KVK has mad interventions during 2008 which was decreasing and adopted more area under rice cultivation replacing the conventional method of rice cultivation during 2009 and 2010. It could be seen that the average size of SRI and ICM plots per household/farmer was increased dramatically from 0.03 ha and 0.14 ha/household in 2009 to 0.097 ha and 0.32 ha/household in 2010, respectively. This might be due to conversion of fallow land to cultivable land under rice although this requires confirmation.

The data revealed that most of the farmers got their SRI (60.97 per cent) and ICM (68.29 per cent) information from the ICAR, KVK, Tura and village committee (Apex Body of NAIP, Component-III) during the implementation of the project in the cluster villages (Table 11.3). When asked "have you introduced SRI/ICM to others?" only few farmers gave the answer yes. Further inquiry showed that those active in promoting SRI/ICM were mostly members of the village committee or member of SHGs/farmers association in the Sibbari cluster villages. Most of the farmers had not spread the information partly because they themselves were not completely sure about the technology, this being their first or second year of adoption. Nevertheless, some expressed their wish to promote SRI/ICM to their friends and relatives in other villages. As the yield increased to 204.57 per cent and 184.93 per cent in SRI and ICM, respectively during 2010 and most of the farmers showed their interest to expand their area under SRI/ICM or to keep their full paddy area under SRI/ICM technology in future, which reflects a favourable opinion to SRI and ICM rice establishment methods.

Table 11.3: Frequency of Sources of Information on SRI, ICM Methods with Integrated Weed Management in Rice Cultivation (N=41)

Sources of Information	*SRI Method*		*ICM Method*	
	Frequency	*Per cent (per cent)*	*Frequency*	*Per cent (per cent)*
Neighbours	3	7.32	2	4.88
Relatives	2	4.88	0	0
Extension personal	1	2.44	0	0
KVK (ICAR),Tura	25	60.97	28	68.29
Village committee	9	21.95	11	26.83
No information	1	2.44	0	0
Total	**41**	**100.00**	**41**	**100.00**

Farmers' Impression on SRI and ICM Technology

SRI/ICM Farmers

As mentioned above, rapid progress was made in the diffusion of SRI/ICM since 2009 in terms of both the number of household/farmers and the area of SRI/ICM fields. This can be ascribed mostly to the advantages that SRI/ICM offers most prominently for saving seeds, nutrients, water, cost etc. According to questionnaire survey, saving seeds(97.56 per cent in SRI and 60.98 per cent in ICM), lower cost(70.73 per cent in SRI and ICM), higher yields(68.29 per cent in SRI and 78.05 per cent in ICM), were tested as first three advantages of SRI/ICM with some other benefit also mentioned such as less fertiliser (68.29 per cent in SRI and 58.54 per cent in ICM), less pest and diseases(53.66 per cent in SRI and 58.54 per cent in ICM), more milling output(60.98 per cent in SRI and 73.17 per cent in ICM) etc. However, about 63.41 per cent of SRI farmers and 39.02 per cent ICM farmers' opinion that these technology had not saved labour for nursery raising, transplanting, weeding etc. (Table 11.4).

Table 11.4: Advantages of SRI and ICM Over Conventional Method Based on Questionnaire Survey

(N=41)

Advantages	*SRI Method*				*ICM Method*			
	Yes		*No*		*Yes*		*No*	
	Frequency	*Per cent*	*Frequency*	*Per cent*	*Frequency*	*Per cent*	*Frequency*	*Per cent*
Higher yield	28	68.29	13	31.71	32	78.05	9	21.95
Saving labour	15	36.59	26	63.41	25	60.98	16	39.02
Saving water	32	78.05	9	21.95	22	53.66	19	46.34
Saving seeds	40	97.56	1	2.44	25	60.98	16	39.02
Less disease and pests	22	53.66	19	46.34	24	58.54	17	41.46
Less pesticides	16	39.02	25	60.98	28	68.29	13	31.71
Lower cost	29	70.73	12	29.27	29	70.73	12	29.27
Less fertiliser	28	68.29	13	31.71	24	58.54	17	41.46
More milling output	25	60.98	16	39.02	30	73.17	11	26.83
Others	28	68.29	13	31.71	25	60.98	16	39.02

Non-SRI/ICM or Conventional Farmers

There were about 52 non-SRI/ICM or conventional farmers in the sample population from Sibbari cluster villages accounting 38.81 per cent of the total sample population of 134 rice farmers. Majority of them were medium or poor farmers. According to the result of questionnaire survey (Table 11.5), the main reason for non adoption of SRI/ICM related to more efforts (90.93 per cent ranked I), faith towards traditional weed control methods(86.54 per cent ranked II), lack of knowledge (67.31 per cent, ranked III) followed by more weeds(63.48 per cent, ranked IV), lack of guidance

(55.77 per cent ranked V) and water management(46.15 per cent). More attention and effort for water management was needed when young seedlings with 1 or 2 tillers were transplanted in to the rice field, to keep them from being flooded under SRI and ICM method which is not a problem for conventional method with older and stronger seedlings. The water management constraint was a real one because all the farmers were dependent on rainfall and no other irrigation sources are available except stream diversion. In some cases also, water management problem arise because the paddies are not well-levelled.

Table 11.5: Reason for Non-Adoption of SRI, ICM Methods and Integrated Weed Management (IWM) in Rice Cultivation

(N=52)

Reasons for Non-Adoption	*No. of Respondent**	*Per cent*	*Rank*
More efforts needed in SRI and ICM	47	90.39	I
Do not know how to do SRI, ICM	35	67.31	III
Do not know the merits of SRI, ICM and IWM	22	42.31	
More weeds to deal with	33	63.46	IV
More seeds needed	12	23.08	
Lack of knowledge of IWM	26	50.00	
Lack of guidance	29	55.77	V
Fear of high cost involvement	18	34.62	
Fear of herbicide side effect on crop	25	48.08	
Faith towards traditional weed control methods	45	86.54	II
Adopted mechanical weed control	7	13.46	
Lack of irrigation facilities	24	46.15	
Lack of skill labour	18	34.62	
Unwillingness	6	11.54	

* Recorded multiple responses.

Impacts on Farmers' Livelihoods

Productivity and Income

The average productivity of rice under demonstration in the Sibbari cluster was 57.43 q/ha in SRI and 52.88 q/ha in ICM methods compared to 18.53 q/ha in conventional method which began to extended in more and more farmers and adopt these crop establishment method for higher productivity, income and improve livelihood, though not yet on a large scale. The gross income from rice cultivation was realized in SRI methods (Rs. 57,432/ha) followed by ICM method (Rs.52, 875/ha) and conventional method (Rs. 18,525/ha) with net return of Rs. 36,788, Rs. 31,830 and Rs. 2,319/ha in SRI, ICM and conventional methods, respectively. The higher benefit cost ratio on variable cost was fetched by SRI farmers (2.78) followed by ICM farmer (2.51) and conventional farmers (1.14) which might be due to lower total variable cost, higher yield and income in SRI method compared to ICM and conventional methods (Table 11.6). Similar result was also reported by Haque *et al.* (1997).

Table 11.6: Economic Viability and Resource Utilization in SRI and ICM Over CRC in Rice Cultivation

Resources	*SRI*		*ICM*		*CRC*	
	Value (Rs/ha)	*Per cent Utilized*	*Value (Rs/ha)*	*Per cent Utilized*	*Value (Rs/ha)*	*Per cent Utilized*
Seed	125	0.61	500	2.38	2000	12.34
FYM/vermicompost	2550	12.35	2550	12.12	1500	9.26
Fertilisers	4152	20.11	4785	22.74	875	5.39
Labour	9524	46.13	8835	41.98	7976	49.22
Power tiller	3750	18.17	3750	17.81	3750	23.14
Plant protection chemical	543	2.63	625	2.97	105	0.65
Total variable cost	20644	100.00	21045	100.00	16206	100
Gross return	57432		52875		18525	
Net return	36788		31830		2319	
B:C ratio	2.78		2.51		1.14	

Economic Viability and Resource Utilization

The resources utilized in rice cultivation in the study area of South Garo Hills district of Meghalaya is given in Table 11.6. Since, only the variable cost was important in the short-run in influencing the decision-making of the farmers, only these were considered for deriving the percentage of resources utilized in rice cultivation. The maximum share of resource utilization was accounted for by labour (46.13 and 41.98 per cent) followed by fertiliser (20.11 and 22.74 per cent) and power tiller (18.17 and 17.81 per cent) of total variable cost in SRI and ICM farmer, respectively. However, the maximum share of resource utilization was accounted for by the conventional farmers was in labour (49.22 per cent), power tiller (23.14 per cent) and seeds (12.34 per cent). This is might be due to the fact that the farmers are very poor and unable to afford costly resources for the cultivation of rice. Similar result was also reported by Suresh and Reddy (2006).

Conclusion

It is important to understand the performance of any new technology in the context of farmer's realities, which are always more complicated and more diversified than on any research station. The results from this study showed that the SRI and ICM methods of rice cultivation could save resources like seeds, water, nutrient, soil health and environment, reduced cost, higher yield etc over conventional method. The study further revealed that the main reason for non adoption of SRI/ICM related to more efforts, faith towards traditional weed control methods, lack of knowledge followed by more weeds, lack of guidance and water management. The higher gross income, net return and benefit cost ratio was realized in SRI methods followed by ICM method and conventional method of rice cultivation. The higher benefit cost ratio on variable cost was fetched by SRI farmers (2.78) followed by ICM farmer (2.51) and conventional farmers (1.14). Hence, the resource conservation technology of SRI and ICM along with integrated weed management (IWM) could replace the conventional method for higher productivity and income of rice, which may lead to improvement of livelihood of the farmers of Sibbari cluster of South Garo Hills district of Meghalaya.

Acknowledgement

The authors duly acknowledge the Indian Council of Agricultural Research, New Delhi for funding to undertake the study under the project "*National Agricultural Innovation Project* (*NAIP*)" at Sibbari cluster of South Garo Hills district of Meghalaya, India during 2007-11.

References

Anonymous (2009-10). Government of Meghalaya, *Statistical Hand Book, South Garo Hills, Meghalaya.*

Balasubramanian, V., Sie, M., Hijmans, R.J. and Otsuka, K. (2007). Increasing rice production Sub Saharan Africa: Challenges and opportunities. *Advances in Agronomy* 94: 55-126.

Das, A., Tomar, J.M.S., Ramesh, T., Munda, G.C., Ghosh, P.K. and Patel, D.P. (2010). Productivity and economics of low land rice as influenced by N-fixing tree leaves under mid-altitude subtropical Meghalaya. *Nutrient Cycling in Agro-ecosystems* 87: 9-19.

Haque, M.A., Mustafi, B.A.A. and Sarkar, A.L. (1997). Rice production in absentee owners land in rainfed saline area of Bangladesh. *Economic Affairs,* 42(1): 14-22.

Nandhini, U.S., Alagumani, T. and Shibi, S. (2006). Economic analysis of agriculture in Southern parts of coastal India. *Agricultura Tropica ET Subtropica,* 39(4): 279-284.

Oniah, M.O., Kuye, O.O. and Idiong, I.C. (2008). Efficiency of Resource 1 Use in Small Scale Swamp Rice Production in Obubra Local Government Area of Cross River State, Nigeria. *Middle-East Journal of Scientific Research,* 3(3): 145-148.

Patel, D.P., Das, A., Munda, G.C., Ghosh, P.K., Ngachan, S.V., Kumar, R. and Saha, R. (2008). SRI and ICM Rice Culture for Water Economy and Higher Productivity. *Research Bulletin* No: 68. ICAR Research Complex for NEH Region, Umiam, Meghalaya, pp. 1-28.

Rajendran, R., Ravi, V., Valliappan, K., Nadanasabapathy, T., Jayaraj, T., Ramanathan, S. and Balasubramanian, V. ,2005. Early production of robust seedlings through modified mat nursery for enhancing rice (*Oryza sativa*) productivity and profit. *Indian Journal of Agronomy,* 50(20): 132-136.

Rao, K.S., Annie, Poonam, Ghosh, A., Mohapatia, P.C. and Pandey, M.P. (2007). Present Status Of SRI Research at CRRI, Cuttack, India, 2nd 16 National Symposium on

SRI in India – Progress and Prospects - Papers and Extended Summaries, 3th – 5th 17 October 2007, Agartala, Tripura, India, pp. 35-37.

Satyanarayana, A., Thiyagarajan, T.M. and Uphoff, N. (2006). Opportunities for water saving with higher yield from the system of rice intensification. *Irrigation Science*, Springer-Verlag publication.

Shekhar, J., Mamkotia, B.S. and Dev, S.P. (2009). Productivity and economics of rice (Oryza sativa) in system of rice intensification in North-Western Himalayas. *Indian Journal of Agronomy*, 54(4): 423-427.

Singh, S.P., Kumar, R.M., Sreedevi, B., Krishnamurthy, P. and Subbaiah, S.V., 2007. System of rice intensification and integrated crop management for sustaining rice production under irrigated ecosystem. In: *Second National Symposium on SRI*, 3-5 October, 2007, Agartala, Tripura, India pp. 55-56.

Stoop, W.A., Uphoff, N. and Kassam, A. (2002). A review of agricultural research issues raised by the system of rice intensification (SRI) from Madagascar: opportunities for improving farming systems for resource poor farmers. *Agricultural Systems*, 71: 249-274.

Suresh, A. and Reddy, T.R.K. (2006). Resource Use Efficiency of paddy cultivation in Peech Command Area of Thrissur district of Kerala: An Economic Analysis. *Agricultural Economic Research Review*, 19: 159-171.

Uphoff, N. (2003). SRI - Higher yields with fewer external input ? The system of rice intensification and potential contributions to agricultural sustainability. *International Journal of Agricultural Sustainability*, 1(1): 38-50.

Chapter 12

Impact of Shifting Cultivation on Sustainable Livelihood of Hill Tribes in North Eastern Region of India

J.W. Momin and D.C. Kalita

Department of RDAP,
North-Eastern Hill University, Tura Campus, Meghalaya

ABSTRACT

The study envisages the practices of traditional sifting cultivation and its role on sustainable livelihood of hill tribes of the region. In the North-Eastern Region of India more than two-third of total geographical area is covered by hills. Shifting cultivation (*Jhum* cultivation) is the common practice in the region. About 4.43 lakh families solely depend on shifting cultivation (Ngachan, Mohanty and Pattanayak, 2010). The practice of *jhum* has been undergoing rapid changes particularly in the recent times. It is not a productive cultivation method that affects the livelihood of the hill people. Since the jhum system has a number of demerits, cultivation practices resulted serious adverse implication to the poor in terms of availability of food grains, soil nutrients, availability of timbers, fuel wood and non-timber forest products (NTFP) and cropping intensity. The continuance of *jhum* in the region is closely linked to ecosystem, socio-economic, cultures and land tenure systems of tribal communities. As community owns the lands, the village council or elders divide the *jhum* land among families for their subsistence on a rotational basis. The *jhum* cultivation from the point of view of sustainability of ecological, economical and livelihoods of the tribal people is highlighted in the chapter, examine the role of land use pattern, food security and forest covers as an indicators of the study. The present study is undertaken on the basis of secondary data available in the previous publications and reports of Government Departments. The study attempts to elaborate the present *jhum* cultivation practices and its negative impact on sustainable livelihood of the people of the region.

Keywords: Impact, Shifting cultivation, Sustainable livelihood, Land use.

Introduction

Agriculture is the mainstay of the people of North-Eastern Region of India where shifting cultivation is a predominant practice in the majority of the hilly tracts of the region. Northeast states viz, Manipur, Tripura, Arunachal Pradesh, Nagaland, Meghalaya except Assam have a large chunk of areas under hilly tracts or terrain. Upland rice in the *jhums* through shifting cultivation systems inter-cropped with maize, pulses etc. is the main crops in the region. There are restrictions of scientific cultivation of every crop due to the natural topography or geographical situations in the hill slopes. It is the most widespread farming system whereby annual and short-term perennial crop production is alternated with longer periods of vegetative fallow. The increase in human population in the region has put tremendous pressures on land. The extension of crop lands, for increasing food production, has been directly responsible for the reduction in areas under forests and grass lands. The current practice of shifting cultivation in north-eastern region of India is an extravagant and unscientific form of land use.

Table 12.1: Families and Tribes Involve in Shifting Cultivation

State	*No. of District*	*No. of Tribal Families*	*Ethnic Groups*
Arunachal Pradesh	10	54,000 (16.49)	Aka. Bangro, Dafla. adi, Miniyong, Padam, Miri, Mishimi
Assam	2	58,000 (17.72)	Karbi, Hmar, Miri, Dafla
Manipur	5	70,000 (21.38)	Kuki, Tangkhul, Hmar; Mao, Marang, Kabui, Kacha Naga
Meghalaya	5	52,290 (15.97)	Khasi, Garo, Jaintia
Mizoram	3	50,000 (15.27)	Mizo, Kuki, Hmar; Lakher
Nagaland	7	1,16,046 (35.45)	Sema, Ao. lotha, Konyak, Rengma, Tangkhul
Tripura	3	43,000 (13.13)	Tripuri, Jamatia,Kuki.Garo,Reang
Total	35	3,27,290	38

Source: Ngachan, S.V., Mohanty, A. K. and Pattanayak, A. 2010. "Status Paper on Rice in North-Eastern Region of India." *Rice Knowledge Management Portal*. Hyderabad. pp. 14.

Figure in the brackets indicate the percentages to total.

About 4.43 lakh families belongs to 38 different ethnic tribes solely depend on shifting cultivation covering the area of 386.3 hectares in 35 numbers of districts in 7 states (Ngachan, Mohanty and Pattanayak, 2010). The characteristics of *jhum* cultivation in the region are commonly as follows: (*i*) Cutting and clearing of forest areas and burning of the dried biomass by fire, (*ii*) rotation of *jhum* land every four to seven years, (*iii*) use of human labour as the chief input, (*iv*) non-employment of animals implements or machinery, (*v*) collective ownership of land, (*vi*) reciprocal labour sharing and (*vii*) mixed cropping system. The produce is transported from the *jhum* land to the village by head-loading. *The earlier 15–20 year cycle of shifting cultivation on a particular land has reduced to 2–3 years now. This has resulted in large-scale deforestation, soil and nutrient loss, and invasion by weeds and other species. The indigenous biodiversity has been affected to a large extent.* This has made the land highly unproductive and is alarmingly leading to extensive land degradation and imbalance in the socio–economic setup of the region. Thus, the overall objective of the study is to analyze the negative impact of shifting cultivation on sustainable livelihood of the people and its role on degrading the environment and economy of the region.

Materials and Methods

The present study is based entirely on secondary sources of data. The secondary data in respect of soil nutrients, land use practices, food security and forest covers were collected from various publications. However, an attempt has been made to study the present cultivation practices and its negative impact on sustainable livelihood of the tribal community in the region. Statistical tools like simple percentage and average are employed to analyze the data.

Results and Discussions

Loss of Soil Nutrients

The problem of soil erosion due to shifting cultivation is very serious in the region. The quantum of soil erosion varies from 84 to 170 t ha^{-1} year in bare fallow. The intensity of erosion depends on slope, soil characteristics, crop coverage, depth of soil and rainfall pattern etc. The annual area under shifting cultivation in the region is 3869 km^2, whereas total area affected is 14,660 km^2 and about 443.3 thousand families are involved in the practice. As high as 70.6 and 130.2 t ha^{-1} of annual soil loss has been reported during first and second year of shifting cultivation on a hill having a slope of 70 per cent (Singh and Singh 1978). Annual soil loss from the region is 88.3 million tons from shifting cultivation land, 90.7 million tons from other agricultural land and 422.7 million tons from non-agricultural land. It is resulted that annual loss of N, P, and K is 686.0, 100.2, and 511.0 thousand tons, respectively from shifting cultivation alone (Table 12.2). The soil fertility is on decline as there is limited material to burn and add to the soil. The results showed that at least 20 years time is necessary for rejuvenation of enough vegetation to get optimum available major nutrients for crop support. In 20 years cycle, the available N, P_2O_5 and K_2O increased by 55.1 per cent, 22.8 per cent and 128.4 per cent, respectively over their initial status. The low increase in available P_2O_5 compared to other nutrients was due to the reason that soil was strongly acidic in reaction, most of the phosphorus got fixed as aluminum and iron compounds.

Land Use Pattern

About 1.6 million hectare area is under shifting cultivation in the region. Out of 4.0 million hectare net sown area of the region, roughly 1.3 million hectare suffers from serious soil erosion problem (Table 12.2). The average area available for cropping in the region is only 32.5 per cent of the total geographical area. Net sown area is highest in Assam (34.12 per cent) followed by Tripura (23.48 per cent). Arunachal Pradesh has lowest net sown area in the region. Cropping intensity is highest in Tripura (156.5 per cent) followed by Manipur (152.1 per cent), Mizoram (136.36 per cent) and Assam (123.59 per cent). The general overview of the data reveals that little expansion of the Net Area Sown over the year has taken place. The potential net sown area could be increased if and when the fallow lands are utilized for cultivation purposes. The cultivable waste land of the region is 20.11 per cent of the geographical area a part of which might be progressively utilized for cultivation purpose in the long run. The cropping intensity of the region is 121 per cent. Moreover cropping intensity has also not revealed from almost stationery Gross Cropped Areas. In the past, when the land was in abundance and population sparse, the average rotational cycle of shifting cultivation in the region used to be 25 to 30 years, the land getting enough time for rejuvenation of vegetation. However, with increase in population, the rotational cycle has come down to 2 to 10 years and the land does not get enough time for rejuvenation. The typology of *jhum* along with the cycle, the consecutive cultivation period and the subsequent fallowing period which is known as *jhum* cycle differs from tribe to tribe. The fallowing period is mainly depend upon the density of population of a particular area, availability of suitable hill slopes, composition of the soil and the capacity of the same to recoup its fertility through natural process, the higher the density of population the shorter the *jhum* cycle.

Table 12.2: Annual Area, Fallow Periods and Soil and Nutrient Loss Under Shifting Cultivation

States	*Area ('000) ha*	*Fallow Period (Year)*	*Soil Loss (MT)*	*Nutrient Loss ('000 T)*		
				N	*P*	*K*
Arunachal Pradesh	70.0	3-10	14.5	217	36.6	153
Assam	69.6	2-10	12.3	201	33.4	155
Manipur	90.0	4-7	20.4	76	7.4	63
Meghalaya	53.0	5-7	14.2	62	7.0	48
Mizoram	63.0	3-4	13.0	60	6.9	40
Nagaland	19.0	5-8	8.0	44	5.2	34
Tripura	22.3	4-9	5.9	26	3.7	18
Total	386.8		88.3	686	100.2	511

Source: Sharma and Sharma. 2010. "Socio-economic aspects and impact of land use change on sediment production dynamics in the northeastern region of India." *Land Reclamation.* 42 (1). Warsaw. Poland. p. 213.

Food Security

Food grains constitute the main food items of the entire population. Hence, food grains production requires top priority attention, so as to reach nearer to self-sufficiency in the near future and to catch up with the food grain requirement of the growing population. Shifting cultivation of upland rice is the most important crops in all rural areas of the region where diverse range of land use and off-farm activities including fruit gardens, animal husbandry, collection of non-timber forest products and wage labour are exceptional. During 2000-2001, the annual average growth of production and productivity of rice were 2.95 per cent and 1.68 per cent respectively in the N.E. states against the corresponding all-India rate of 3.6 per cent and 2.68 per cent. Total food grain production increased from 27.19 lakh tons to 47.24 lakh tons while yield increased from 1,002 kg/ha to 1,424 kg/ha during the period. The percentage increase of production of rice was recorded as 18.00 per cent in 2008-09 over 2000-01 with an increasing trend in national rice production from 2000-2001 to 2008-2009, the increase being 12.78 per cent in 2008-09 over 2000-01. The productivity of rice was highest in Tripura (2586 kg/ha) followed by Manipur (2357 kg/ha) and Nagaland (1994 kg/ha) during 2008-09. Assam registered highest production (41.74 lakh MT) in 2008-09 sharing 62.18 per cent of the total NE rice production (67.18 lakh MT) in 2008-09. However, the productivity was much less against the national average of 2178 kg/ha of 2008-09. The share of NE India to total national production of rice was 6.96 per cent in 2008-09, respectively. The average productivity is resulted decreasing due to loss of fertility in the region (Table 12.4). Since food productivity is highly dependent on the spatial and seasonal changes in water availability, the food security cannot be ensured for all times unless suitable measures are undertaken to minimize the effect of runoff on soil erosion. The region has a food grains deficit of about 2.5 million tons and the deficit gap is widening year after year (Table 12.5).

The data on food grain production and requirement indicate that there is a deficit existing in all the states of the region varying from 10 per cent for Tripura to 69 per cent for Mizoram, except Nagaland having some surplus food grains (Table 12.5). In cereals alone there is about 19 per cent deficiency in the region (Singh *et al.*, 2010). The region as a whole is deficit of about 3.14 mt of food grains (25 per cent deficiency). The projected food grain demand for the region is 10 mt and 12.5 mt for the year 2020 and 2030, respectively. In order to make the region self sufficient in food grain production, the productivity of all the food crops has to be increased from the present low level with effective utilization of natural resources and wide variabilities in climatic conditions. Problems of soil erosion, lack of technical skills, and mixed cropping which produces lower yields owing favourable climatic condition are the major concern for food security.

Changes in Forest Covers

Shifting cultivation causes loss of flora and fauna which includes precious species of tree plants, shrubs, medicinal plants and minor forest products. Forest cover in the region is below the national norm of recommended for hilly areas. This is because a sizable proportion of the forest area is reportedly under shifting cultivation resulting in depletion of the forest cover. Total forest cover in the region is 14.2 million ha, which is about 66.5 per cent of the geographical area as against the national average of 19.39 per cent. The rate of deforestation is about one per cent per year (Table 12.6) and it is emphasized that deforestation in this context refers to the annual change (increase) in the area supporting annual crops compared to overall forest

Table 12.3: Land Use Pattern in North East India

State	Land Use ('000 ha)									
	Reporting Area	Forest	Non-Available for Cultivation	Permanent Pastures and Other Grazing Land	Land Under Misc. Tree Crops and Grooves	Cultivable Wasteland	Fallow Lands Other than Current Fallows	Current Fallow	Net Area Sown	Total
Arunachal Pradesh	5659	5154	65	18	37	67	73	37	209	8,374
Assam	7850	1950	2512	160	209	77	60	127	2753	7,844
Manipur	1951	1693	27	1	6	1	0	0	224	2,233
Meghalaya	2227	942	228	0	158	450	169	68	213	2243
Mizoram	1946	1594	134	5	10	5	82	24	92	2180
Nagaland	1595	863	75	0	120	57	76	82	322	1658
Tripura	1049	606	134	0	27	1	1	1	280	1049
Total	22,277	12,802	3175	184	567	658	461	339	4093	25,581

Source: Land Use Statistic, Ministry of Agriculture, GOI, 2009. p. 15.

Table 12.4: Area (A), Production (P) and Productivity (Y) of Upland Rice Under Shifting Cultivation

States	*2000-2001*			*2008-2009*			*Increase/ Decrease of Production in 2008-2009 Over 2000-2001 (per cent)*
	Area (Lakh Ha)	*Production (Lakh MT)*	*Yield (Kg/ Ha)*	*Area (Lakh Ha)*	*Production (Lakh MT)*	*Yield (Kg/ Ha)*	
Arunachal Pradesh	1.19	1.26	1059	1.26	2.19	1293	73.80
Assam	26.75	39.99	1495	24.84	41.74	1614	4.37
Manipur	1.95	4.53	2320	1.68	5.79	2357	28.09
Meghalaya	1.08	1.91	1772	1.08	3.10	1886	60.32
Mizoram	0.57	1.09	1913	0.52	0.79	885	–27.52
Nagaland	1.46	2.00	1411	1.73	3.95	1994	97.5
Tripura	2.56	5.88	2297	2.42	9.30	2586	58.16
NER Total	35.56	56.66	1752	33.53	66.86	1802	18.00
All India	449.72	855.00	1901	455.37	964.31	2178	12.78

Source: Ngachan, S. V., Mohanty, A. K. and Pattanayak, A. 2010. "Status Paper on Rice in North-Eastern Region of India." *Rice Knowledge Management Portal.* Hyderabad. pp. 18-19.

Table 12.5: Food Grains Situation in the Northeastern Region (Requirements for 2011 Population)

States	*Food Grains (Million Tons)*			
	Production (000' t)	*Requirement (000' t)*	*Deficit (000' t)*	*Deficit (per cent)*
Arunachal Pradesh	250.4	268.1	–17.7	–6
Assam	3714.2	6043.1	–2328.9	–38
Manipur	401.3	527.7	–126.4	–23
Meghalaya	233.7	574.7	–341.0	–59
Mizoram	65.2	211.5	–146.4	–69
Nagaland	491.5	384.0	–107.5	28
Tripura	637.6	711.7	–74.2	–10
Total Food Grain	5793.9	8720.8	–3142.1	–25.28

Source: Singh, *et al.*, 2010. *Food Security and Climate Change in Himalayan Region: Status and Strategies for Livelihood Improvement.* ICAR Publication: New Delhi. p. 12.

area. The region has complex environmental gradient involving changes in physiography, forest structure, climate, geology and natural disturbances. An analysis of the forest cover statistics of the region reveals some unexpected and contradictory trend (Table 12.6). During the period 1999-2001, the region witnessed an increase in forest cover of 2374 Km^2 or annual growth of 1.28 per cent. Increase in forest cover in the states of Assam (7.26) and Tripura (9.34) in 1991-2001 and Meghalaya (3.96 per cent) and Tripura (9.34 per cent) in 2001-2003 are the main drivers of growth during this period, while Arunachal Pradesh, Mizoram and Nagaland also witnessed considerable loss in the forest cover during that period (Table 12.7). The growth of forest in the states of Assam and Tripura was maximum than any other states in the region. However, during the period from 2001-2005, there is maximum increase in forest cover of 3950 Km^2 with an annual growth rate of 0.46 per cent. Thus the individual state level data provide a clearer indication on the areas where exactly the forest cover decreases over the period. The overall interpretation of regional forest statistics suggest the drivers of loss of forest cover in terms of threats and pressures being enhanced by erratic changes in local climatic factors due to global change

Table 12.6: Change of Forest Covers in North Eastern Region of India

State	*Total Geographical Area*	*Forest Covers (Km²)*			
		1999	*2001*	*2003*	*2005*
Arunachal Pradesh	83,743	68,847	68,045	68,019	67,777
Assam	78,438	23,688	27,714	27,826	27,645
Manipur	22,327	17,384	16,926	17,219	17,086
Meghalaya	22,429	15,633	15,584	16,389	16,988
Mizoram	21,081	18,338	17,494	18,430	18,684
Nagaland	16,579	14,164	13,345	13,609	13,719
Tripura	10,486	5,745	7,065	8,093	8,155
Total	255083	163799	166173	169585	170054

Source: Forest Survey of India, Ministry of Environment and Forest, GOI, 2009. p. 3.

Table 12.7: Percentage of Annual Change in Forest Cover

State	*Change of Forest Covers (per cent)*		
	1999-2001	*2001-2003*	*2003-2005*
Arunachal Pradesh	–0.59	–0.26	0.06
Assam	7.26	0.04	–0.16
Manipur	–1.35	0.96	–0.51
Meghalaya	–0.16	3.96	0.19
Mizoram	–2.41	2.3	0.27
Nagaland	–3.07	2.39	–1.08
Tripura	9.34	6.51	0.20
Average Total	1.28	2.27	–0.14

Source: Forest Survey of India, MoEF, GOI, 2009. p. 4.

Impact of Work Force in Economic Growth

More than 57.2 per cent of the workers in the region are engaged in agriculture or allied sectors and the rest are marginal or non-workers (Table 12.8). There exists a wide variation in labours output in the region where household industries, processing, etc. and agricultural labourers are negligible. It is indicated that the region with high proportion of cultivators are involved in *jhum* cultivation with less contribution to production. Due to maximum work load adversely involved in *jhum* operation, the people cannot spare their time for other activities like household industries, processing, etc. which would productively contribute for growth.

Table 12.8: Economic Classification of Population in Northeastern Region (per cent)

State	*Cultivators*	*Agricultural Labourers*	*Household Industry, Processing, etc.*	*Marginal Workers/ Non-Workers*
Arunachal Pradesh	60.4	5.1	0.2	34.3
Assam	50.9	12.1	0.8	36.1
Manipur	61.8	6.7	5.8	25.7
Meghalaya	55.3	12.3	0.4	31.8
Mizoram	61.3	3.3	1.0	34.4
Nagaland	72.6	1.4	0.4	25.6
Tripura	38.1	23.9	1.4	36.6
Average Total	57.2	2.25	1.43	32.07

Source: Sharma and Sharma. 2010. "Socio-economic aspects and impact of land use change on sediment production dynamics in the northeastern region of India." *Land Reclamation*. 42 (1). Warsaw. Poland. p. 214.

Conclusions

The economies of hill tribes solely depend on traditional shifting cultivation where the system does not give sufficient yield to sustain their livelihood. This is due to increase in demographic pressure and reduction in the shifting cycle period. The system appears to be relatively sustainable when the fallow period is fairly long. The long-term success of this traditional cultivation depends upon the recovery of soil fertility during the fallow interval. Decline of forest covers is one of the major concern in the region where the growth of forest covers are in decreasing trends. From the above study, it can be noted that the shifting cultivation alone is responsible for deforestation and environmental degradation. Lack of control over land and forests in the hills by the Government Administrations and the acknowledged power of traditional village councils and headmen, and the assertion by the hill tribes of their rights over local natural resources are the main reason for failure in growth of economy. Despite all these problems, the traditional agricultural system is well adapted to the environment and the people's knowledge of growing cereals and tree crops enabled them to maintain an ecological balance. Nonetheless, very recently, it is realized the crisis situation of the cultivation system and efforts are being made to motivate the *jhumias* for permanent cultivation like commercial agroforestry/homegarden, cash crops garden, etc. So, innovations built on integration of merits of traditional and modern farming systems could offer more effective means of addressing the problems of cultivation system than a strategy aiming for replacement of traditional agriculture by the modern one. Proper land use planning to achieve efficient intensive use of land resources is recommended in the region to ensure intensive utilization of land, prevent its deterioration, create wide spread productive employment and reduce disparity. Thus, the practice needs to be replaced with sustainable, socially acceptable and eco-friendly land use systems.

References

Bujarbarua, P. and Baruah, S. (2009). "Vulnerability of fragile forest ecosystem of North East India in context with the global climate change: An ecological projection." *Climate Change: Global Risks, Challenges and Decisions.* IOP Publishing: Philadelphia, USA. pp. 1-4.

Chauhan, B.S. (2001). *Shifting Cultivation in Perspective.* Publication of Nagaland University. pp. 3-67.

Das *et al.* (2006). "Traditional Agriculture Practices in Meghalaya, North-East India". *Indian Journal of Traditional Knowledge.* Vol. 5 (1). pp.17-18.

Deb, B.J. and Ray, B.D. (2006). *"Changing Agricultural Scenario in North-East India".* New Delhi: Concept Publishing Company.

Kumar, P. and Mittal, S. (2006). "Agricultural Productivity Trends in India: Sustainability Issues". *Agricultural Economics Research Review.* Vol. 19. pp. 71-88.

Kumar, A., Marcot, B. G. and Roy, P.S. (2006). "Spatial Patterns and Process for Shifting Cultivation Landscape in Garo Hills, India." *Landscape Ecology.* New Delhi. pp. 2-8.

Mishra, A.K. and Misra, J.P. (2007). "Sustainable Development of Agriculture in North-Eastern India: A Quest for more Economical and Resources Sustainable Alternatives". *ENVIS Bulletin.* Vol. 14 (2). pp. 2-44.

Murtem, G., Sinha, G. N. and Dopum, J. (2008). "*Jhumias* View on Shifting Cultivation in Arunachal Pradesh." *Bulletin of Arunachal Forest Research.* 24 (1 and 2). Arunachal Pradesh. pp. 35-40.

Ngachan, S.V., Mohanty, A.K. and Pattanayak, A. (2010). "Status Paper on Rice in North-Eastern Region of India." *Rice Knowledge Management Portal.* Hyderabad. pp. 14-22.

Seidenberg, C., Mertz, O. and Kias, M.B. (2003). "Fallow, labour and livelihood in shifting cultivation: implications for deforestation in northern Lao PDR." *Danish Journal of Geography.* 103(2). pp. 71-79.

Sharma, U.C. and Sharma, V. (2010). "Socio-economic aspects and impact of land use change on sediment production dynamics in the northeastern region of India." *Land Reclamation.* 42 (1). Warsaw. Poland. pp. 210-217.

Sillitoe, Paul (2008). "Shifting Cultivation and Sustainable Development of North-Eastern India: Tradition in Transition". *Mountain Research and Development*, 28 (1), pp. 88-92.

Singh, A.K. *et al.* (2010). *Food Security and Climate Change in Himalayan Region: Status and Strategies for Livelihood Improvement.* ICAR Publication: New Delhi. pp. 9-18.

http://www.icarneh.ernet.in/publication/Vision per cent 202025, per cent 20 ICARNEH Region.

Chapter 13

Price Spread of Rice in Assam

D.C. Kalita[1]*, B.C. Bhowmick[2], C. Hazarika[2] and R.S. Saikia[2]

[1]Lecturer, Department of agricultural Economics, Nagaland University, SASRD, Campus Medziphema – 797 106
[2]Department of Agricultural Economics, AAU, Jorhat – 13

ABSTRACT

The present study was undertaken in Assam to examine the marketing channels, marketing margins, marketing costs and price spread of wheat. The study was based on primary data collected through personal interview method of 300 samples from three size groups being selected by multi-stage stratified random sampling procedure. Thus the study relates to the sample size of 82, 95, and 123 in small, medium and large size respectively. The data pertain to the year 1997-98. The results showed that there were eight marketing channels of paddy-rice marketing in Assam. The marketing cost showed a positive relationship with the length of the marketing channel being highest (Rs. 405.35 per quintal) in channel III. Marketing margin was found to be highest (Rs. 347.80 per quintal) in channel IV. While the producer's share in consumer's rupee was estimated to be highest (87.42 per cent) in channel V and the index of marketing efficiency was also highest in channel V (6.95).

Keywords: Marketing channel, Marketing cost, Marketing margin, Price spread, Intermediaries.

Introduction

Rice is the major cereal in India contributing about 47 per cent to the total cereal production in the economy and occupies about 23.29 per cent of the total cultivated

---

* *Corresponding Author:* E-mail: durgack@yahoo.com

area during 1995-96 (Anomymous, 1997). In Assam also, rice is the major cereal crop covering 70.50 per cent of the gross cropped area. During the past decades the productivity of rice has increased significantly and this increased production ultimately lead to the increase in marketed surplus and market arrivals. This increased production of rice resulted in the increased arrivals in the market and thereby increase the trade in the state. The market arrivals of rice produce increased from 47914 quintal in 1985-86 to 170519 quintals in 1994-95 in the primary market as against 75793 quintals in 1985-86 to 75793 quintals in 1994-95 in the secondary market in Assam (Anonymous, 1996). This increase of market arrivals needed improvements of marketing efficiency of the existing marketing system and the reduction of marketing costs and margins of the intermediaries involved in the marketing system and increase of producers share in consumers rupee are the indicators of increased marketing efficiency. Acharya and Agarwal (1999) defined price spread as the difference between the price paid by the consumer and the price received by the producers for an equivalent quantity of farm produce. The study of marketing margins and costs are important as they reveal many facets of marketing and the price structure as well as the efficiency of the marketing systems. The study of price spread helps the farmers to know the price structure involved in the marketing process and thereby can help the farmer in increasing the efficiency of marketing systems. This type of study also helps the policy maker to take appropriate price policy. Therefore, the study of marketing costs, margin and price spread in marketing through different channels is important for producers, consumers and ultimately policy makers of the nation. This study also helps in comparison of price paid by the consumers and price received by producers. Hence in this paper an attempt is made to study price spread of paddy-rice in Assam.

Methodology

The study was conducted in all the six agro-climatic zones of Assam covering a total of 300 sample farmers from three size groups *viz.*, 82 small (1.01-2.00 ha), 95 medium (2.01-4.00 ha) and 123 large (above 4.00 ha) farmers following the multistage stratified random sampling method. Marketing costs was calculated by estimating the costs incurred in the process of marketing the rice. The cost incurred after the harvesting of the crop till it reaches the consumers hand constitute the marketing cost that includes transportation cost, handling cost, storage cost, market fees, weighing charges, labour charges for packing, loading and unloading charges etc. While marketing margin at any stages of marketing was estimated by using the following formula.

$$Mm_i = Sp_i - (Pp_i + Mc_i)$$

where,

Mm_i = Marketing margin of the i th middleman

Sp_i = Selling price of the i th middleman

Pp_i = Purchase price of the i th middleman

Mc_i = Marketing cost incurred by the i th middleman

Price spread was calculated as the difference between the price paid by the consumers and the price received by the producers. In order to examine the efficiency of the marketing system, producer's share in consumer's rupee was calculated as follows:

$$\text{Producer's share in consumer's rupee} = \frac{\text{Producers price}}{\text{Consumers price}} \times 100$$

Similarly, the share of total marketing costs and the total marketing margins were also estimated. The marketing efficiency of various channels was calculated by using the Shepherd's formula

$$ME = V/I - I$$

where,

ME = Index of marketing efficiency

V = Value of goods sold (consumers price)

I = Total marketing costs

Results and Discussion

Marketing Channels for Paddy

The study revealed that there are eight major marketing channels of paddy in Assam. These eight channels were as follows.

Channel I: Producer – itinerate trader – primary wholesaler – processor-cum-wholesaler – Retailer – consumer

Channel II: Producer – primary wholesaler –commission agent –cooperative agent –processor-cum-wholesaler – secondary wholesaler – retailer – consumer.

Channel III: Producer –Market Fariah – primary wholesaler –wholesaler – processor-cum-wholesaler – secondary wholesaler – retailer – consumer

Channel IV: Producer – commission agent – processor-cum-wholesaler – secondary wholesaler – retailer – consumer

Channel V: Producer – processor-cum-wholesaler – retailer – consumer

Channel VI: Producer – consumer

Channel VII: Producer – village level wholesaler – processor-cum-wholesaler – retailer – consumer

Channel VIII: Producer – Government agent – retailer – consumer

Table 13.1 represents the marketing cost incurred by different intermediaries in the paddy-rice marketing channel. From the table the highest marketing cost was observed in channel-I, the amount being Rs.405.35 per quintal. It was mainly because of the large number of intermediaries involved in that channel. Mahapatra, *et al.* (1998) also reported that the marketing cost varies according to the length of the

Table 13.1: Marketing Cost of Intermediaries in Paddy Rice Marketing in Different Marketing Channels

Intermediaries	*Channel I*	*Channel II*	*Channel III*	*Channel IV*	*Channel V*	*Channel VI*	*Channel VII*	*Channel VIII*
Producers	52.8 (8.09)	28.9 (9.14)	30.15 (8.43)	35.25 (11.04)	33.5 (9.73)	49.3 (14.95)	–	39.2 (100.00)
Itinerant trader	28.7 (7.08)	–	–	–	–	–	–	–
Petty dealer	–	–	25.35 (7.09)	–	–	–	–	–
Village level wholesaler	53.2 (13.12)	61.4 (19.42)	–	–	–	–	–	–
Primary wholesaler	37.8 (9.33)	–	34.55 (9.66)	–	32.5 (9.44)	–	–	–
Wholesaler	41.4 (10.21)	–	27.1 (7.58)	–	–	–	–	–
Commission agent	–	–	–	38.5 (12.06)	34.2 (9.93)	–	–	–
Processor	195.4 (10.21)	208.3 (65.88)	205.2 (57.39)	225.3 (70.58)	210.5 (61.13)	261.05 (79.04)	–	–
Secondary wholesaler	–	–	15.9 (4.45)	–	18.2 (5.29)	–	–	–
Govt. agent	–	–	–	–	–	–	122.3 (69.92)	–
Retailer	16.05 (3.96)	17.55 (5.55)	19.25 (5.38)	20.15 (6.31)	15.45 (4.49)	19.5 (5.91)	20.9 (11.61)	–
Total marketing cost	405.35 (100.00)	316.15 (100.00)	357.5 (100.00)	319.2 (100.00)	344.35 (100.00)	329.85 (100.00)	180.05 (100.00)	39.2 (100.00)

Figures in parentheses indicate percentages to total cost.

Table 13.2: Marketing Margin of Intermediaries in Paddy Rice Marketing in Different Marketing Channels

Intermediaries	*Channel I*	*Channel II*	*Channel III*	*Channel IV*	*Channel V*	*Channel VI*	*Channel VII*
Itinerant trader	42.45 (13.96)	–	–	–	–	–	–
Petty dealer	–	–	64.1 (18.43)	–	–	–	–
Village level wholesaler	29.3 (9.63)	91.45 (28.47)	–	–	21.5 (6.43)	–	–
Primary wholesaler	21.55 (7.09)	–	17.85 (5.13)	–	–	–	–
Whole saler	24.7 (8.12)	–	12.9 (3.71)	–	–	–	–
Commission agent	–	–	–	22.25 (8.50)	25.05 (7.49)	–	–
Processor	139.3 (45.81)	171.45 (53.50)	121.75 (35.01)	191.05 (72.98)	175.65 (52.53)	189.15 (77.19)	–
Secondary wholesaler	–	–	85.95 (24.71)	–	69.65 (20.81)	–	–
Govt. agent	–	–	–	–	–	–	127.15 (44.84)
Retailer	46.8 (15.39)	57.9 (18.03)	45.25 (13.01)	48.5 (18.53)	42.5 (12.71)	55.9 (22.81)	156.4 (55.16)
Total marketing cost	304.1 (100.00)	321.2 (100.00)	347.8 (100.00)	261.8 (100.00)	334.35 (100.00)	245.05 (100.00)	283.55 (100.00)

Figures in parentheses indicate percentages to total marketing margin.

Table 13.3: Price Spread Analysis in Paddy Rice Marketing System for Different Marketing Channels

Items	*Channel I*	*Channel II*	*Channel III*	*Channel IV*	*Channel V*	*Channel VI*	*Channel VII*	*Channel VIII*
Consumers price (Rs./qt)	974.15	972.95	975.10	925.25	1003.20	971.25	942.55	311.70
Total marketing cost (Rs./qt)	405.35	316.15	357.50	319.20	344.35	329.85	180.05	39.20
Total marketing margin (Rs./qt)	304.10	321.20	347.80	261.80	334.35	245.05	283.55	–
Producer's share in consumer's rupee	27.16	34.57	27.67	37.54	32.62	33.60	37.67	87.42
Index of marketing efficiency	1.40	2.08	1.73	1.90	1.91	1.94	3.12	6.95

distribution channel. The lowest marketing cost of Rs. 39.20 per qt was observed in channel-VIII, where producer directly sold their product to the consumer. The marketing cost was also very low (Rs.180.05 per qt) in channel-VII, where producer sold their commodity through Govt. agent. It is also evident from the table that the processor incurred the maximum amount of marketing costs in all the channels where they are involved. The cost incurred by processor ranged between 48.21 per cent to 79.14 per cent of total marketing costs in various marketing channels of paddy-rice marketing. This might be due to the fact that the processor undertakes the processing operation before he sells the commodity to the next intermediaries. Because of the high amount of the processing charges the marketing cost incurred by the processor becomes very high.

The total marketing margin was highest in channel-III (Rs.347.80 per qt) (Table 13.2). This might be due to the large number of intermediaries associated with this channel, the number of intermediaries being six. Like the marketing cost, the share of marketing margin was also highest for processor. The percentage of marketing margin of the processor to the total margin ranged between 35.01 to 77.19 in various channels. The share of marketing margin was very less for primary wholesaler, wholesaler and commission agent where the percentage was found to be below 10.00 per cent.

The producer's share in consumers rupee was found to be highest (87.42 per cent) in channel VIII, where no intermediaries are involved. This indicated that, out of the total money paid by the consumer the producer in channel VII receives 87.42 per cent. In other channels, the producer's share in consumer's rupee ranged between 27.16 per cent to 37.67 per cent for various channels of paddy-rice marketing (Table 13.3). The low producer's share was observed due to high amount of marketing cost and margins enjoyed by the processor. The index of marketing efficiency was also highest in channel VIII, where direct sale of commodities took place. It is also evident from the table that the index of marketing efficiency depends upon the number of intermediaries involved in the channel. In the channel VII, the index is 3.12, where only two intermediaries were involved. The efficiency was very less (1.40) in channel-I, where large number of intermediaries were involved. So, among all these channels, channel II, VII and VIII were highly efficient. But from the effectiveness point, channel-II, V and III were very much effective dealing 47.02, 20.16 and 12.41 per cent of total product.

Conclusion

From the above analysis we can infer that the total marketing cost is directly related to the number of intermediaries involved in that channel. Larger the number higher will be the cost. The marketing margin was also found to have a direct relationship with the number of intermediaries. The marketing efficiency index was found to have an inverse relationship with the number of intermediaries. Finally, it can be concluded that when the length of the channel becomes very longer, the marketing cost and margin tend to rise and thereby reduces the efficiency of the marketing channel. It was also evident from the above analysis that the effective channels are not necessarily being efficient and vice-versa.

References

Acharya, S.S. and Agarwal, N.L. (1999). *Agricultural Marketing in India*, 3rd Edition. Oxford and IBH Publishing Co. Pvt. Ltd., New Delhi, pp. 316-335.

Anonymous (1996). *Basic Statistics for Agriculture*, Govt. of Assam, Guwahati.

Mahapatra, S.C., Mishra, S.N. and Pradhan (1998). Marketing Strategies of Rice in Orissa: A Case Study. *The Bihar Journal of Agricultural Marketing*, 6(1): 30-37.

Chapter 14

An Economic Analysis of Maize in Assam

D.C. Kalita[1], B.C. Bhowmick[2], C. Hazarika[2] and R.S. Saikia[2]*

[1]Lecturer, Department of agricultural Economics, Nagaland University, SASRD, Campus Medziphema – 797 106
[2]Department of Agricultural Economics, AAU, Jorhat – 13

ABSTRACT

The study was conducted in all the six agro-climatic zones of Assam covering a total of 300 sample farmers from three size groups *viz.*, 82 small (1.01-2.00 ha), 95 medium (2.01-4.00 ha) and 123 large (above 4.00 ha) farmers following the multistage stratified random sampling method. The study revealed that there are three important marketing channels of Maize through which the sample farmers sold their marketed surplus. Among the three major channels, channel I (Producer – petty dealer – primary wholesaler – wholesaler – miller-cum-wholesaler – Retailer – consumer) was found to have incurred highest amount marketing of marketing cost (Rs. 252.45 per qt) and the channel III (Producer – Govt. agent – Retailer – consumer) has incurred lowest amount of marketing cost (Rs. 116.55 per qt). The producer's share in consumers rupee was 44.08, 45.69 and 50.18 per cent for channel-I, II and III respectively. The effectiveness of channel-I and II were very high transacting 52.73 and 42.60 per cent of total produce through these channels respectively by the maize farmers of Assam.

KeyWords: Maize, Marketing cost, Marketing margin, Price spread.

* *Corresponding Author:* E-mail: durgack@yahoo.com

Introduction

The production of maize in Assam increased from 1135 tones in 1951-52 to 14721 tones in 1998-99. This increased production of maize resulted in the increased arrivals in the market and thereby increase the trade in the state. The market arrivals of rice produce increased from 2208 quintal in 1985-86 to 5483 quintals in 1994-95 in the primary market as against 2237 quintals in 1985-86 to 4761 quintals in 1994-95 in the secondary market in Assam (Anonymous, 1996). This increase of market arrivals needed improvements of marketing efficiency of the existing marketing system and the reduction of marketing costs and margins of the intermediaries involved in the marketing system and increase of producer's share in consumer's rupee are the indicators of increased marketing efficiency. Acharya and Agarwal (1999) defined price spread as the difference between the prices paid by the consumer and the price received by the producers for an equivalent quantity of farm produce. The study of marketing margins and costs are important as they reveal many facets of marketing and the price structure as well as the efficiency of the marketing systems. The study of price spread helps the farmers to know the price structure involved in the marketing process and thereby can help the farmer in increasing the efficiency of marketing systems. This type of study also helps the policy maker to take appropriate price policy. Therefore, the study of marketing costs, margin and price spread in marketing through different channels is important for producers, consumers and ultimately policy makers of the nation. This study also helps in comparison of price paid by the consumers and price received by producers. Hence in this paper an attempt is made to study the economic analysis of marketing of maize in Assam.

Methodology

The study was conducted in all the six agro-climatic zones of Assam covering a total of 300 sample farmers from three size groups *viz.*, 82 small (1.01-2.00 ha), 95 medium (2.01-4.00 ha) and 123 large (above 4.00 ha) farmers following the multistage stratified random sampling method. Marketing cots was calculated by estimating the costs incurred in the process of marketing the maize. The cost incurred after the harvesting of the crop till it reaches the consumers hand constitute the marketing cost that includes transportation cost, handling cost, storage cost, market fees, weighing charges, labour charges for packing, loading and unloading charges etc. While marketing margin at any stages of marketing was estimated by using the following formula.

$$Mmi = Spi - (Ppi + Mci)$$

where,

Mmi = Marketing margin of the ith middleman

Spi = Selling price of the ith middleman

Ppi = Purchase price of the ith middleman

Mci = Marketing cost incurred by the ith middleman

Price spread was calculated as the difference between the price paid by the consumers and the price received by the producers. In order to examine the efficiency of the marketing system, producer's share in consumer's rupee was calculated as follows:

$$\text{Producer's share in consumer's rupee} = \frac{\text{Producers price}}{\text{Consumers price}} \times 100$$

Similarly, the share of total marketing costs and the total marketing margins were also estimated. The marketing efficiency of various channels was calculated by using the Shepherd's formula

$$ME = V/I - I$$

where,

ME = Index of marketing efficiency

V = Value of goods sold (consumers price)

I = Total marketing costs

Results and Discussion

Marketing Channels for Maize

There are three important marketing channels of Maize through which the sample farmers sold their marketed surplus. The channels are as follows.

Channel I : Producer – petty dealer – primary wholesaler – wholesaler – miller-cum-wholesaler – Retailer – consumer

Channel II: Producer – primary wholesaler – wholesaler – miller-cum-wholesaler – Retailer – consumer

Channel III: Producer – Govt. agent – Retailer – consumer.

Marketing Cost

The marketing cost incurred by various intermediaries in the marketing of maize through different channels is presented in Table 14.1. Among the three major channels, channel I was found to have incurred highest amount marketing of marketing cost (Rs. 252.45 per qt) and the channel III has incurred lowest amount of marketing cost (Rs. 116.55 per qt). From the table, it was clearly evident that the marketing cost is directly related to the number of intermediaries involved in the channel. Mahapatra, *et al.* (1998) also reported that the marketing cost varies according to the length of the distribution channel. Longer the channel, higher will be the total marketing cost. Like other cereals, in maize marketing also the processor incurred the highest percentage of total marketing cost (47.12 per cent in channel I and 53.30 per cent in channel II).

Marketing Margin

The total marketing margin was found to be the highest (Rs. 317.30 per qt) in channel III where Govt. agents are involved. The lowest margin was observed in channel-II (Rs. 266.75 per qt), where processor and retailer earned 47.31 per cent and 28.40 per cent respectively (Table 14.2).

Table 14.1: Marketing Cost of Intermediaries in Maize Marketing in Different Marketing Channels

Intermediaries	*Channel I*	*Channel II*	*Channel III*
Producers	33.5 (13.27)	35.25 (14.38)	38.4 (32.95)
Petty dealer	15.25 (6.04)	–	–
Primary wholesaler	19.4 (7.69)	31.05 (12.66)	–
Whole saler	18.55 (7.35)	16.4 (6.69)	–
Processor	118.95 (47.12)	130.7 (53.30)	–
Secondary wholesaler	21.3 (8.44)	–	–
Govt. agent	–	–	55.4 (47.33)
Retailer	25.5 (10.10)	31.8 (12.97)	22.75 (19.52)
Total marketing cost	252.45 (100.00)	245.2 (100.00)	116.55 (100.00)

Figures in parentheses indicate percentages to total cost.

Table 14.2: Marketing Margin of Intermediaries in Maize Marketing in Different Marketing Channels

Intermediaries	*Channel I*	*Channel II*	*Channel III*
Petty dealer	27.2 (9.83)	–	–
Primary wholesaler	36.3 (13.12)	35.5 (13.31)	–
Whole saler	31.1 (11.24)	29.3 (10.98)	–
Processor	94.9 (34.30)	126.2 (47.31)	–
Secondary wholesaler	17.9 (6.47)	–	–
Govt. agent	–	–	152.5 (48.06)
Retailer	69.25 (25.03)	75.75 (28.40)	164.8 (51.94)
Total marketing margin	276.65 (100.00)	266.75 (100.00)	317.3 (100.00)

Figures in parentheses indicate percentages to total marketing margin.

Price Spread

Table 14.3 represents the price-spread analysis for maize marketing channel in Assam. The producer's share in consumers rupee was 44.08, 45.69 and 50.18 per cent for channel-I, II and III respectively. Similarly, the marketing index was also 2.75, 2.84 and 6.47 for channel I, II and III respectively. The effectiveness of channel-I and II were very high transacting 52.73 and 42.60 per cent of total produce through these channels respectively by the maize farmers of Assam.

Table 14.3: Price Spread Analysis in Maize Marketing System for Different Marketing Channels

Items	*Channel I*	*Channel II*	*Channel III*
Consumers price (Rs./qt)	946.20	942.65	870.80
Total marketing cost (Rs./qt)	252.45	245.20	176.55
Total marketing margin (Rs./qt)	276.65	266.75	317.30
Producer's share in consumer's rupee	44.08	45.69	50.218
Index of marketing efficiency	2.75	2.84	6.47

Conclusion

From the above analysis we can infer that the total marketing cost is directly related to the number of intermediaries involved in that channel. Larger the number higher will be the cost. The marketing margin was also found to have a direct relationship with the number of intermediaries. The marketing efficiency index was found to have an inverse relationship with the number of intermediaries. Finally, it can be concluded that when the length of the channel becomes very longer, the marketing cost and margin tend to rise and thereby reduces the efficiency of the marketing channel. It was also evident from the above analysis that the effective channels are not necessarily be efficient and vice-versa.

References

Acharya, S.S. and Agarwal, N.L. (1999). *Agricultural Marketing in India*, 3rd Edition. Oxford and IBH Publishing Co. Pvt. Ltd., New Delhi, pp. 316-335.

Anonymous (1995). *Basic Statistics for Agriculture*, Govt. of Assam, Guwahati.

Mahapatra, S.C., Mishra, S.N. and Pradhan (1998). Marketing Strategies of Rice in Orissa: A Case Study. *The Bihar Journal of Agricultural Marketing*, 6(1): 30-37.

Chapter 15

Homestead Agroforestry: Practice for Sustainable Livelihood

Kishor Kr. Roy

SMS, Assam Agricultural University, Akbarpur, Karimganj – 788 710, Assam

ABSTRACT

The present study is an attempt to study the Homestead Agroforestry: Practice for Sustainable Livelihood. The study revealed that Combination of Multipurpose trees with food crops on croplands provides important environmental benefits in terms of both ecological benefits and specific on-site benefits. Agroforestry systems on croplands and farmlands yield significant economic benefit to the farmer, the community and the nation as a whole. Besides these, Improvement in rural living standards from sustained employment and higher incomes, Improvement in nutrition and health due to increased quality and diversity of food outputs, and Stabilization and improvement of upland communities through elimination of the need to shift sites from farm activity.

Keywords: Agroforestry, Sustainable livelihood, Social benefit, Economic benefit.

Introduction

Land use options that increase livelihood security and reduce vulnerability to climate and environmental change are the need of the hour. Traditional resource management adaptation such as agroforestry systems may potentially provide option for improvement in livelihood through simultaneous production of food, fodder, and firewood as well as mitigation of the impact of climate change. A livelihood is a means of deriving a just and dignified living by the individuals, family, and the society. It comprises of the assets available to household *viz.* human, financial, physical, natural and social capital, the activities, and the access to these mediated by institution and social relations that together determine the living gained by the society, household or individuals. A livelihood can be sustainable when it cope with and recover from stresses and shocks, and maintain or enhance its capabilities and assets both now and for the future, while not undermining the natural capital. Steps to promote and mainstreaming of adaptation into livelihood may potentially deliver better results when combined with adaptive management of natural resources and agro ecosystems. There is need for intensified conservation effort as well as growing products and generating services in agro ecosystems. Tree-growing in combination with agriculture (agroforestry system) as well as numerous vegetation management regime in cultural landscape (ethonoforestry system), including individual farms, watershed and regional landscape can be integrated to take advantage of the service provided by adjacent natural, semi-natural or restored ecosystems. (Pandey, 2007)

Agroforestry has been defined as "a suitable management system for land that increases overall production, combines agricultural crops, tree crops and forest plants and or animals simultaneously or sequentially and applies management practices that are compatible with the cultural pattern of local population" (Govt. of India 2001). Moreover, World Agroforestry Center (ICRAF) has defined, "Agroforestry is a dynamic ecologically based, natural resource management system that through the integration of trees in farm and rangeland, diversify and sustains smallholder production for increased social, economic and environmental benefits."

Agroforestry, the combination of woody perennials with crops and/or animals on the same unit of land management is an age-old practice in India. Being a traditional land use system is able to satisfy a large diversity of socio-economic needs in a sustainable way and in many different agro-ecological conditions. Agroforestry is found from east to west and north to south India through hundreds of systems and an equivalent number of management practices of multipurpose trees, shrubs, herbs and palms. In general on the basis of nature of components the following are common agroforestry systems prevailing in different agro-ecological regions of India:

1. Agri-silviculture (trees + crops)
2. Agri-silvipasture (trees + crops + pasture or animals)
3. Silvi-pasture (trees + pasture/animals)
4. Agri-silvi-horticulture (trees + fruit trees + crops)
5. Agri-horticulture (fruit trees + crops)
6. Horti-pasture (fruit trees + pasture or animals)
7. Boundary plantation (tree on boundary + crops)
8. Alley cropping (hedges + crops)
9. Block plantation (block of tree + block of crops)
10. Silvi-olericulture (tree + vegetables)
11. Horti-olericulture (fruit tree + vegetables)
12. Energy plantation (trees + crops during initial years)

13. Forage forestry (forage trees + pasture)
14. Shelter-belts (trees + crops)
15. Wind-breaks (trees + crops)
16. Live fence (shrubs and under- trees on boundary)
17. Silvi or Horti-sericulture (trees or fruit trees + sericulture)
18. Horti-apiculture (fruit trees + honeybee)
19. Aqua-forestry (trees + fishes)
20. Homestead (multiple combination of trees, fruit trees, vegetable etc)

These agroforestry systems are practiced throughout India by the people at large. But unfortunately, they are little known in terms of functioning and dynamics. Also their descriptive knowledge, the economic benefit and environmental importance of associated trees and the system as a whole is rarely been assessed. From a research and development point of view, little progress has been made in the study and promotion of agroforestry and other tree-based systems, or in giving this domain of study a comprehensive understanding, at the interface between forestry, agriculture, livestock rearing, and social sciences (Depommeir, 2003).

Benefits from Agroforestry

Environmental Benefits

Combination of Multipurpose trees with food crops on croplands provides important environmental benefits in terms of both ecological benefits and specific on-site benefits. The general ecological benefits include:

1. Reduction of pressure on natural forest,
2. Efficient recycling of nutrients by deep rooted trees on the side,
3. Better protection of ecological systems,
4. Reduction of surface run-off, nutrient leaching and soil erosion through impending effect of tree roots and stems of these processes,
5. Improvement of microclimate such as lowering soil surface temperature and reduction of evaporation of soil moisture through a combination of mulching and shading,
6. Increment of soil nutrient through addition and decomposition of litter-fall, and
7. Improvement of soil structure through the constant addition of organic matter from decomposed litter.

Economic Benefit

Agroforestry systems on croplands and farmlands yield significant economic benefit to the farmer, the community and the nation as a whole. Such economic benefits may include:

1. Increment in the output of food, fuel wood, fodder, fertilizer, timber,

2. Reduction in incidence of total crop failure, common to single cropping, or monoculture system, and
3. Increase in level of farm incomes due to improved and sustained productivity.

Social Benefit

Besides economic benefits, the social benefits occur because of the sustainable nature of these types of systems. These benefits include-

1. Improvement in rural living standards from sustained employment and higher incomes,
2. Improvement in nutrition and health due to increased quality and diversity of food outputs, and
3. Stabilization and improvement of upland communities through elimination of the need to shift sites from farm activity.

Home-Gardens/Homesteads Cultivation

Home gardens can be defined as land use systems involving deliberate management of multipurpose trees and shrubs in intimate association with annual and perennial agricultural crops and invariably, livestock within the compounds of individual house, the whole tree-crop-animal unit being intensively managed by family labour. Such systems are essentially man made and reflect the wisdom of the traditional culture and ecological knowledge that have evolved over the years. Such valuable traditional ecological knowledge systems are based on strong socio-cultural and traditional belief, confounded by the economic status of the people. In this context, if designed and managed appropriately, the traditional home gardens based largely on indigenous knowledge systems harbouring a wide diversity of local crops of their social and traditional significance can be the most promising and ecologically feasible option for the farming community. (Tangjang and Arunachalam, 2009)

The practice of traditional home garden or homestead is an old-age practice in coastal states particularly in Kerala, Tripura, Assam, North-eastern states and parts of West Bengal and Andaman and Nicobar Islands. Although the home-gardens appear to be a mixture of trees, shrubs and herbs, a certain general pattern seems to exist. There is wide variation in the intensity of trees, species and crops based on the size of holding, needs of the people residing in homesteads and micro-climate. Coconut, arecanut, guava, mango, citrus, tamarind, jack-fruit, papaya, banana, moringa, sesbania, custardapple and many multipurpose trees are the major trees found grown in home-gardens. (Tewari, 2008)

Domestic animals and poultry are the main components of homesteads; therefore, sometimes forages like stylo (*Slylosanthes guianensis/S. hamata*), guinea-grass, Guatemala, napier and others are also grown frequently. The components are so intimately mixed in horizontal and vertical strata as well as in time that a complex interaction exists among soil, plants, other aboveground and below ground components, nutrition and environmental factors. There is critical competition both for light and nutrition. The holder chooses his crops and crop combinations based on his wisdom, needs and perceptions acquired over generations of experience.

Many multipurpose trees having productive and protective functions are integrated into the system in different spatial and temporal arrangements. Most of these trees closely interact with agricultural crops when grown. These trees include teak (*Tectona grandis*), jack-trees (*Artocarpus spp.*), *Casuarina equisetifolia, Mangifera india, Ceiba pentendra, Leucaena leucocephala, Grevillea robusta, Bambusa arundinacea, Erythrina variegata* and *Gliricidia sepium* (both good support to black pepper). *Thespesia populnea* is common in low lying homesteads and the wood is commonly used for agricultural implements. Mangroves form an essential part of homesteads of backwater areas in lowlands. These are commonly used as fuelwood. Coconut and pandanus (*Pandanus tectorius*) can be commonly seen near canals and backwaters. Palmyrah palm (*Borassus flabellifer*) is common multipurpose palm in coastal Andhra Pradesh. Fish and shrimp culture in backwater channels and in association with mangroves is the main activity in homestead. Arecanut palm along with betel vine/black peeper is being traditionally grown in the N. E. states particularly Assam and Meghalaya. In recent years tea() has taken significant proportion in homestead gardens particularly in upper Assam. Different species of Bamboo are also being grown around the boundaries of homestead as natural fence. Thus, multitude of crop species in the homesteads helps to satisfy primary needs of the farmer such as food, fuel, fodder, timber and cash. This in spite of high intensity of cropping also helps to conserve fertility by nutrient cycling or organic manuring or mulching and increased microbial activity in the rhizosphere of crops. Among social benefits are-high family labour utilization and risk minimization. This system also helps in checking soil erosion due to high intensity of vegetation cover, environmental health and conserving biodiversity.

Homestead Agroforestry and Biodiversity

It is well known that when agricultural crops replace forest, biodiversity is affected to a great extent. However, the situation is different in homestead/homegardens since this is grown in valleys surrounding human habitat without disturbing natural forests. Complex, traditional, intensively managed homestead agroforestry systems prevailing in the country provide options for ethnic societies to use and conserve plant biodiversity (Agnihotri *et al.*, 2004). This homegardens have been described as 'living genebanks', where an abundance of germplasm, in the form of indigenous varieties, obsolete cultivars, landraces and some rare species, as well as introduced species thrive in proximity, and have been conserved through generations (Palni *et al.*, 1998, Agelet *et al.*, 2000). The farmers maintain a variety of economically important plant species in homestead agroforestry systems. Tynsong and Tiwary (2010), investigated plant species composition of arecanut agroforests of south Meghalaya and encountered 160 plants, which included 83 tree species, 22 shrubs species, 41 herb species and 14 climber species. Further, they concluded that this agroforestry system serve as home for many economically important plant species, harbour rich biodiversity and mimic the natural forests both in structural composition as well as ecological and economic functions.

Alam and Masum (2005) studied the diversity of plants, livestock and fish species, species richness in the offshore island of Bangladesh and identified a total of 142 plant species belonging to 61 families. Marked variation in species richness and diversity was also recorded in the homestead of different farm categories. The highest types of species (108) were found in the large farm category whereas the lowest types of species (71) were found in the small farm category. Various macro- and micro-environmental factors of the homestead, needs and choices of the family influence the distribution of the plant species. That is why the species composition varies from one location to another and from one farm category to another (Alam and Masum 2005).

In a study conducted by Das and Das (2005), a total of 122 trees and shrubs have been identified in the homegardens of Barak valley of Assam. They observed distinct variations in size, diversity and composition within the small holder farmers (<1 ha). With increase in holding size, more variations in species composition were encountered. In larger homegardens, the land is demarcated into more micro zones or management zones and larger areas allotted to arecanut cultivation. Well defined dense zones were also encountered in many such homegardens. The estimated plant density was 1535 ha^{-1} and basal cover was 33.86 m^2ha^{-1}. The woody plant density in this zone is much higher than the values of 238-319 ha^{-1} reported by Kumar *et al.* (1994) in Kerala homegardens. The higher plant density and the resultant higher basal area can be attributed to the presence of a larger number of arecanut trees and also the inclusion of bamboo species in the area. Pandey *et al.*, 2006 in their study at Andaman and Nicobar Islands reported that Arecanut contributed maximum (54 to 76 per cent) followed by coconut to the density in both home - forest - garden as well as home garden.

A number of factors such as socio-economic status, market integration, land holding size, etc. affect the diversity in the homegardens. The homegarden size and diversity were found to be related to the socio economic conditions of the families that maintain them. Poor families with no or less paddy land holdings had smaller homegardens and therefore, less diversity (Das and Das 2005).

Homegardens exhibit complex structure, both vertically and horizontally. The vertical structure of homegardens is composed of 3-4 canopy layers %the emergent layer, the canopy, the understory, the shrub and the herb layer. In a study by Das and Das (2005) it was found that the emergent layer had a height of 15 m or more and was composed of multipurpose tree species. The canopy layer was between 10 and 15 m with dominant species such as arecanut, jackfruit, bamboo etc. The understory layer had the height of 5-10 m, the shrub layer had the height of 1-5 m and the herb layer was less than 1 m. The herb layer was mainly composed of vegetables, ornamentals and medicinal species. The shade provided by the upper layer support large number of shade loving climbers like *Piper nigrum* and herbs like *Alocasia macrorrhiza.* Perera *and Rajapakse* (1991), in their study of Kandyan homegardens in Sri Lanka, recorded four vertical layers. From the ground layer comprising herbaceous food crops, forage, medicinal and other crops to the upper canopy of fast-growing multipurpose trees, the gradient of light and relative humidity creates different niches, enabling various species group to exploit them.

Areca nut constituted the most important component of larger homegardens. Because of its commercial importance, areca farming is practiced in most homegardens represented as betel zones. It is important to note that a tendency was observed among the larger homegarden owners (< 1 ha) to allot more land, especially portions from the dense zones to areca nut cultivation. Even though diversity in such homegardens is remarkable, there is a chance that in future such commercialization of areca nut may reduce species diversity of homegardens. In an investigation on the weed intensity in arecanut based agroforestry models at Karnataka (Girish *et al.*, 2003) found that sole crop of arecanut not only recorded maximum dry weight of weeds but also weed diversity.

Many traditional crop species having medicinal properties were found generally in the home gardens as traditional practice. In total, 17 such medicinal plant species were reported to be used by the local people to cure different ailments in Arunachal Pradesh. (Tangjang and Arunachalam, 2009).

Homestead Agroforestry and Avian Diversity

Homestead agroforestry having host of perennial agricultural crops like arecanut, coffee, pepper, cocoa etc. serve as conservation sites for birds when grown near and under the shade of forest trees. In fact innumerable studies in India have indicated that agricultural fields/plantations located adjacent to natural forest support a rich biodiversity of bird fauna.

Agricultural areas planted with arecanut palm in Western Ghats of India was reported to retain 90 per cent of the bird species from the adjacent forest The fundamental reasons for retention of such rich avian biodiversity was due to high vertical structural complexity of the forest intact, providing varied homes for many different bird species. The researchers found that arecanut possessed even more conservation importance than retaining bird biodiversity. The arecanut based agroforestry system sustains a rich avifauna compared with many other agricultural areas cleared from tropical forests because of the preservation of the vertical forest system (Jeremy 2008).

Arecanut based agroforestry system in western ghat of India are of high conservation value which contained a statistically identical set of bird as that contained within production forest, harbouring such flagships of Asian conservation as the Great Hornbill and Malabar Gray Hornbill (Ranganathan *et al.*, 2010).

Economic Contribution of Homestead Agroforestry System

Home gardens play an important role in the rural economy of the Himalayan region. Nautiyal *et al.* (1998) described their importance in monetary terms; the highest per unit area annual output was obtained from simultaneous agroforestry systems (₹25370; ₹35=US$1 in 1998) followed by home gardens (₹18200) and sequential agroforestry (₹9426). Complex, traditional, intensively managed home gardens still thrive in many parts of India. They provide options for ethnic societies to use and conserve plant biodiversity. They comprise many tree species (underutilized fruit and nut crops or those serving other functions), medicinal plants, and host species for beneficial insects and birds that support integrated pest management, crop wild relatives, traditional landraces and cultivars of numerous crop species (Maikhuri *et al.*, 1997; Borthakur *et al.*, 1998).

Arecanut based agroforestry has been reported to provide non-staple food (*e.g.* fruits), timber and other house construction materials, biomass energy, fodder, organic manure for the fields, medicinal plants, edible oils, spices, and even material for clothing (Kothari, 1999). It is also reported that forty-four per cent (44 per cent) of farmers were using agroforestry produces for their self consumption and remaining proportion of farmers were using for self consumption, marketing, medicinal and decoration purposes. Fifty-six per cent (56.34 per cent) of farmers regularly marketing the homestead agroforestry produces, which gave on an average ₹13,725 (USD 276.52) as additional income per year per family. This income was generated from an average land area of 200 square meters without any external input and using only family labour (Saravan, 2010).

In a study conducted by Rahman *et al.*, 2005 at Gazipur sadar Upazila of Bangladesh reported average annual value of gross production of homestead forest is ₹10267.00 per household, out of which ₹4014.00 is used by the household and ₹6205.50 is sold which is 39.1 per cent and 60.9 per cent respectively. The overall contribution of homestead forest income to average household income is 11.8 per cent. Large households received the highest forest income of ₹17869.50, whereas marginal households received the minimum forest income of ₹3028.50. The findings further indicate that homestead forest production, own consumption and income increase with increasing of landholding size class. Total forest gross production per household per year is ₹10267.00 which turns out to be ₹93055.50 per ha, which is 1.2 times as high as that of a monoculture of jackfruit, 1.6 times that of a monoculture of mango, and 5.5 times that of HYV paddy cultivation. These findings suggest that homestead forestry is a more profitable option than other land-use systems, and the diverse species composition of homestead forests leads to higher gross production on a per hectare basis than other monocultures. The fact that overall income from homestead forests of '6205.50 is 11.8 per cent of average annual household income, and households sold a higher proportion of the homestead forest products (60.9 per cent) than own consumption (39.1 per cent), reveals the importance of homestead forestry as a supplementary cash income generating activity in the study area. The results of the study also reveals that homestead forests contribute to the household economy by providing subsistence cash income, employment opportunities and various household materials.

References

Agnihotri, R.K., Sharma, S., Joshi, M., and Palni, L.M.S. (2004). Crop Diversity in the Homegardens of the Kumaun Region of central Himalaya, India. *Plant Genetic Resources-Newsletter* 138: 23-28.

Agelet, A., Bonet, M.A. and Valles, J. (2000). Home gardens and their role as a main source of Medicinal Plants in mountain region of Catalonia (Iberian Peninsula). *Economic Botany* 54 (3): 295-303.

Alam, M.S. and Masum, K.M. (2005). Status of Homestead Biodiversity in the Offshore Island of Bangladesh. *Research Journal of Agriculture and Biological Sciences* 1 (3): 246-253.

Borthakur, S.K., Sarma, T.R., Nath, K.K. and Deka, P. (1998). The house gardens of Assam: a traditional Indian experience of management and conservation of biodiversity– I. *Ethnobotany* 10: 32–37.

Das, Tapasi and Das, A.K. (2005). Inventorying plant biodiversity in homegardens: A case study in Barak Valley, Assam, North East India. *Current Science* 89(1):155-163.

Depommier, D. (2003). The tree behind the forest: ecological and economic importance of traditional agroforestry systems and multiple uses of trees in India. *Tropical Ecology*, 44(1): 63-71.

Girish, B.S., Shivanna, H. and Girisha, H.V. (2003). Studies on weed intensity in existing Arecanut based Agroforestry models. *Karnataka Journals of Agricultural Science*. 16(2): 260-264.

Jeremy, H. (2008). Rainforest agriculture preserves bird biodiversity in India. *www.mongabay.com.*

Kumar, B.M., George, S.J. and Chinnamani, S. (1994). Diversity, Structure and standing stock of wood in the home gardens of Kerela in peninsular India. *Agricultural Systems* 25: 243-264.

Maikhuri, R.K., Semwal, R.L., Nautiyal, S. and Saxena, K.G. (1997). Eroding traditional crop diversity imperils the sustainability of agricultural system in the Central Himalaya. *Current Science* 73: 777–782.

Nautiyal, S., Maikhuri, R.K., Semwal, R.L., Rao, K.S. and Saxena, K.G. (1998). Agroforestry systems in the rural landscape–a case study in Garhwal Himalaya, India. *Agroforestry Systems* 3: 151–165

Pandey, C.B., Kanaklata, Venkatesh, A. and Medhi, R.P. (2006). Diversity and species structure of home gardens in South Andaman. *Tropical Ecology*, 47(2): 251-258.

Pandey, D.P. (2007). Multifunctional agroforestry systems in India. *Current Science* 92 (4): 455-463.

Perera, A.H. and Rajapakse, R.M.N. (1991). A Baseline Study of Kandyan Forest gardens of Sri Lanka: Structure, Composition and Utilization. *Forest Ecology and Management* 45: 269-280.

Rahman, M.M., Furukawa, Y., Kawata, I., Rahman, M.M. and Alam, M. (2005). Homestead forest resources and their role in household economy: A Case Study in the Villages of Gazipur Sadar Upazila of Central Bangladesh. *Small-scale Forest Economics, Management and Policy*, 4(3): 359-376

Ranganathan, J., Krishnaswami, J. and Anand, M.O. (2010). Landscape level effects on avifauna within tropical agricultural in the Western Ghats: Insights for management and conservation. *Biological Conservation*, 143: 2909-2917

Saravan, R. (2010). Traditional knowledge, innovation Systems and democracy for sustainable Agriculture: a case study on adi tribes of Eastern Himalayas of North-East India. *Innovation and Sustainable Development in Agriculture and Food*, 1: 28-30.

Tangjang, S. and Arunachalam, A. (2009). Role of traditional home garden systems in Northeast India. *Indian Journel of Traditional Knowledge* 8(1): 47-50.

Tewari, S.K. (2008). Agroforestry: A boon for agriculture and forestry. *Farm Forestry (Agroforestry)*, Oxford.

Tynsong, H. and Tiwari, B.K. (2010). Diversity of plant species in arecanut agroforests of South Meghalaya, North-East India. *Journal of Forestry Research*, 21(3): 281-286.

Chapter 16

Sustainable Development of Small Ruminants through Feed Resource Management in Eastern India

Lalmohan Mandal

Professor, Department of Animal Science and Ex-Dean, Faculty of Agriculture, Bidhan Chandra Krishi Viswavidyalaya Mohanpur – 741 252, Nadia, W.B

ABSTRACT

The present study is an attempt to study the Sustainable Development of Small Ruminants through Feed Resource Management in Eastern India. The study is mainly based on secondary data. The study revealed that to augment the feed resources (green) for rearing of sheep and goats, various Central and State Govt. schemes under feed and fodder development activities are being implemented to bridge the gap between demand and supply; but those schemes are generally biased towards large ruminants. Special emphasis is to be given on the requirements of these small ruminants. Surplus green fodder production, if any, during monsoon period may be utilized in the lean period by preserving these fodder in the form of hay Most of the states of Eastern India are deficient in fodder. There exists a large gap between the requirements of feed and fodder and their availability. Short fall of fodder production might be due to non availability of fodder land (1.08 per cent only in W.B) and also acute shortage of availability of quality fodder seed.

Keywords: Sustainable development, Small ruminants, Management.

Introduction

The livestock sector has a significant potential for round the year employment generation particularly in rural areas. This provides subsidiary source of livelihood to the people leaving below the poverty line due to lack of sufficient agricultural land to sustain, particularly in the drought prone, hilly, tribal and other remote areas, where crop production on its own may not be capable of engaging them fully.

According to 50th National Sample Survey, (NSSO 1998), the employment in the Animal Husbandry Sector was 9.8 million (provisional) in the principal status and another 8.6 million(provisional) in the subsidiary status. However, in case of landless laborers and marginal farmers, rearing of large ruminants specially the high yielders are mostly not possible due to shrinkage of common property resources(grazing land) and financial constraints but farming of small ruminants has more or less overcome these problems and can contribute additional income for the family. Among the small ruminants, goat in the true sense is called as poor men's cow due to its tremendous economic importance in contributing milk, meat and ultimately the household nutrition security and livelihood to the downtrodden people. Census data revealed that India accounts for 20 per cent of the World's goat population with annual growth rate of 1.6 per cent (Department of Animal Husbandry and Dairying, 2005). During the period from 1951 to 1992, there has been a growth of 31.7 per cent in the cattle population and 92.2 per cent in buffalo population whereas the population of sheep increased by 30 per cent and those of goat and pigs increased by about 145 per cent and 190 per cent respectively(Press Information Bureau, 2005). The reason for high growth rate include low cost and handy technology to the landless laborers, marginal farmers and industrial workers' particularly less volume of feed is required than cattle, more resistant to diseases, easily manageable by child or female and easy market accessibility.

About 5 million households in the Country are engaged in the rearing of small ruminants (sheep, goats and rabbits) and other allied activities.

The need of protein feed is known to every one- as a source of animal protein; meat, milk and eggs are mostly consumed. Animal proteins are not only rich in essential amino acids but they are also properly balanced. Out of meat and eggs- different types of meat play a major role to meet the requirement of protein. Meat production in India has been increased by 30 per cent in between 1990-2000, while, the consumption of meat has been increased by 2.7 per cent (FAO, 2005). Around 39 per cent of goat is slaughtered annually in India, which contributes about 8.02 per cent of total meat produced in India (Deptt. of Animal Husbandry and Dairying, 2005). The production of meat is presented in Table 16.1.

Out of the states in Eastern India, the state of West Bengal is richly endowed with various livestock and poultry. Despite having the numbers, production of livestock products does not commensurate with the internal requirement of state as per ICMR norms. Per capita availability and consumption norms for Animal Husbandry products in 2007-08 in the state of West Bengal have been presented in Table 16.2.

Table 16.1: Production of Meat in West Bengal (in thousand tons)

Year	Mutton	Goat Meat (Chevon)	Pork	Bovine (Cattle and Buffalo)	Poultry
1990-91	4.0	91.5	31.4	533.4	35.4
2000-01	16.2	133.3	31.6	124.4	132.7
2001-02	16.3	135.2	32.4	123.6	134.6
2002-03	15.4	138.9	34.1	123.2	138.4
2003-04	15.7	144.3	33.7	125.8	143.5
2004-05	16.2	148.3	31.2	132.5	147.0
2005-06	16.6	152.2	31.8	135.5	151.0
2006-07	17.0	156.3	32.8	139.3	155.2
2007-08 (P)	17.7	161.5	33.7	140.7	151.4

P: Provisional.

Source: Directorate of Animal Resources and Animal Health, Govt. of West Bengal.

Table 16.2: Per Capita Availability and Consumption Norms for Animal Husbandry Products in 2007-08

Item	Unit	Per Capita Nutritive Requirement as per ICMR	Per Capita Nutritive Requirement as per State Norms	Actual Availability	Per cent of Availability as per ICMR	Per cent of Availability as per State Norms
Milk	Gm/day	220	180	128	58	71
Egg	No./annum	183	100	35	19	35
Meat	Gm/day	60	30	16	27	53

Source: Department of Animal Resources Development, Govt. of West Bengal.

The state of West Bengal has however, made significant progress in the development of animal resources. During 1976-77, production of milk was 10.64 lakh MT whereas it shot up to 40.77 lakhs MT during 2007-08; during the same period corresponding figures for production of meat were 1.29 lakh MT and 5.14 lakh MT and for production of eggs was 676 million and 3057 million respectively. The corresponding internal requirements for 2007-08 as per ICMR norm had been 57.47 lakh MT, 9.58 lakh MT and 8747 million. But despite such progress, the state is still facing a number of challenges in augmenting productivity of livestock and poultry birds for bridging the wide gap that exist between requirement and availability of livestock products. Requirement and production of major Livestock products in 2007-08 for state of West Bengal has been presented in Table 16.3.

During the last 30 years, the state of West Bengal has registered more growth in number of all animals, except buffalo, than the average growth of the country. The state ranks third in the production of meat but it is far behind in mitigating the availability of meat for the people of this state as per ICMR norms. Productivity of the meat producing animals is very low. One goat contributes 7 kg of meat, a sheep 7.7 kg, a pig 26.9 kg and a chicken 1 kg. The state wise distribution of sheep in Eastern India is presented in Table 16.4.

Table 16.3: Requirement and Production of Major Livestock Products in 2007-08

Item	*Unit*	*Requirement*	*Production*	*Annual Shortfall*
Milk	Lakh MT	57.47	40.77	16.70
Egg	Million No.	8747	3057	5690
Meat	Lakh MT	9.58	5.14	4.44

Source: Department of Animal Resources Development, Govt. of West Bengal.

It is evident from Table 16.4 that total sheep in India is 7,15,58,232 and that in the major states of Eastern India are 3,53,799 in Assam; 2,18,235 in Bihar; 4,83,336 in Jharkhand; 18,18,309 in Orissa and 15,77,193 in West Bengal. The state wise distribution of goats in Eastern India is presented in Table 16.5.

The data in Table 16.5 shows the goat population as 14,05,37,405 in India and that are 43,19,773 in Assam; 1,01,67,009 in Bihar; 65,91,592 in Jharkhand ; 71,27,038 in Orissa and 1,50,69,286 in West Bengal.

Sheep mainly thrive on grazing. In spite of a number of sheep development activities in operation in different states of the Country, sheep rearing still continues to be a nomadic/backward proposition and thus mostly concerned to poor and landless people. For scanty suitable grazing lands in most of the states, the shepherds keep on migrating their flocks over extensive areas within or even in the neighboring states. Generally, two types of migratory flocks are observed:-

1. Truly nomadic flocks with no fixed centers but following seasonal migratory routes to grazing areas.
2. Flocks on the fallow land but following definite migratory routes to season pastures.

Top feeding by lopping tree branches and shaking of pods is also common.

India possesses the second largest goat population in the world. In the prevailing socio-economic conditions in the country, where per capita land holding is hardly 0.2 ha, goat rearing becomes an inseparable component of mixed farming system. Goat farming has been recommended as the best choice for the rural people in developing countries because of low investment, wide adaptability, high fertility and fecundity, low feed and management needs, high feed conversion efficiency, quick pay-off and low risk involved. It is backbone of the economy of small and landless farmers in India. It is an insurance against crop failure and provides alternate source of livelihood to the farmers all year round.

The controversy over goats is on damage it causes to the environment, predominantly due to its browsing nature. On one hand, the goat is accused as the major cause of deforestation and soil erosion, and on the other hand, it is claimed as

Table 16.4: State-wise Details of Sheep in Eastern India (Sheep Indigenous)

State/Union Territory	Male			Female			Total Indigenous	Total Sheep
	Under 6 Months	6 Months and Above	Total	Under 6 Months	6 Months and above	Total		
Arunachal Pradesh	2,407	3,963	6,370	3,304	5,855	9,159	15529	19880
Assam	72,366	67,985	1,40,351	87,042	1,24,380	2,11,422	3,51,773	3,53,799
Bihar	31,278	47,868	79,146	44,360	84,191	1,28,551	2,07,697	2,18,235
Jharkhand	81,840	1,09,651	1,91,491	98,619	1,90,434	2,89,053	4,80,544	4,83,336
Manipur	1,804	1,975	3,775	2,010	2,416	4,426	8,205	8,696
Meghalaya	3,929	4,025	7,954	4,869	7,976	12,845	20,799	21,041
Mizoram	182	222	404	177	199	376	780	974
Nagaland	494	564	1,058	588	691	1,279	2,337	3,649
Odisha	4,35,833	4,65,025	9,00,858	3,99,903	5,17,444	9,17,347	18,18,205	18,18,309
Sikkim	762	987	1,749	299	464	763	2,512	2,536
Tripura	1	3,682	3,683	0	2	2	3,685	3685
West Bengal	2,90,348	2,85,217	5,75,565	3,77,674	6,12,604	9,90,278	15,65,843	15,77,193
All India Total	**84,91,794**	**82,38,590**	**1,67,30,384**	**1,43,84,054**	**3,67,14,252**	**5,10,98,690**	**6,78,28,690**	**7,15,58,232**

Source: 18th All India Livestock Census-2007.

Table 16.5: State Wise Details of Goats in Eastern India-2007

State/Union Territory	Male			Female			Total Goats
	Under 1 year	1 Year and Above	Total	Under 1 year	1 Year and Above	Total	
Arunachal Pradesh	65,356	56,091	1,21,447	70,831	99,843	1,70,674	2,92,121
Assam	8,90,880	7,41,271	16,32,151	8,51,652	18,35,970	26,87,622	43,19,773
Bihar	19,16,479	15,27,635	34,44,114	20,70,997	46,51,898	67,22,895	1,01,67,009
Jharkhand	10,99,170	12,02,908	23,02,078	11,48,585	31,40,929	42,89,514	65,91,592
Manipur	9,911	10,938	20,849	9,929	19,799	29,728	50,577
Meghalaya	70,502	64,839	1,35,341	76,419	1,53,723	2,30,142	3,65,483
Mizoram	3,175	3,255	6,430	2,931	6,349	9,280	15,710
Nagaland	35,797	38,297	74,094	31,445	72,533	1,03,978	1,78,072
Orissa	15,87,906	15,97,540	31,85,446	11,74,764	27,66,828	39,41,592	71,27,038
Sikkim	13,876	24,436	38,312	14,550	39,133	53,683	91,995
Tripura	1,14,672	73,597	1,88,269	1,33,309	3,11,474	4,44,783	6,33,052
West Bengal	29,86,307	29,17,746	59,04,053	35,58,406	56,06,827	91,65,233	1,50,69,286
All India Total	**2,19,54,935**	**1,88,38,515**	**4,07,93,450**	**2,70,92,150**	**7,26,51,805**	**9,97,43,955**	**14,05,37,405**

Source: 18th All India Livestock Census-2007.

a useful animal for poor people and is responsible for cleaning the bushes and making the land worthy of cultivation.

In our country, goats are reared by both men and women with diverse working and professional background. The production systems are mainly:

1. *Tethering*: It is a convenient means of rearing goats from the stand point of control, minimum labor input and utilization of feed in situ. Sometimes combining tethering with grazing up to 5 goats at a time, led by ropes held by women and children are also in practice.
2. *Extensive system*: This involves low carrying capacity in situations where land is marginal and in plentiful. It is characterized by low rainfall and browse plants. System is followed by nomadic people during winter months when crop residues, are available.
3. *Intensive system*: Goats are mostly fed in confinement with limited access to land which require high labor and cash inputs. Cultivated grasses and agro-industrial by products are fed in situ.
4. *Semi intensive system*: It is practiced in most of the situations. The advantages of this system are increased fertility of land via the returns of dung and urine, control of waste herbage growth, reduced fertilizer usage, increased crop yields and greater economic returns.
5. *The goat industry* in India has yet to be firmly laid down on scientific uses. The minimum goat unit may consist of one goat and the maximum could go to a few hundreds under range management. Goat farming in the country is mainly based on "zero input". However, large scale sheep and goat farms are also maintained by Govt. and other concerns where small ruminants are reared in intensive system with feeding of some amount of concentrate mixture- but this involves cost also. For this purpose, ICAR has prescribed the nutrient requirements of small ruminants (both sheep and goat) for different physiological conditions. Different nutrient requirements for adult sheep and goats are presented in Tables 16.6 and 16.7.

However, this system is practically not practiced by our villagers where mostly indigenous breeds of sheep and goats are reared and also due to ignorance, less availability of feed resources and high cost in- volvement. Hence attempts are to be made to increase grazing lands; availability of tree leaves and greens from forests *i.e.* silvipasture, shrubs and tree as source of fodder so that small and marginal farmers may be benefitted. Feed requirement of goats is generally met up mostly by grazing for 6-8 hours a day on common property resources on the bunds of canals, rivers and ponds or nearby forests. Goat is also kept on grazing after the crop is harvested. During other seasons, it is stall fed with local grasses collected from bunds or grazing areas or fodder trees as available in villages. Very little concentrate like broken rice or maize or wheat or pulse chunies is given along with rice bran and mustard cake.

The system of raising sheep and goats for meat purpose under grazing with supplementation, although cost effective for the producer, has not been largely adapted by the farmers not only owing to their economic back ground but also for their age old traditional practices. It is however, realized with progressive shrinkage of grazing land and increase in stocking density on such land, most small ruminants have to be raised under grazing with supplementation in future. In addition to grazing on range, limited amount of concentrate will provide better finishing weight at 6 and 9 months of age (at slaughter age) in meat type sheep and goat respectively. Table 16.8 shows the district-wise area under forests by legal status in West Bengal. The consumable leaves of these forests along with grasses may also serve as sources of green roughages for sheep and goats, if permitted.

Table 16.6: Daily Nutrient Requirements for Maintenance of Adult Sheep

Body Weight (kg)	*DM (g)*	*DM as per cent of Body Weight*	*Energy*		*DCP (g)*	*Ca (g)*	*P (g)*
			TDN (g)	*ME (Mcal)*			
20	575	2.8	258	0.93	38	1.1	0.8
25	678	2.7	305	1.10	33	1.3	0.9
30	775	2.6	350	1.26	38	1.5	1.0
35	873	2.5	393	1.42	43	1.7	1.2
40	964	2.4	434	1.56	48	1.9	1.3
45	1,055	2.3	475	1.85	53	2.3	1.5
50	1,140	2.3	513	1.85	56	2.3	1.5
55	1,225	2.2	551	1.99	60	2.4	1.6
60	1,310	2.2	588	2.12	65	2.6	1.7

Mineral requirements are based on NRC (1985).

Source: Nutrient Requirements of Livestock (1998) ICAR; by S.K.Ranjhan, P: 11.

Table 16.7: Daily Nutrient Requirements for Maintenance of Adult Goats

Body Weight (kg)	*DM (g)*	*DM as per cent of Body Weight*	*Energy*		*DCP (g)*	*Ca (g)*	*P (g)*
			TDN (g)	*ME (Mcal)*			
15	500	3.3	240	0.86	23	1.1	0.7
20	615	3.1	295	1.07	29	1.3	0.9
25	730	2.9	350	1.26	34	1.6	1.1
30	830	2.8	400	1.45	39	1.8	1.2
35	940	2.7	450	1.62	44	2.1	1.4
40	1040	2.6	500	1. 79	48	2.3	1.5
45	1125	2.5	540	1.96	53	2.5	1.7
50	1,230	2.4	590	2.12	57	2.2	1.8
55	1315	2.4	630	2.28	62	2.9	1.9
60	1,410	2.3	67 5	2.43	66	3.1	2.1

Source: Nutrient Requirements of Livestock (1998) ICAR; by S.K.Ranjhan. P: 15.

Table 16.8: District-wise Area Under Forests by Legal Status in West Bengal (in Sq.mt)

District	*Reserved Forests*	*Protected Forests*	*Unclassed State Forests and Others*	*Total Area*
Darjeeling	1115	–	89	1204
Jalpaiguri	1483	217	90	1790
Cooch Behar	–	42	15	57
Bankura	80	1311	91	1482
Purba Midnapur	8	1166	535	1709
Paschim Midnapur				
Burdwan	3	192	82	277
Purulia	112	729	35	876
Birbhum	8	54	97	159
Hooghly	3	–	–	3
Nadia	5	3	5	13
Murshidabad	1	7	–	8
Malda	8	5	7	20
Uttar Dinjapur	8 4 6 18			
Dakshin Dinajpur				
24-Parganas(N)	43	–	–	43
24-Parganas(S)	4177	42	1	4220
Howrah	–	–	–	–
Kolkata	–	–	–	–
Total	**7054 (59.38 per cent)**	**3772 (31.75 per cent)**	**1053 (8.87 per cent)**	**11879 (100 per cent)**
All India	**4,23,311**	**2,17,245**	**1,27,881**	**7,68,437**

Source: State Forest Report, 2004-05.

Data on all other Eastern Indian States are less available. However data from field conditions from Bihar are presented in Table 16.9.

Data shows that in field conditions also, twining rate is 44.5 per cent which is higher than single kidding and the growth rate is also satisfactory; thus making goat rearing a profitable one.

Use of common property resources in the livestock economy in the region of Eastern Himalayas have been presented in Table 16.10 which shows 20 per cent of the household are habituated in grazing livestock and 11 per cent are habituated in collecting fodder resources.

Table 16.9: Production and Reproduction Performances of Goat Under Field Conditions of Bihar

Parameters	*Performance*
Adult Weight(male)	15 kg
Adult Weight(female)	12 kg
Age at 1st Oestrous	8-9 months
Age at 1st kidding	422 days
Kidding interval	250 days
Average litter size	2.1
Service per conception	1.6
Lactation yield	56 kg
Lactation length	80-120 days
Single kidding	32.3 per cent
Twining	44.5 per cent
Triplicate kidding	23.2 per cent
Growth rate	30g/day

Source: Unpublished project report, ICAR-RCER, Patna 2006.

Table 16.10: Use of Common Property Resources in Livestock Economy

Agroclimatic Zones	*Per cent Household*		
	Possessing Livestock	*Grazing Livestock*	*Collecting Fodder*
Eastern Himalayas	57	20	11

Source: Compiled from NSSO. 1999.

The data on pattern of land utilization under permanent pasture and other grazing lands (statewise 1990-91) in Eastern Indian states have been presented in Table 16.11 which shows it is highest in Odisha, (711,000 hectares) followed by Assam (184,000 hectres and the than United Bihar(146,000 hectares.

Directorate of Economics and Statistics; Ministry of Agriculture

The data on livestock, feed and fodder position of West Bengal have been presented in Table-12 which shows that the total requirement of green fodder is 460.53 lakh MT; the availability is 110.76 lakh MT and there is a deficit of 349.77 lakh MT. Efforts are to be made to bridge this gap between the requirement and the availability.

Table 16.11. Pattern of Land Utilization (State-wise 1990-91) Under Permanent Pasture and Other Grazing Lands in Eastern India States (in thousand hectares)

Name of the State	*Area*
Arunachal Pradesh	—
Assam	184
Manipur	—
Meghalaya	17
Mizoram	4
Nagaland	–
Odisha	711
Sikkim	69
Tripura	–
United Bihar	146
West Bengal	7

Source: Agricultural Statistics at a Glance (1993).

Table 16.12: Position of Livestock, Feed and Fodder in West Bengal

Total Livestock Population	
Jersey Cross	22, 36,178
Holstein Cross	1, 83,039
Other Cross bred Cattle	1, 22,927
Indigenous Cattle	6, 46,815
Bengal Non-descriptive indigenous Cattle	1, 58, 99,310
Indigenous breed of Buffalo	2, 14,872
Non-descriptive Buffalo	5, 48,894
Total Sheep	15, 77,193
Bengal Goat	1, 48, 34,718
Other breeds of Goats	2, 34,568

Source: All India Livestock Census 2007.

Total Requirement of Feeds and Fodder (in Lakh MT)	
Dry Fodder	189.66
Green Fodder	460.53
Concentrates	127.78
Total Availability of Feeds and Fodder (in Lakh MT)	
Dry Fodder	141.79
Green Fodder	110.76
Concentrates	15.41

Contd...

Table 16.12–Contd...

Deficit of Feeds and Fodder (in Lakh MT)	
Dry Fodder	47.87
Green Fodder	349.77
Concentrate	112.37
Total Land available for cultivation (in thousand hectare)	**5295.77**
Total Area under Fodder Cultivation (in Thousand hectare)	**139.87**

Source: Statistical Abstract- 2008, W.B.

Conclusion

1. To augment the feed resources (green) for rearing of sheep and goats, various Central and State Govt. schemes under feed and fodder development activities are being implemented to bridge the gap between demand and supply; but those schemes are generally biased towards large ruminants. Special emphasis is to be given on the requirements of these small ruminants.
2. Common property resources such as canal and pond embankment, roadside, side of railway tracks etc.,may be used for production of greens(forages) involving local bodies and self help groups.
3. Surplus green fodder production, if any, during monsoon period may be utilized in the lean period by preserving these fodders in the form of hay (as sheep and goat do not prefer silage).
4. Mono cropped lands must be utilized for fodder cultivation.
5. Cultivable waste lands are to be brought under fodder cultivation by awareness programme among the farmers.
6. Short duration fodder crops may be cultivated among two major crops.
7. Marshy land unsuitable for cultivation of main crops, ditches, periphery of the ponds and canal may be utilized for cultivation of perennial fodder crops like para grass and hybrid napier.
8. Roadsides, periphery of the cultivated plots and lands those are normally not cultivated may be used for planting fodder trees like subabul, babla etc.
9. Local bodies, NGO's, SHG's should be involved to develop pasture and grazing lands on waste lands.
10. The main constraint that stands in the way of effective growth of this sector is that all livestock are mostly indigenous and low productive.
11. Most of the states of Eastern India are deficient in fodder. There exixts a large gap between the requirements of feed and fodder and their availability. Short fall of fodder production might be due to non availability of fodder land (1.08 per cent only in W.B) and also acute shortage of availability of quality fodder seed.

12. District specific low cost concentrate mixture should be formulated based on locally available feed ingredients because feed cost alone is 70 per cent of the total rearing cost.
13. Health of animals should be monitored strictly including regular deworming and dipping for proper and full use of nutrients by the animals without any waste.
14. Animal Husbandry extension strategies should be strengthened and followed to increase the small ruminant feed resources

References

All India Livestock Census (18th) (2007). Govt. of India.

Banerjee, G.C. (1998). *A Textbook of Animal Husbandry*. Published by Oxford and IBH Publishing Co Pvt. Ltd. New Delhi.

Biswas, S. and Mukherjee, S. (2011). Understanding common property resources and environmental degradation in the vulnerable states of India- A Perspective. IEMCON,2011. Organized in collaboration with IEEE, India. Proceedings pages: 465-470.

Department of Animal Husbandry and Dairying (2005). Basic Animal Husbandry Statistics. Department of Animal Husbandry and Dairying. Govt. of India, New Delhi, India http://dahd.nic.in/stat files/ BAHs 2006 per cent web per cent 20 web.pdf.

Dey, A., Barari, S.K. and Yadav, B.P.S. (2007). Goat production scenario in Bihar, India Livestock Research for Rural Development 19 (9).

Directorate of Animal Resources and Animal Health, Govt. of W.B. (2011). Status Report on Fodder Development activities in the state of West Bengal. Presented to the Regional Co-ordination Committee meeting held on 20.7.2011 at R.S.F.P.D., Kalyani.

Economic Review (2008-2009). Government of West Bengal.

FAO (2005). Livestock sector Brief: India. Food and Agriculture Organization, Rome. http://www.fao.org/AG/AGA info/resources/cn/publications/sector.briefs/isb IND.pdf.

Kaushish, S.K. (2010). *A Textbook of Animal Husbandry*. Published by Oxford and IBH Publishing Co Pvt L td. New Delhi.

NSSO (1998). Ownership of Livestock,cultivation of selected crops and consumption levels. Report No.424 5th quinquernnia survey of consumer expenditure, NSS 50th Round National Sample Survey Organization, Department of Statistics, GOI.

Press Information Bureau (2005). www.pib.nic.in/feature/fery98/fed798/PIBF2907981.html.

Ranjhan, S.K. (1998). *Nutrient Requirements of Livestock*. Published by ICAR.

Chapter 17

Therapeutic Value of Milk in Human Nutrition

B.K. Mishra

Department of RDAP, North Eastern Hill University, Tura campus, Tura, Meghalaya – 794 002

ABSTRACT

Milk and milk products are rich in proteins and peptides which play a major role in development of the immune system in mammalian offspring. Immunotropic properties of these compounds prompted investigators to search for their utility in prevention and therapy of various disorders in humans. The following constituents of milk are of particular interest: a) Lactoferrin (LF)–exhibits antibacterial, antifungal, antiviral, antiparasitic and antitumor activities. It is protective with regard to intestinal epithelium, promotes bone growth and accelerates recovery of the immune system function in immunocompromised animal; b) A Proline-Rich Polypeptide (PRP) shows a variety of immunotropic functions, including promotion of T-cell maturation and inhibition of autoimmune disorders. PRP was recently found to improve or stabilize the Instrumental Activity of Daily Living status in Alzheimer's disease patients. c) Casein–has been protective in experimental bacteremia by eliciting myelopoiesis. Casein hydrolyzates were also protective in diabetic animals, reduced the tumor growth and diminished colicky symptoms in infants. Casein-derived peptides have been found to have antihypertensive effects. Glycomacropeptide (GMP)–a peptide derived from kappa casein, exhibits antibacterial and antithrombotic activities. d) Alpha lactalbumin (LA)–demonstrates antiviral, antitumor and anti-stress properties. LA-enriched diets were anxiolytic, lowered blood pressure in rats, prevented diarrhea and led to a better weight gain in malnourished children. e) Lysozyme–is effective in treatment of periodentitis and prevention of tooth decay. Milk enriched in lysozyme was used in feeding premature infants suffering from concomitant diseases and lactoperoxidase–shows antibacterial properties. In conclusion, milk-derived proteins and peptides are bio-accessible and safe for the prevention and treatment of numerous disorders in humans. This paper tries to focus on nutritional and therapeutic potential of milk and dairy origin components for better utilization as human food to cure several diseases and improve the health status of human beings.

Keywords: Milk, Dairy products, Nutritional, Therapeutic, Utilization.

Introduction

The effect of diet is only one factor that affects human health. Influences of gender, genetics, and lifestyle patterns such as exercise and smoking are significant factors affecting the health of an individual. There is a considerable amount of variation among individuals and results from different studies on the same nutrient may be contradictory, and thus confusing to the consumer. Human health topics covered below are the relationship between milk and *allergies, arthritis, cancer, conjugated linoleic acid* (*CLA*), *coronary heart disease, dental caries, diabetics, hypertension, lactose intolerance, osteoporosis and bone health, probiotics,* and *raw milk consumption*.

Milk and Allergies

A food allergy is an immune response to proteins resulting in inflammatory symptoms involving the skin, respiratory system, and gastrointestinal tract. Some people think they are allergic to milk because they have symptoms of bloating, cramps, flatulence, and diarrhoea after consumption of dairy products. This is not an allergic response but is a condition known as lactose malabsorption.

Allergic responses have different mechanisms, and some are mediated by immunoglobulin, specifically IgE. The major food proteins that trigger allergic reactions are milk, eggs, peanuts, tree nuts, fish, shellfish, soy, and wheat. Prevalence of milk allergy in the population is about 2 per cent in children and 1 per cent in adults. Onset of milk allergy may occur within about 1 yr of age and most children outgrow this allergy by six years of age.

There are two main categories of *proteins* in milk, the caseins and whey proteins, and there are multiple proteins within each category. Although the level of sensitivity to individual milk proteins varies, to date all the milk proteins have been shown to contribute to milk allergy. Human reactivity to the diverse types of proteins in milk also suggests that people allergic to cow's milk are also likely to be allergic to goat and buffalo milk, although more research is needed to confirm this. Most therapy recommendations for milk allergies are to remove dairy products from the diet during the allergic period. Because many children outgrow milk allergies, dairy products can be reintroduced into the diet so that children may receive the nutritional benefits from consuming dairy products. Typical *pasteurization conditions* used for milk do not affect the protein content or structure and, therefore, have no effect on its allergen city.

Milk and Arthritis

Arthritis is characterized by damaged joints and leads to physical disabilities. There are many forms of the disease caused by degeneration of the joint (osteoarthritis), autoimmune diseases and inflammation (rheumatoid, psoriatic arthritis), infection (septic arthritis), and others.

It has been suggested that consumption of raw milk helps relieve arthritis. However, there are no studies in the literature that report an effect of drinking milk and relieving arthritis. There are currently no scientifically supported dietary recommendations to help relieve arthritis aside from maintaining a heavy weight and balanced diet.

Milk and Cancer

Cancer is a disease characterized by malignant growth in the body. Different cancers have different causes, modes of progression, treatments, and possible risk factors. The area of cancer research is very active and there is a considerable amount of the data in the literature. Because of the variation in cancer types, individual metabolism, and factors evaluated in any given study there may be conflicting results published between studies on the same topic. Factors affecting the risk of cancer include genetics, lifestyle choices such as smoking and exercise, diet and other environmental causes.

Diets high in fat and calories and low in fiber and fruits and vegetables appear to increase the risk of some cancers. Diets high in dairy were thought to increase the risk of breast cancer because of the high fat content of some dairy products. However, there are conflicting research studies on the association between high fat (any type) intake and breast cancer. It used to be thought that dairy product consumption increased the risk of cancer, but we now know that there are several components in milk that are protective against cancer: *calcium, vitamin D, conjugated linoleic acid (CLA), sphingomyelin,* and *whey proteins.*

Calcium and vitamin D may reduce the risk of colon cancer. The protective effect of calcium and vitamin D was observed in numerous human studies in several countries. The amount of calcium consumed in most of these studies was at least 700 mg/d from either food sources or supplements. Sphingomyelin and whey proteins were shown to inhibit colon cancer cell growth in laboratory settings.

Calcium and vitamin D consumption have also been shown to have a protective effect against breast cancer in humans. Conjugated linoleic acid (CLA) is a very effective inhibitor in breast cancer development, as well as inhibiting skin, stomach, and colon cancers in laboratory settings. Whey proteins have been shown to inhibit breast cancer in laboratory settings. Some reports suggest that insulin-like growth factor (IGF-1) and growth hormones (GH) found in milk may promote breast cancer. However, the amount of IGF consumed in milk is very small compared to the amount produced by the human body, and bovine GH is not biologically active in humans. Therefore, these compounds do not contribute to breast cancer in humans.

Milk and Conjugated Linoleic Acid (CLA)

Conjugated linoleic acid (CLA) is a fatty acid found in *milk fat* that has a specific chemical arrangement and unique health benefits. There is a growing understanding that different fatty acids are metabolized differently and contribute differently to human health, both positively and negatively. In general, *trans* fatty acids are considered bad for health and contribute to the risk of coronary heart disease. There is a movement in the U.S. and other developed countries to minimize consumption of *trans* fatty acids and their presence in the food supply. However, the *trans* fatty acids of concern are those that are generated during partial hydrogenation of vegetable oils to create desirable functional properties for use in margarines, baked goods, and many other processed foods. There is a difference in the health effects of *trans* fatty acids obtained by industrial processing of vegetable oil and trans fatty acids that naturally occur in ruminant animal fat in milk and meat.

The amount of CLA needed by humans to obtain these benefits currently is not known. The content of CLA in milk varies based on many factors including feed and cow-to-cow variation. In general, milk from cows on pasture has a higher CLA content than milk from cows fed on rations, although the ration diet may be supplemented to increase CLA content. CLA is found in the milk fat, therefore, dairy products that have higher fat content will have higher CLA content.

Milk and Coronary Heart Disease

Coronary heart disease (CHD) is the leading cause of death in developed countries. Coronary heart disease is characterized by the accumulation of plaque inside the artery walls which narrows the arteries and reduces the flow of blood and oxygen to the muscle. There are many risk factors that contribute to CHD. Some risk factors cannot be modified such as being male, getting older, and a family history of CHD, while some risk factors can be modified such as smoking, high blood pressure, obesity, lack of exercise, and high blood cholesterol levels. The positive correlation of high blood cholesterol levels and increased risk of CHD is well-documented.

Total blood cholesterol consists of several components that have different effects on CHD. A high level of low density lipoprotein (LDL) has the strongest effect of increasing the risk of CHD, whereas a high level of high density lipoprotein (HDL) has a protective effect against CHD. Increased blood triglyceride, or tricylglycerol, levels also increase the risk of CHD, the blood cholesterol profile of an individual can be influenced by diet and thus modified, but there is also a strong genetic influence that cannot be modified. High intakes of total fat, saturated fat, and to some extent dietary cholesterol are associated with increased total and LDL cholesterol.

Milk fat is considered to be a highly saturated fat, with approximately 65 per cent of the fatty acids being saturated. Milk fat typically contains 10 per cent of fatty acids that are 10 carbons or under, 4 per cent lauric acid (12 carbons), 10 per cent myristic acid (14 carbons), 27 per cent palmitic acid (16 carbons), and 14 per cent stearic acid (18 carbons). Milk fat contains approximately 30 per cent monosaturated and 5 per cent polysaturated fatty acids. With respect to blood cholesterol, if only the palmitic, myristic and lauric acids are considered then 41 per cent of milk fat contributes to increased blood cholesterol rather than the total 65 per cent saturated fatty acids. Consumption of nonfat or lowfat dairy products is recommended for cardiovascular health.

Milk and Dental Caries

Dental caries occur when the enamel on the teeth loses its mineral content (decentralization) resulting in a pitted surface. Saliva protects teeth from caries by maintaining a balanced pH (approx.7) in the mouth and by a continuous depositing of calcium and phosphorus on the enamel caries occurs when demineralization is greater than mineralization of the enamel. Fermentation of sugars by the bacteria in the mouth results in the formation of acids which then lower the pH (to 5.5) of the mouth and allows for loss from tooth enamel. It has long been recognized that sugar consumption contributes to dental caries.

Milk contains the sugars lactose, but lactose may not have as much effect on dental caries as other sugars. Consumption of milk products, particularly cheese, has been shown in research studies to have a protective effect against dental caries. Although the mechanism by which cheese protects against dental caries is still not clear, milk proteins, calcium and phosphorus all contribute to this effect. Milk protein contains a large amount of calcium and phosphorus, thus increasing the concentration of these minerals in the mouth during the remineralization process. It is also thought that consumption of milk and cheese may have a buffering effect on mouth pH that prevents the pH from dropping to the conditions favourable for demineralization to occur (pH 5.5). Pasteurization of milk does not effect the function of milk protein, calcium, or phosphorus.

Milk and Diabetes

Diabetes is a disease of carbohydrate metabolism. Type I diabetes is caused by an inability to produce insulin and is controlled by insulin injections. Type I diabetes is often associated with childhood onset. Type II diabetes is caused by an inability to respond to insulin that typically occurs in adults, and can be controlled through diet and exercise.

There were some studies that suggested that milk consumption contributed to childhood diabetes. However, these data have not been confirmed in larger studies, and more definitive research on the effects of dairy product consumption on Type I diabetes is needed. Several studies have evaluated the effect of whole diets on diabetes and conclude that people with higher dairy consumption have a lower incidence of Type II diabetes. Although the mechanism of this action is still unknown, it is possible that *calcium*, which also protects against *hypertension*, is active in protecting against diabetes.

Milk and Hypertension

Hypertension is characterized by high blood pressure. High blood pressure increases the risk of coronary heart disease, stroke, and kidney disease. Factors that contribute to high blood pressure include obesity, a sedentary life style, and high intake of sodium in some individuals.

Dietary factors that have been studied and shown to reduce blood pressure are adequate intake of *calcium, potassium*, and *magnesium*. Calcium and potassium were identified as important nutrients for the reduction of blood pressure in the 1980s. Milk is a source of *calcium, potassium*, and *magnesium*. Lowfat dairy products have been used successfully in dietary intervention studies as a combined source of these minerals to reduce blood pressure.

Angiotensin-I-converting enzyme (ACE) is an enzyme in the body that regulates fluid and electrolyte balance which affects blood pressure. Inhibition of ACE activity helps control blood pressure. Recent studies have identified bioactive peptides from milk casein and whey *proteins* that inhibit ACE activity and reduce blood pressure. Milk proteins must be broken down into peptides (short amino acid chains) before they are bioactive, and this is accomplished by enzymes in the digestive tract or during the fermentation of milk.

Milk and Lactose Intolerance

Problems with lactose digestion result in bloating, abdominal cramps, flatulence, and loose stools. Many people consider these symptoms to reflect an intolerance to lactose, but there are several concepts surrounding lactose intolerance that need to be clarified. *Lactose* is a sugar that is unique to milk and requires the enzyme lactase (B-galactosidase) for digestion. All humans, with a very small number of exceptions, are born with the ability to generate adequate amounts of the enzyme lactase in their digestive system. Milk is the only food newborns receive so their bodies must be able to digest lactose in order to obtain the energy required for growth and development. As humans age, the liability to produce the enzyme lactase decreases, a condition called *lactase nonpersistence*. Most people, even into adulthood, maintain some lactase activity in their gut. People that experience the symptoms described above after eating products typically have exceeded the activity of the lactase available in their system and this is called *lactose malabsorption*. The term *lactose intolerance* is used to describe the above symptoms in response to a defined amount of lactose consumed, usually tested in a clinical setting.

Another clarification that needs to be made is that fresh milk does not contain lactase. Lactase may be present in dairy products, but it comes from lactic acid bacteria that are either added specifically to milk for fermentation or through airborne or other contamination. It also should be noted that *pasteurization* does not affect lactose, and pasteurized milk is neither more nor less digestible, nor has a different lactose content than raw milk.

Individuals vary in their ability to digest lactose, the severity of their symptoms, and their perception of their symptoms. There is a greater incidence of lactose malabsorption in adults from countries that traditionally do not have a strong dairy industry. The prevalence of lactose malabsorption is very high (almost 100 per cent) in Asian countries, above 50 per cent in South America and Africa, and is approximately 53 per cent among Mexican Americans and 80 per cent among African American.

The issue of lactose intolerance and maldigestion is a very emotional one. Although many people consider themselves to be lactose intolerant, they are often able to digest a limited quantity of lactose. Some people are more sensitive to the discomfort experienced with bloating and abdominal cramps and may perceive the symptoms to be more severe. Symptoms incorrectly attributed to poor digestion of dairy products may be caused by other bowel disorders. Many people may consider themselves lactose intolerant and therefore avoid dairy products, when in fact they could include dairy products in their diets. Long term avoidance of dairy products, due to perceived lactose malabsorption, can result in decreased consumption of calcium, potassium, magnesium and other minerals and vitamins that can lead to problems with bone health, hypertension and other disorders.

There are several strategic available for people with lactose maldigestion to allow them to enjoy dairy products with minimal discomfort. Consuming small quantities of dairy products at a time will help to keep the lactose load manageable for the enzymes available. Consuming dairy products with a meal helps because the other in the meal will prolong gastric emptying, which means the stomach empties slower and therefore the amount of lactose reaching the small intestine is spread out over a longer period of time, thereby not overloading the lactase enzymes. Continued consumption of small quantities of dairy products will not increase lactase activity, but the environment of the digestion system will begin to adapt to the presence of lactose and reduce unpleasant symptoms. Fermented dairy foods contain lactic acid bacteria that have lactase present, providing additional active enzymes to assist with human digestion. Fermented dairy products usually have less lactose present because the lactose has been partially used by the bacteria to produce the desirable flavours and textures of products like yogurt and cheese. There are also lactose-reduced dairy products on the market. Lactase enzyme can also be taken in pill form prior to consuming dairy products to ease digestion.

Milk and Osteoporosis and Bone Health

Bone is approximately 50 per cent calcium phosphate. Bone serves as a reservoir for calcium in the body. Blood levels of calcium are tightly regulated to remain at a constant level for proper body function. Bone is constantly undergoing a remodeling process where minerals are removed from bone and added to bone. If dietary intake is insufficient to replace the minerals then bone loss occurs. Bone remodeling is a complex process that involves hormones, protein, calcium, phosphate, vitamin D, and other vitamins and minerals. There is a large variation among individuals with respect to peak bone mass, bone loss, and fracture rates at different sites on the body. Many studies have been conducted to understand this complex behaviour, and our knowledge about these relationships continues to change.

An undisputed fact is that accumulation of maximum peak bone mass during childhood and early adulthood is the best protection against osteoporosis. Adequate calcium intake during this time is critical. Osteoporosis can be considered a disease of the elderly but conditions are set in childhood and early adulthood. Adequate calcium intake throughout life provides the minerals needed for bone remodeling and slows or reduces overall bone loss. The most important dietary source of calcium is dairy products. Calcium also may be obtained from other sources, such as leafy greens, fortified products, and supplements. However, fortified dairy products have the benefit of providing vitamin D, which regulates calcium absorbtion into bones, and other important components of bone metabolism such as phosphorus, protein, magnesium, and zinc. Dairy products contain a large amount of calcium per serving than that other foods. Many studies have shown that people with higher calcium consumption have a lower fracture rate, and often the calcium comes from diary products. The importance of dairy as a source of calcium has led to dietary recommendations of three servings per day.

Milk and Probiotics

Probiotics are defined as live bacteria that provide a health benefits when consumed in adequate amounts. Probiotics bacteria can be consumed as powders. Capsules, or as part of foods, which may be called "functioval foods." Many probiotic bactaria are Lactic acid bactaria, meaning that they use Lactose as an energy source, and hence are associated with dairy products. Yogurt is often associated with probiotics because the bacterial culture used to ferment into yogurt are probiotic cultures, and it is a nutritionally complete food for the maintenance of other probiotics that are added for further health benefits. Most yogurt is made with milk that is pasteurized before fermentation and, therefore, the bacteria are alive the probiotic organisms. An important goal of the use of probiotics is that the bacteria are alive at consumption. Some yogurts may be the pasteurized after fermentation and this would inactive the probiotics organisms. An important goal of the use of probiotics is that the organisms reach their target location in the body alive and in sufficient numbers to be effective. However, the number of bacteria needed for health benefits is unknown in many cases, because the literature on the survival and persistence of organisms in the human body is spare and further research is needed.

The majority of the benefits of the benefits of probiotics are seen in the digestive system. Probiotics improve the general health of the gastrointestinal tract, reduce the symptoms of lactose intolerance, help in the treatment of diarrhoea, inflammatory bowel disease, and irritable bowel syndrome, and help prevent colon cancer. Probiotics also enhance the immune system and help reduce some allergic reactions. There is also an indication that probiotics may play a role in inhibiting Heliobacter pylori infections which cause ulcers, and in maintaining vaginal health.

Many different bacteria have been identified as probiotics. However, the benefits provided are often specific to the general type of bacterial and may be specific even to individual strains within a bacterial type. Streptococus thermophilus and Lactobacillus bulgaricus are the primary organisms used in yogurt production and produce lactase which helps with lactose digestion in people with lactose malabsorption (intolerance). Lactobacillus specified such as L. acidophilus, L. bulgaricus, L. cacei, L. rhamnosus, L. reuteri, and Streptococcus thermophilus and Sacchromyces boulardii reduce the duration and seveity of diarrhea associated with rotavirus and antibiotics use in children, and traveler's diarrhea. Lactobacillus and Bifidobacteria may help to treat the abdominal pain, flatulence and irregulars bowl movements associated with irritable bowel syndrome, and the inflammation associated with inflammatory bowel diseases such as ulcerative colitis and Cohn's disease. The mechanisms behind these benefits are as varied as the disease, and in many cases are still unknown. Probiotics may help by reestablishing the balance of the intestinal microflora and stimulating other enzyme activity, producing antimicrobial substances such as bacteriocins and organic acids, changing the pH of the gastrointestinal tract, producing lactase and stimulating other enzyme activity, producing antimicrobial substances such as bacterocins and organics acids, changing the permeability of the intestinal lining, and stimulating the immune system and antibody responses.

Raw Milk Consumption

Raw milk is often consumed by milk producers on their farms because of convenience and in a preference for the raw milk. Although not scientifically proven, people that grow up drinking raw milk often do not have adverse health reactions to raw milk consumption. However, there are many instances of farm visitors an others getting ill from raw milk and products made from raw milk. Raw milk may be used to make cheese that has unique flavors, but raw milk cheese are required by law to be aged for 60 days to reduce the likelihood of illness from disease causing organisms (pathogens) that may be present in the raw milk.

The risk in drinking raw milk is real, and those choose to do so should make informed decisions about the risks and benefits of drinking raw milk. Pathogens come from the farm environment and may be present in raw milk, even from healthy cows and in milk produces under sanitary conditions. While adults in good health may consume raw milk without adverse affects, that same milk may cause severe illness in the young, elderly, and people with compromised immune systems. A particular concern is the consumption of raw milk by children who are not able to make the risk –benefits decision for them. There have been several outbreaks of E. coli 0157:H7 in children who drank raw milk that resulted in hemolytic uremic syndrome, which causes kidney failure and may lead to life long problems or death. These illnesses could have been prevented.

There is a movement to consume raw milk because of perceived health benefits of raw milk compare to pasteurized milk. It is suggested that raw milk is easier to digest, particularly for people that are lactose intolerant, because enzymes present in raw milk are inactivated by pasteurization. However, as discussed above in the section on lactose intolerance, there is no lactase present in fresh milk. Lactase may be present in lactic acid bacterial that are added purposefully to milk for fermentation or from airborne or other contamination. Enzymes used to digest food are found in the mouth, stomach, and primarily in the intestines of human-enzymes present in the food are found in insignificant role in their digestion.

It is suggested that the pasteurization process changes the nutritional properties of milk fat, proteins, lactose, vitamins, and minerals. Typical pasteurization conditions used for milk do not affect the nutritional or functional properties of milk fat, protein, lactose, minerals, and fat soluble vitamins, but there is a small reduction in the quantity of heat sensitive water-soluble vitamins. It should be noted here that milk is an insignificant source of vitamins C, so the small reduction that occurs during pasteurization is not important in the overall diet. In some products, like shelf-stable canned milk, the heat treatments need to be more severe in order to sterilize the milk and there may be changes in the nutritional qualities of these products.

The Importance of Milk and Dairy Products as Part of a Healthy Balance Diet

Milk and dairy products are particularly good sources of many nutrients and provide a quick and easy way of supplying these nutrients to the diet within relatively few calories.

Milk, cheese and yogurt are all rich in the following beneficial nutrients

- ☆ Calcium- for healthy bones and teeth
- ☆ Phosphorous-for energy release
- ☆ Magnesium-for muscles function
- ☆ Protein- for growth and repair
- ☆ Vitamin B12-for production of healthy cells
- ☆ Vitamin A-for good eyesight and immune function
- ☆ Zinc-for immune function
- ☆ Ribofflavin- for healthy skin
- ☆ Folate- for production of healthy cell
- ☆ Vitamin C- for formation of healthy connective tissues.

1 glass of milk alone can make a significant contribution to the daily recommendation intake of many important nutrients for all age groups.

Table 17.1: Contribution of 1 Glass of Milk to Daily Nutrient Requirements

Nutrient	*Amount Present in a Glass (200ml) Semi Skimmed Milk*	*Contribution to Recommended Daily Amount for Females 11-18 Years (per cent)*	*Contribution to Recommended Daily Amount for Males 11-18 Years (per cent)*
Protein (g)	7.2	13	16
Calcium (mg)	247	25	31
Phosphorus (mg)	194	25	31
Magnesium (mg)	23	8	8
Sodium (mg)	89	6	6
Potassium (mg)	321	9	9
Chloride (mg)	179	7	7
Iron (mg)	0.04	0.4	0.3
Zinc (mg)	0.8	8	11
Copper (mg)	Trace	–	–
Selenium (ug)	2	3	3
Iodine(ug)	62	44	44
Vitamin A(ug)	39	6	7
Thiamin(mg)	0	0	0
Riboflavin (mg)	0.5	38	45
Niacin (mg)	0.2	1	1.4
Vitamin B6(ug)	0.12	8	10
Vitamin B12(ug)	1.9	127	127
Folate (ug)	19	10	10
Vitamin C(mg)	4	10	10
Vitamin D (ug)	Trace	–	–

It is highly recommended that milk and dairy foods should be consumed as part of a healthy balanced diet and that ideally 3 portions should be consumed per day. 3-A-Day includes 200 ml/a glass of milk, 150g or a pot yogurt and 30 g or a matchbox sized piece of cheese. Portions sizes will vary for different age groups but will provide each group with the daily recommended calcium requirement.

The human body can be thought of as an engine that releases the energy present in the foods that it digest. This energy is utilized partly for the mechanical work performed by the muscles and in . . a distance of one meter by a force of one Newton. The relatively higher levels of energy in human nutrition are more like to be measured in kilojoules (1kilojoule= 10^3) joules or megajouls (1 megajoule = 10^6 joules). One kilocalorie is equivalent to 4.184 kilojoules.

This nutrient pack drink is given to patient even during critical stage. Various milk products such as curd, butter milk, ghee, cheese, paneer, khoya, rabri etc are used commonly in our food preparations. This time tested nutritious drink is been criticized by few vocal people to the extent that milk is equated to poison.

Conclusion

Dairy sector in India has the major responsibility of providing milk to growing population, ensuring animal protein in their diets and thus ensuring better health. Among farm animals dairy cows and buffaloes are major producers of valuables human foods such as milk land meat and thus plays vital role in improving the nutritional status of people especially in villages. Dairying and crop products are mutually integrated since both are interdependent and benefited each other. Milk is a nearly complete food supporting better growth rate and is very important in the diet for growing children, sick and elderly. Major group of farmers involved in dairying are small farmers and agricultural Labourers. Hence it is an occupation of poor land landless people providing them a means for "landless agriculture". Dairying also plays a vital role in improving the nutritional status of their families.

References

William H. Bowen and Ruth A. Lawrence (2005). "Comparison of the Carcinogenicity of Cola, Honey, Cattle Milk, Human Milk, and Sucrose", *Pediatrics* 116 (4): 921–6. Doi,10.1542/peds.2004-2462. PMID 16199702

Hemme and Otte (2010). "Status and Prospects for Smallholder Milk Production: A Global Perspective". Food and Agriculture Organization of the United Nations

Basnet, S., Schneider, M., Gazit, A., Mander, G., Doctor, A. (April 2010). "Fresh Goat's Milk for Infants: Myths and Realities–A Review", *Pediatrics* 125 (4): e973–977.doi:10.1542/peds.2009-1906. PMID 20231186

Goff, Douglas (2010). "Introduction to Dairy Science and Technology: Milk History, Consumption, Production, and Composition". Dairy Science and Technology, University of Guelph

Bewley, Elizabeth (24 June 2010). "Dairy farmers tackle big coops". Burlington, Vermont: Burlington Free Press. pp. 8B

Vorbach C., Capecchi M.R., Penninger J.M. (2006). "Evolution of the mammary gland from the innate immune system, Bioessays 28 (6): 606–616.Doi: 10.1002/bies.20423. PMID 16700061

Goldman A.S. (2002). Evolution of the mammary gland defense system and the ontogeny of the immune system" *Journal of Mammary Gland Biology and* Neoplasia 7 (7): 277–289. PMID 12751892

Vigne, D., Helmer, J.D. (2007). "Was milk a "secondary product" in the Old World Neolithisation process? Its role in the domestication of cattle, sheep and goats".Anthropozoologica 42 (2): 9–40.

Chapter 18

Concept, Need and Scope of Sustainable Agriculture in North Eastern India

Md. Mokidul Islam*

Subject Matter Specialist (Agronomy)
ICAR Research Complex for NEH Region
Krishi Vigyan Kendra, Tura, Sangsanggre,
P.O: Dobasipara – 794 005, West Garo Hills (Meghalaya)

ABSTRACT

The biodiversity in North- Eastern Indian states remain unexplored to utilize the resources properly for development of agrarian economy and to feed the millions of the region. The region has about 72 per cent area under hill ecosystem with inaccessible terrain, fragility, marginality, excessive sloppy land with rolling topography, rich biodiversity, unique ethnicity and socio-ecological set up. The agricultural system of North Eastern Hill Region of India is in Complex-Diverse-Risk Prone (CDRP) ecosystem which leads to low productivity at subsistence level. Sustainable, high yield agriculture must be at the top of India's agenda for food security, nutritional security, and environmental safety. Maximum economic yield goals can be achieved on irrigated land by using improved genetics, yield targeted site specific nutrient management, and improved crop husbandry practices. The region as a whole is having a deficit of 1.6 million tonnes of food grains. In order to make the region self sufficient in food grain production, the productivity of all the food crops has to be increased sustainable way from the present level. The sustainability in agriculture is possible to attain in the north-eastern India with proper soil and water conservation measures to prevent degradation of soil productivity, efficient use of limited irrigation water, diversification of

---

* E-mail: mislam01d@yahoo.co.in

crops (including livestock), resource conservation technologies like system of rice intensification (SRI), integrated crop management(ICM), aerobic rice cultivation, organic farming, precision farming, conservation agriculture, crop rotations, intercropping, integrated plant nutrient management system(IPNMS), integrated weed management(IWM), integrated pest management (IPM) and disease management(IDM)etc. Hence, the sustainable agriculture in north- eastern region of India is environmentally feasible, socially acceptable and economically viable for improvement of livelihood of the farming community.

Keywords: Biodiversity, Complex-diverse-risk prone, Diversification, Sustainability.

Introduction

Agriculture has changed dramatically, especially since the end of World War II. Food and fibre productivity soared due to new technologies, mechanization, increased chemical use, specialization and government policies that favoured maximizing production. These changes allowed fewer farmers with reduced labour demands to produce the majority of the food and fibre in the globe. Although these changes have had many positive effects and reduced many risks in farming, there have also been significant costs. Prominent among these are topsoil depletion, groundwater contamination, the decline of family farms, continued neglect of the living and working conditions for farm labourers, increasing costs of production, and the disintegration of economic and social conditions in rural communities. A growing movement has emerged during the past two decades to question the role of the agricultural establishment in promoting practices that contribute to these social problems. Today this movement for sustainable agriculture is garnering increasing support and acceptance within mainstream agriculture. Not only does sustainable agriculture address many environmental and social concerns, but it offers innovative and economically viable opportunities for growers, labourers, consumers, policymakers and many others in the entire food system. Because the concept of sustainable agriculture is still evolving, we intend this essay not as a definitive or final statement, but as an invitation to continue the dialogue.

Sustainable, high yield agriculture must be at the top of India's agenda for food security, nutritional security, and environmental safety. Maximum economic yield goals can be achieved on irrigated land by using improved genetics, yield targeted site specific nutrient management, and improved crop husbandry practices. Attention must be paid to quantities of nutrients being removed by crops and quantities of nutrients supplied by all. To minimize India's current annual negative nutrient balance of 8 to 10 M t of $N+P_2O_5+K_2O$, increased use of fertilizers along with possible use of organic manures and bio-fertilizers would be essential and inevitable. Organics, which can supply a portion of the P and K along with the secondary and micronutrients required by crops can help offset the negative nutrient balance and slow down nutrient depletion processes; however it cannot meet the nutrient requirements of crops. Proponents of strictly organic farming systems tend to overlook three important factors: (a) the adequacy of supply of organic materials on a national basis, (b) the nutrient content and rate of supply to growing crops, and (c) the high labour costs to collect and apply organic materials. The importance of organics in improving soil physical and biological properties cannot be denied, their efficient use to the extent possible should, therefore, be promoted and integrated nutrient management be practiced.

Sustainability of Indian agriculture to maintain food self-sufficiency will depend on the 'high input-high output' principle. The 'low input-high output' concept is merely a dream, and adherence to this invalid view would prove fatal for food security and nutritional security. In the Indian context, this is more true now than ever before because of emerging demands for horticultural, floricultural, and plantation crop products. The founding father of the green revolution in India, Dr. Norman Borlaug, has rightly stated that but for the use of chemical fertilizers, India and China would have needed two to three times more land under cereals to meet food needs of 1991 if they used the technology of 1960..and not increase the fertilizer input to sustain its present level of production.

The North Eastern India at a Glance

The North Eastern Region of India comprising of Assam, Arunachal Pradesh, Meghalaya, Manipur, Mizoram, Nagaland, Sikkim and Tripura. The region lies between 21°50′ and 29°34′ N latitude and 85°34′ and 97°50′E longitude. The region has a population of more that 39 million and geographical area of 26.2 million hectare, which is 3.85 per cent and 8 per cent of the population and area of the country, respectively. The region has about 72 per cent area under hill ecosystem with inaccessible terrain, fragility, marginality, excessive sloppy land with rolling topography, rich biodiversity, unique ethnicity and socio-ecological set up. The north eastern region of India received high rainfall and therefore clothed with diverse and dense vegetation. Beyond the transitional pre-glacial region with increasing altitude, there appears the greater Himalayan region devoid of significant vegetative cover. Rocky surface, alpine vegetation and snow-capped high peaks dominate the physical landscape of the region. The varied physiological features and altitudinal differences gives rise to varied types of climate ranging from near tropical to temperate and alpine. The annual rainfall in the region is received mainly from South-West monsoon from middle of May and continued till October. On an average, NE region of India receives about 2450 mm of rainfall. The Cherrapunji-Mawsynram range receives rainfall as high as 11,500mm annually. The region shows great variation in temperature regime too. The temperature varies from 15 to 32° C in summer and 0 to 26°C in winter.

The NE region of India is a biodiversity hot spot and harbours the largest number of endemics and species then anywhere in the country. The agricultural practices of the north-eastern region of India are broadly of two types *viz.*,(*i*) settled cultivation practices in plains, valleys, foot hills and terraced slopes and (*ii*)shifting cultivation in hill areas of all the NE states with the exception of Sikkim, where settled cultivation is practiced on terraces. Agricultural operations are carried out up to 3600 m altitude and on slope up to 60 per cent. Cropping system in the region is predominantly rice based except Sikkim, where maize based systems. The NE India, more than 80 per cent gross cropped area is under food grains and rice occupies 89.8 per cent of the total food grain area and 93.2 per cent of the total food grain production. The agriculture in the region is subsistence level. The region as a whole is having a deficit of 1.6 million tonnes of food grains. The cropping intensity on an average in the region is 120 per cent and low cropping intensity is mainly due to lack of irrigation facilities and insufficient soil moisture during winter. For inducing sustainability in agriculture in the NE region of India, the cropping intensity is required to be increased substantially through rain water harvesting during rainy season as well as creating more irrigation facilities and introducing new land use systems. Diversification through farming system approach involving cattle, pig, poultry, duck, fish and other livestock can compensate the requirement to some extent to the non-vegetarian people of north-eastern India. The climatic conditions of the region are ideally suited for cultivation of fruits (pine apple, orange etc.), vegetables, spices (ginger, turmeric, large cardamom etc), multipurpose trees, plantation crops (tea, coffee, rubber etc), orchids, flowers etc.

Status of Food Grain Production in NE India

The biodiversity in NE states remain unexplored to utilize the resources properly for development of agrarian economy and to feed the millions of the region. The data on food grain production and requirement indicated that there is a deficit existing in all the NE states varying from 0.013 to 1.16 million tonnes (Table 18.1). The projected food grain demand for NE India is 12.19, 15.24, 16.76 and 27.23 million tonnes for the year 2011,2021, 2025 and 2050, respectively (Table 18.2). In order to make the region self sufficient in food grain production, the productivity of all the food crops has to be increased sustainable way from the present low level.

Table 18.1: Food Grain Production, Requirement and Deficit (million tonne) in NE States (2001)

States	*Production*	*Requirement*	*Deficit*
Arunachal Pradesh	0.203	0.218	0.015
Assam	4.167	5.328	1.611
Manipur	0.378	0.478	0.100
Meghalaya	0.203	0.260	0.057
Mizoram	0.124	0.178	0.054
Nagaland	0.277	0.398	0.121
Sikkim	0.091	0.104	0.013
Tripura	0.523	0.638	0.115
Total NE India	**5.966**	**7.602**	**1.636**

Source: Sharma and Datta (2006).

Table 18.2: Demand of Food Grain (million tonne) in NE Region of India

Trend	*2011*	*2021*	*2025*	*2050*
High (250 kg/capita/annum)	12.19	15.24	16.76	27.23
Low (200 kg/capita/annum)	8.97	10.32	10.94	15.04

Source: Sharma and Datta (2006).

Components of Sustainable Agriculture

Major component of the sustainable agriculture systems are:

- ✩ Soil and water conservation to prevent degradation of soil productivity and lengthening crop growing season for optimum crop productivity
- ✩ Efficient use of limited irrigation water without leading to problems of soil salinity, alkalinity and high ground water table
- ✩ Crop rotations that mitigate weed. Diseases and insects-pests problems, increase soil productivity and minimise soil erosion
- ✩ Integrated nutrient management that reduces the need for chemical fertilisers, improves the soil health and minimise environmental pollution by conjunctive use of organics, inorganics and bio-fertilizers
- ✩ Integrated pest management that reduces the need for agrochemicals by crop rotation, weather monitoring, use of resistant cultivars, planting time and biological pest control and
- ✩ Management systems to control weeds by preventive measures, tillage, timely intercultivation and crop rotation to improve plant health

Concept Themes of Sustainable Agriculture

The concept of sustainability basically implies a characteristic of a system, a programme or a resource to last intact forever. Currently, there are two distinct concepts of sustainability in vogue. One, the economist's worldview of sustainability is concerned about the long-term constancy of economic output, income or consumption. Two, the ecologist's and biologist's concept of sustainability relates to long-term preservation of biosphere, that is, the sustenance of human populations and biodiversity conservation in a given geographical area/region, endowed with limited natural resources (FAO 1989, Bartelmus 1997). Economic sustainability implies the maintenance/constancy of produced capital and natural capital (natural resource and environment) used in the production of goods and services. However, ecological sustainability can be defined in terms of compliance with the carrying capacity limit of natural systems. Carrying capacity is usually measured by the number of people a natural system/area/watershed can sustain indefinitely, or for a specified time period, at a particular standard of living. On the other hand, sustainable development defined as 'development that meets the needs of the present without compromising the ability of future generations to meet their own needs. Therefore, the WCED or the Brundtland Commission in its 1987 report, "Our Common Future", emphasised the key role of agricultural sustainability as the basis of sustainable development (WCED, 1987).

Sustainable agriculture integrates three main goals–environmental health, economic profitability, and social and economic equity. A variety of philosophies, policies and practices have contributed to these goals. People in many different capacities, from farmers to consumers, have shared this vision and contributed to it. Despite the diversity of people and perspectives, the following themes commonly weave through definitions of sustainable agriculture.

Sustainability rests on the principle that we must meet the needs of the present without compromising the ability of future generations to meet their own needs. Therefore, *stewardship of both natural and human resources* is of prime importance. Stewardship of human resources includes consideration of social responsibilities such as working and living conditions of labourers, the needs of rural communities, and consumer health and safety both in the present and the future. Stewardship of land and natural resources involves maintaining or enhancing this vital resource base for the long term.

A *systems perspective* is essential to understanding sustainability. The system is envisioned in its broadest sense, from the individual farm, to the local ecosystem, *and* to communities affected by this farming system both locally and globally. An emphasis on the system allows a larger and more thorough view of the consequences of farming practices on both human communities and the environment. A systems approach gives us the tools to explore the interconnections between farming and other aspects of our environment.

A systems approach also implies *interdisciplinary efforts in research and education*. This requires not only the input of researchers from various disciplines, but also farmers, farm workers, consumers, policymakers and others.

Making the transition to sustainable agriculture is a process. For farmers, the transition to sustainable agriculture normally requires a series of small, realistic steps. Family economics and personal goals influence how fast or how far participants can go in the transition. It is important to realize that each small decision can make a difference and contribute to advancing the entire system further on the "sustainable agriculture continuum." The key to moving forward is the will to take the next step.

Finally, it is important to point out that *reaching toward the goal of sustainable agriculture is the responsibility of all participants in the system*, including farmers, labourers, policymakers, researchers, retailers, and consumers. Each group has its own part to play, its own unique contribution to make to strengthen the sustainable agriculture community.

The remainder of this document considers specific strategies for realizing these broad themes or goals. They represent a range of potential ideas for individuals committed to interpreting the vision of sustainable agriculture within their own circumstances. The strategies are grouped according to three separate though related areas of concern:

- ✰ Farming and Natural Resources,
- ✰ Plant and Animal Production Practices, and
- ✰ Economic, Social and Political Context.

Farming and Natural Resources

Water

When the production of food and fibre degrades the natural resource base, the ability of future generations to produce and flourish decreases. The decline of ancient civilizations in Mesopotamia, the Mediterranean region, Pre-Columbian southwest U.S. and Central America is believed to have been strongly influenced by natural resource degradation from non-sustainable farming and forestry practices. Water is the principal resource that has helped agriculture and society to prosper, and it has been a major limiting factor when mismanaged.

The north –eastern region of India is rich in water resources, both the surface and ground water. But poor management has done great damage than deriving any benefit of significance from this valuable natural wealth. The region receives about 510 km^2 of water annually as rainfall and surface water resources have been estimated as 1487.65 km^2 of water. The region ha s ground water potential of 25.31 km^2. Inspite of high potential, there is scarcity of even drinking water during winter months, envisage the urgent need for improving water harvesting measures.

Water Supply and Use

In some areas of the country, extensive water storage and transfer systems have been established, which has allowed crop production to expand to very arid regions. In drought years, limited surface water supplies have prompted overdraft of groundwater and consequent intrusion of salt water, or permanent collapse of aquifers. Periodic droughts, some lasting up to 50 years, have occurred in California for example.

Several steps should be taken to develop drought-resistant farming systems even in "normal" years, including both policy and management actions:

1. Improving water conservation and storage measures,
2. Providing incentives for selection of drought-tolerant crop species,
3. Using reduced-volume irrigation systems,
4. Managing crops to reduce water loss, or
5. Not planting at all.

Water Quality

The most important issues related to water quality involve salinization and contamination of ground and surface waters by pesticides, nitrates and selenium. Salinity has become a problem wherever water of even relatively low salt content is used on shallow soils in arid regions and/or where the water table is near the root zone of crops. Tile drainage can remove the water and salts, but the disposal of the salts and other contaminants may negatively affect the environment depending upon where they are deposited.

Temporary solutions include the use of salt-tolerant crops, low-volume irrigation, and various management techniques to minimize the effects of salts on crops. In the long-term, some farmland may need to be removed from production or converted to other uses. Other uses include conversion of row crop land to production of drought-tolerant forages, the restoration of wildlife habitat or the use of agroforestry to minimize the impacts of salinity and high water tables.

Wildlife

Another way in which agriculture affects water resources is through the destruction of riparian habitats within watersheds. The conversion of wild habitat to agricultural land reduces fish and wildlife through erosion and sedimentation, the effects of pesticides, removal of riparian plants, and the diversion of water. The plant diversity in and around both riparian and agricultural areas should be maintained in order to support a diversity of wildlife. This diversity will enhance natural ecosystems and could aid in agricultural pest management.

Energy

Modern agriculture is heavily dependent on non-renewable energy sources, especially petroleum. The continued use of these energy sources cannot be sustained indefinitely, yet to abruptly abandon our reliance on them would be economically catastrophic. However, a sudden cut-off in energy supply would be equally disruptive. In sustainable agricultural systems, there is reduced reliance on non-renewable energy sources and a substitution of renewable sources or labour to the extent that is economically feasible.

Air

Many agricultural activities affect air quality. These include smoke from agricultural burning; dust from tillage, traffic and harvest; pesticide drift from spraying; and nitrous oxide emissions from the use of nitrogen fertilizer. Options to improve air quality include incorporating crop residue into the soil, using appropriate levels of tillage, and planting wind breaks, cover crops or strips of native perennial grasses to reduce dust.

Soil

Soil erosion continues to be a serious threat to our continued ability to produce adequate food. Numerous practices have been developed to keep soil in place, which include reducing or eliminating tillage, managing irrigation to reduce runoff, and keeping the soil covered with plants or mulch.

The soil of the NE region of India is mostly rich in organic matter and nitrogen content except in shifting cultivation areas, low to medium in phosphorus and medium to high in potassium. However, the soils of the sub-tropical region *viz.*, Tripura and other foot hills are deficit in organic matter and phosphorus. More than 90 per cent of soils of the region are acidic in nature having mild to severe acidity, attributed to the leaching of bases due to heavy rainfall. Application of well decomposed organic manures is important for P availability in these soils. The soils of the region can be broadly classified in five orders *viz., inceptisols, entisols, alfisols, ultisols and mollisols.*

Plant Production Practices

Sustainable production practices involve a variety of approaches. Specific strategies must take into account topography, soil characteristics, climate, pests, local availability of inputs and the individual grower's goals. Despite the site-specific and individual nature of sustainable agriculture, several general principles can be applied to help growers select appropriate management practices:

- ✰ Selection of species and varieties that are well suited to the site and to conditions on the farm;
- ✰ Diversification of crops (including livestock) and cultural practices to enhance the biological and economic stability of the farm;
- ✰ Management of the soil to enhance and protect soil quality;
- ✰ Efficient and humane use of inputs; and
- ✰ Consideration of farmers' goals and lifestyle choices.

Selection of Site, Species and Variety

Preventive strategies, adopted early, can reduce inputs and help establish a sustainable production system. When possible, pest-resistant crops should be selected which are tolerant of existing soil or site conditions. When site selection is an option, factors such as soil type and depth, previous crop history, and location (*e.g.* climate, topography) should be taken into account before planting.

Diversity

Diversified farms are usually more economically and ecologically resilient. While monoculture farming has advantages in terms of efficiency and ease of management, the loss of the crop in any one year could put a farm out of business and/or seriously disrupt the stability of a community dependent on that crop. By growing a variety of crops, farmers spread economic risk and are less susceptible to radical price fluctuations.

Properly managed, diversity can also buffer a farm in a biological sense. For example, in annual cropping systems, crop rotation can be used to suppress weeds, pathogens and insect pests. Also, cover crops can have stabilizing effects on the agroecosystem by holding soil and nutrients in place, conserving soil moisture with mowed or standing dead mulches, and by increasing the water infiltration rate and soil water holding capacity.

Cover crops in orchards and vineyards can buffer the system against pest infestations by increasing beneficial arthropod populations and can therefore reduce the need for chemical inputs. Using a variety of cover crops is also important in order to protect against the failure of a particular species to grow and to attract and sustain a wide range of beneficial arthropods.

Optimum diversity may be obtained by integrating both crops and livestock in the same farming operation. This was the common practice for centuries until the mid-1900s when technology, government policy and economics compelled farms to become more specialized.

Mixed crop and livestock operations have several advantages. First, growing row crops only on more level land and pasture or forages on steeper slopes will reduce soil erosion. Second, pasture and forage crops in rotation enhance soil quality and reduce erosion; livestock manure, in turn, contributes to soil fertility. Third, livestock can buffer the negative impacts of low rainfall periods by consuming crop residue that in "plant only" systems would have been considered crop failures.

Soil Management

A common philosophy among sustainable agriculture practitioners is that a "healthy" soil is a key component of sustainability; that is, a healthy soil will produce healthy crop plants that have optimum vigor and are less susceptible to pests. While many crops have key pests that attack even the healthiest of plants, proper soil, water and nutrient management can help prevent some pest problems brought on by crop stress or nutrient imbalance.

In sustainable systems, the soil is viewed as a fragile and living medium that must be protected and nurtured to ensure its long-term productivity and stability. Methods to protect and enhance the productivity of the soil include using cover crops, compost and/or manures, reducing tillage, avoiding traffic on wet soils, and maintaining soil cover with plants and/or mulches.

Efficient Use of Inputs

Many inputs and practices used by conventional farmers are also used in sustainable agriculture. Sustainable farmers, however, maximize reliance on natural, renewable, and on-farm inputs. Equally important are the environmental, social, and economic impacts of a particular strategy.

Converting to sustainable practices does not mean simple input substitution. Frequently, it substitutes enhanced management and scientific knowledge for conventional inputs, especially chemical inputs that harm the environment on farms and in rural communities. The goal is to develop efficient, biological systems which do not need high levels of material inputs.

Growers frequently ask if synthetic chemicals are appropriate in a sustainable farming system. Sustainable approaches are those that are the least toxic and least energy intensive, and yet maintain productivity and profitability. Preventive strategies and other alternatives should be employed before using chemical inputs from any source. However, there may be situations where the use of synthetic chemicals would be more "sustainable" than a strictly nonchemical approach or an approach using toxic "organic" chemicals.

Development of an integrated plant nutrient system (IPNS) in NE India involving an approximate mix of organics, biological nitrogen fixation, phosphate solubilising microbes and need based chemical fertilisers would be crucial for the sustainability of production and soil as the resource base for it. Popularization of integrated plant nutrient supply by incorporating all components of nutrient sources including legume in rotations and intercropping system is the need of the hour. Keeping in view, the contribution of component wise inputs for sustainable agriculture through integrated plant nutrient system and their contribution has been worked out by Reddy (1999) is given in Table 18.3.

Consideration of Farmer Goals and Lifestyle Choices

Management decisions should reflect not only environmental and broad social considerations, but also individual goals and lifestyle choices. For example, adoption of some technologies or practices that promise profitability may also require such intensive management that one's lifestyle actually deteriorates. Management decisions that promote sustainability nourish the environment, the community *and* the individual.

Table 18.3: Component of Integrated Nutrient Management System (INMS) and their Contribution in Sustainable Agriculture

Component	*Scope*	*Equivalent Under Favourable Conditions*	*Factors*
Fertilisers	All crops	One unit of 60:30:30=10-15 units of wheat, 10-17 units of rice	Variety, weather and management
FYM	To the extent available	One unit of FYM= 3 kg fertiliser for single crop and 5 kg for double cropping	Composition and quality
Compost	Where available	1 t = 3.5 kg N for rice	Composition and quality
Crop residues	Where available	1 t = 3.5 kg N in *kharif* rice, 2 kg N in *rabi* rice	Composition, quality and interval between application
Green manure	Where available	45-50 days *Sesbania*= 30-60 kg N/ha for rice	P application and economics
Rhizobium	Legumes	Equivalent to 10-40 kg N/ha	Quality of inoculums, P application and time of incorporation
BGA	Rice	10 kg BGA = 15-25 kg N/ha	Suitable culture, temperature, moisture and P application
Azolla	Low land rice	10 t fresh *azolla* = 20-25 kg N/ha	P, moisture, pH, temperature, incorporation

Source: Reddy (1999).

Animal Production Practices

In the early part of this century, most farms integrated both crop and livestock operations. Indeed, the two were highly complementary both biologically and economically. The current picture has changed quite drastically since then. Crop and animal producers now are still dependent on one another to some degree, but the integration now most commonly takes place at a higher level–*between* farmers, through intermediaries, rather than *within* the farm itself. This is the result of a trend toward separation and specialization of crop and animal production systems. Despite this trend, there are still many farmers, particularly in the Midwest and North-eastern U.S. that integrate crop and animal systems–either on dairy farms, or with range cattle, sheep or hog operations.

Even with the growing specialization of livestock and crop producers, many of the principles outlined in the crop production section apply to both groups. The actual management practices will, of course, be quite different. Some of the specific points that livestock producers need to address are listed below.

Management Planning

The mobility of the stock, daily feeding, health concerns, breeding operations, seasonal feed and forage sources, and complex marketing are sources of this complexity. Therefore, a successful ranch plan should include enterprise calendars of operations, stock flows, forage flows, labour needs, herd production records and land use plans to give the manager control and a means of monitoring progress toward goals.

Animal Selection

Including livestock in the farming system increases the complexity of biological and economic relationships. The animal enterprise must be appropriate for the farm or ranch resources. Farm capabilities and constraints such as feed and forage sources, landscape, climate and skill of the manager must be considered in selecting which animals to produce.

For example, ruminant animals can be raised on a variety of feed sources including range and pasture, cultivated forage, cover crops, shrubs, weeds, and crop residues. There is a wide range of breeds available in each of the major ruminant species, *i.e.*, cattle, sheep and goats. Hardier breeds that, in general, have lower growth and milk production potential, are better adapted to less favourable environments with sparse or highly seasonal forage growth

Animal Nutrition

Feed costs are the largest single variable cost in any livestock operation. While most of the feed may come from other enterprises on the ranch, some purchased feed is usually imported from off the farm. Feed costs can be kept to a minimum by monitoring animal condition and performance and understanding seasonal variations in feed and forage quality on the farm. Determining the optimal use of farm-generated by-products is an important challenge of diversified farming.

Reproduction

Use of quality germplasm to improve herd performance is another key to sustainability. In combination with good genetic stock, adapting the reproduction season to fit the climate and sources of feed and forage reduce health problems and feed costs.

Herd Health

Animal health greatly influences reproductive success and weight gains, two key aspects of successful livestock production. Unhealthy stock waste feed and require additional labour. A herd health program is critical to sustainable livestock production.

Grazing Management

Most adverse environmental impacts associated with grazing can be prevented or mitigated with proper grazing management. First, the number of stock per unit area (stocking rate) must be correct for the landscape and the forage sources. There will need to be compromises between the convenience of tilling large, unfenced fields and the fencing needs of livestock operations. Use of modern, temporary fencing may provide one practical solution to this dilemma. Second, the long term carrying capacity and the stocking rate must take into account short and long-term droughts. Finally, the manager must achieve sufficient control to reduce overuse in some areas while other areas go unused.

Confined Livestock Production

Animal health and waste management are key issues in confined livestock operations. The moral and ethical debate taking place today regarding animal welfare is particularly intense for confined livestock production systems. The issues raised in this debate need to be addressed.

Confinement livestock production is increasingly a source of surface and ground water pollutants, particularly where there are large numbers of animals per unit area. Expensive waste management facilities are now a necessary cost of confined production systems. Waste is a problem of almost all operations and must be managed with respect to both the environment and the quality of life in nearby communities. Livestock production systems that disperse stock in pastures so the wastes are not concentrated and do not overwhelm natural nutrient cycling processes have become a subject of renewed interest.

The Economic, Social and Political Context

In addition to strategies for preserving natural resources and changing production practices, sustainable agriculture requires a commitment to changing public policies, economic institutions, and social values. Strategies for change must take into account the complex, reciprocal and ever-changing relationship between agricultural production and the broader society.

The "food system" extends far beyond the farm and involves the interaction of individuals and institutions with contrasting and often competing goals including farmers, researchers, input suppliers, farm workers, unions, farm advisors, processors, retailers, consumers, and policymakers. Relationships among these actors shift over time as new technologies spawn economic, social and political changes.

A wide diversity of strategies and approaches are necessary to create a more sustainable food system. These will range from specific and concentrated efforts to alter specific policies or practices, to the longer-term tasks of reforming key institutions, rethinking economic priorities, and challenging widely-held social values. Areas of concern where change is most needed include the following:

Food and Agricultural Policy

Existing federal, state and local government policies often impede the goals of sustainable agriculture. New policies are needed to simultaneously promote environmental health, economic profitability, and social and economic equity. For example, tax and credit policies could be modified to encourage a diverse and decentralized system of family farms rather than corporate concentration and absentee ownership. Government and land grant university research policies could be modified to emphasize the development of sustainable alternatives. Marketing orders and cosmetic standards could be amended to encourage reduced pesticide use. Coalitions must be created to address these policy concerns at the local, regional, and national level.

Land Use

Conversion of agricultural land to urban uses is a particular concern in many areas, as rapid growth and escalating land values threaten farming on prime soils. Existing farmland conversion patterns often discourage farmers from adopting sustainable practices and a long-term perspective on the value of land. At the same time, the close proximity of newly developed residential areas to farms is increasing the public demand for environmentally safe farming practices. Educating land use planners and decision-makers about sustainable agriculture is an important priority. The land use pattern for hill slopes as an alternative to shifting cultivation areas in NE states are given in Table 18.4. For non arable lands like B1 class lands suitable for pasture and fodder crops, livestock enterprises like goat rearing, milch cows etc are to be taken up with pasture grasses like guinea grass, congosignal, broom grass etc. On terraces in the foot hills, the crops like maize, groundnut, ginger, rice bean, sweet potato in *kharif* season and mustard, radish, pea, cabbage, cauliflower and turnip in *rabi* season can be grown. In the middle portion, horticultural plants like mandarin, guava, Assam lemon and pine apple can be grown on contour bunds. Based on the studies have been made so far at ICAR Research Complex for NEH Region, mixed land use with agri-horti-silvipastoral system with appropriate topo sequence of cropping has been found to be one of the most sustainable means to provide an alternative farming system to replace *Jhuming* (Table 18.5 and 18.6).

Table 18.4: Land Use Pattern for Hill Slopes as an Alternative to Shifting Cultivation

Slope	*Portion of Total Area (Approx.)*	*Land Use*	*Conservation Measures*
Top portion	1/3rd	Agroforestry	Contour bunds
Mid portion	1/3rd	Horti-pastoral	Half moon terracing for horticultural plants and contour bunds
Lower portion	1/3rd	Agriculture	Bench terracing

Source: Sarangi and Satapathy (2007).

Table 18.5: Economics of Agri-Horti-Silvipastoral Land Use System in NEH Region

Components	*Gross Income (Rs/ha)*	*Net Return (Rs/ha)*	*Per cent of Total*
Agriculture	7945	3655	18.06
Horticulture, Fruits	11640	7328	36.22
Vegetable	8147	5428	26.83
Silvipasture	4877	3822	18.89
Total	32609	20233	100.00

Source: Sarangi and Satapathy (2007).

Table 18.6: Economics of Micro Watershed Based Land Use System

Land use	*Input (Rs)*	*Output (Rs)*	*Profit (Rs)*	*Ratio*
Dairy farming	34,107	65,688	31,581	1.93
Agropastoral	33,938	64,787	30,849	1.91
Agri-horti-silvipasture	33,015	68,118	35,103	2.06
Silvi-horti- system	15,345	18,588	3,243	1.21

Source: Annual Report, ICAR Research Complex for NEH Region.

Modified Land Capability Classification for NEH Region

Land capability classification is widely employed as the basis for land use planning; land is graded into a number of capability classes usually I-VIII on the basis of its inherent limitations of erosion, wetness, soil, climate and so on. Class I-IV are arable- suitable for rainfed arable use; class V is applied to special situations such as wet valley floors whilst class VI to VII are non arable land considered suitable for grazing, forestry etc. According to these classifications, in North Eastern Himalayas there is no class I and II land, little class III land and some class VI and class VII lands (to be categorized as non arable land classes) are used for cultivation in the region and therefore, the concept comes into conflict with the current land use. Areas of sloping land are already being used to grow subsistence food crops, and families and sometimes whole communities are dependent upon the produce. It would be socially undesirable to attempt to change this situation. Owing to the typical soil and climatic conditions in the region, a modified land capability classification as mentioned below have been tried out (Table 18.7) in the North Eastern Region of India.

Table 18.7: Modified Land Capability Classification for North Eastern Hills (NEH)

Class A–1	Slopes, 0–5 per cent	Suitable for cultivation without special soil conservation measures.
Class A–2	Slopes 6–50 per cent	Suitable for cultivation with special conservation measures
Class B–1	Slopes 51–100 per cent	With shallow soil depth (less than 1.75 mm), suitable for pasture and fodder.
Class B–2	Slopes 51–100 per cent	Having soil depth of more than 1.75 mm, suitable for orchards, cash crops, plantation crops etc.
Class C	Slopes over 100 per cent	Suitable for forest.

Source: Borthakur (1992).

Labour

Generally, the conditions of agricultural labour are far below accepted social standards and legal protections in other forms of employment. Policies and programs are needed to address this problem, working toward socially just and safe employment that provides adequate wages, working conditions, health benefits, and chances for economic stability. The needs of migrant labour for year-around employment and adequate housing are a particularly crucial problem needing immediate attention

Rural Community Development

Rural communities in California are currently characterized by economic and environmental deterioration. Many are among the poorest locations in the nation. The reasons for the decline are complex, but changes in farm structure have played a significant role. Sustainable agriculture presents an opportunity to rethink the importance of family farms and rural communities.

Economic development policies are needed that encourage more diversified agricultural production on family farms as a foundation for healthy economies in rural communities. In combination with other strategies, sustainable agriculture practices and policies can help foster community institutions that meet employment, educational, health, cultural and spiritual needs.

Consumers and the Food System

Consumers can play a critical role in creating a sustainable food system. Through their purchases, they send strong messages to producers, retailers and others in the system about what they think are important. Food cost and nutritional quality have always influenced consumer choices. The challenge now is to find strategies that broaden consumer perspectives, so that environmental quality, resource use, and social equity issues are also considered in shopping decisions.

At the same time, new policies and institutions must be created to enable producers using sustainable practices to market their goods to a wider public. Coalitions organized around improving the food system are one specific method of creating a dialogue among consumers, retailers, producers and others. These coalitions or other public forums can be important vehicles for clarifying issues, suggesting new policies, increasing mutual trust, and encouraging a long-term view of food production, distribution and consumption.

The north-eastern India has great scope to bring sustainability in agriculture through proper soil and water conservation measures to prevent degradation of soil productivity, efficient use of limited irrigation water, diversification of crops (including livestock), rain water harvesting though watershed management approach, resource conservation technologies like system of rice intensification (SRI), integrated crop management(ICM), aerobic rice cultivation, organic farming, precision farming, conservation agriculture, crop rotations, intercropping, integrated plant nutrient management system(IPNMS), integrated weed management(IWM), integrated pest management (IPM) and disease management(IDM)etc.

Conclusion

The agriculture in north-eastern region of India is a complex diverse and risk prone (CDRP) production system mostly rainfed, monocropped, low input based and at subsistence level. Use of local varieties, low consumption of chemicals, low moisture retention capacity of upland soil, low temperature, no drainage in valley land and lack of irrigation facilities, low sunshine hours along with traditional management practices have resulted into low productivity of agricultural and horticultural crops which leads to low cropping intensity in the region. The population of the region has already crossed 40 million marks and futuristic increase in demographic pressure and urbanization are critical issues in terms of food, nutrition, environment and employment opportunities, thus putting tremendous pressure on shrinking natural resources and threatening the resource base itself. The NE region of India is still deficit in food grains and the productivity of most of the crops is far below than the national average. There is a potential and scope for at least two fold increase in productivity of crops in the region. The NE India is bestowed with rich natural resources of soil, water, biodiversity hot spot and varied agro-climate which can be utilise for sustaining the agriculture and allied productivity for the improvement of livelihood of the people without deteriorating the environment and natural resource base. To bring the sustainability in agriculture farming system, the cropping intensity is required to be increased substantially through rain water harvesting, watershed approach, integrated farming system, diversification of crops (including livestock), creation of more irrigation facilities and introduction of new land use systems including the resource conservation techniques like system of rice intensification, integrated crop management, organic farming, precision farming, conservation agriculture etc. Hence, the sustainable agriculture in north- eastern region of India is environmentally feasible, socially acceptable and economically viable for improvement of livelihood of the farming community.

References

Abrol, I.P. and Gupta, R.K.. (1996). Sustainable high production systems: Research issues and approaches. Souvenir, 2nd International Crop Science Congress, Nov. 17-24, 1996, New Delhi, India.

Annual Report (2001-2006). ICAR Research Complex for NEH Region, Umiam Meghalaya

Bertelmus, Peter (1997). 'Whither Economics? From Optimality to sustainability'. *Environment and Development Economics*, 2(3):326-27.

Borthakur, D.N. (1992). Agriculture of the North Eastern Region with special reference to Hill Agriculture. BEE CEE Prakashan, Guwahati.

Food and Agriculture Organisation (FAO) (1989). *The State of Food and Agriculture*1989, Rome: Food and Agriculture Organisation, *FAO Agriculture Series No*. 22: 65.

Katyal, J.C. (1997). Sustainability: Relevance of integrated concept. Survey of Indian Agriculture. The Hindu, Chennai, India.

Panwar, A.S., Ngachan, S.V., Munda, G.C. and Das, Anup (2007). Sustainable crop production for food sufficiency in NE India. ICAR Research Complex for NEH Region, Umiam, Meghalaya, pp. 1-471.

Reddy, S.R. (1999). *Principles of Agronomy*. Kalyani Publishers, New Delhi, pp. 433-447.

Sarangi, S.K. and Satapathy, K.K. (2007). Crop production through farming system approach. In: *Sustainable Crop Production for Food Sufficiency in NE India*, (Eds.) A.S. Panwar, S.V. Ngachan, G.C. Munda and Anup Das. ICAR Research Complex for NEH Region, Umiam, Meghalaya, pp. 116-137.

Sharma, U.C. and Datta, M. (2006). Physiography and Resources. *In*: *Soil and their Management in North East India*, (Eds.) U.C. Sharma, M. Datta and J.S. Samra. ICAR Research Complex for NEH Region, Umiam, Meghalaya, pp. 35-62.

World Commission on Environment and Development (WCED) (1987). *Our Common Future. Report of World Commission on Environment and Development*. New Delhi Oxford University Press.

Chapter 19

Enhancing and Sustaining the Productivity of *Jhum* Rice in Manipur

***B.B. Panda*[1] *and Mokidul Islam*[2]**

[1]Senior scientist, Central Rice Research Institute, Cuttack – 753006, Odisha

[2]SMS (Agronomy), Krishi Vigyan Kendra, Tura, West Garo Hills, Meghalaya

ABSTRACT

Jhum rice is grown in five hill districts of Manipur under temperate sub-tropical climatic conditions covering an average area of 65,560 ha with the production of 57,770 metric tonnes and productivity of 701 kg/ha. Low temperature during reproductive phase, high rainfall and humidity and low sunshine hours are the characteristics feature of the growing season, Lack of suitable HYVs for *jhum* rice, Lack of improved crop management practices for *jhum* rice, Lack or no plant nutrient supplementation for higher yield of *jhum* rice, Remoteness of *jhum* rice areas, Lack of infrastructure facilities are some of the constraints faced in enhancing the rice productivity in *jhum* areas in the state of Manipur. The study revealed that there is a need for developing and using high yielding, short duration, drought resistant *jhum rice* varieties with tolerance to low temperature in high altitudes. The introduction of modern method of terraces cultivation in *jhum* rice areas in lower slopes of hills by practicing line sowing/planting and use of small implements can reduce the pressure in *jhum* areas and *jhum* cycle can be maintained, if not increased

Keywords: Jhum rice, Productivity, Constraints, Status.

Introduction

Jhuming is the most common form and main traditional system of cultivation among the tribal peoples of different hill districts of Manipur. *Jhuming* is the integrated farming system involving agroforestry, agriculture and strong social organization on the part of the communities. This system is believed to be originated in the Neolithic period around 7000 BC (Sharma, 1976). The system is labour intensive, requiring very little capital and inputs, the labour being provided by the entire family. The main crop in *Jhum* field is paddy. In this traditional system of farming, so far, there is no report of application of chemical or FYM. The crops are grown as rainfed without tilling the land and harvesting started from July onwards. *Jhum* is the main sources to generate livelihood among various tribes of Manipur and is closely linked with socio-cultural life of the people. So, the people are reluctant to adopt any alternate system of farming. Now, with the fact that *Jhum* causes heavy damage to the surface soil by washing the fertile top soil exposing the rocks, result in the depletion of forest cover due to continuous cutting of forest trees, leading to the loss of the biodiversity.

The hill districts of Manipur are under temperate sub-tropical climate at the average altitude of 3000 M above MSL total hill area of 2.05 lakh ha, out of which only 0.64 lakh ha are cultivated. The deficiency of rice is becoming more and more acute in recent years in this areas because of the rising population on the fixed area of cultivated land. Rice grown under *jhuming* is direct seeded and mainly on hill slopes where slopes are sometimes more than 50 per cent and tilla lands. A long duration Japonica type of rice varieties are grown in *jhum* lands which are glutinous and bold seeded. Frequent moisture stress is a common phenomenon in the *jhum* ecosystem where soil depth varies from shallow in hilltops to somewhat deeper towards lower reaches.

Area, Production and Productivity of *Jhum* Rice

Jhum rice is grown in five hill districts of Manipur under temperate sub-tropical climatic conditions covering an average area of 65,560 ha with the production of 57,770 metric tonnes and productivity of 701 kg/ha. *Jhum* rice covers about 31.49 per cent area of the total rice of the state producing only 13.35 per cent of the total production (Table 19.1). The productivity of *jhum* rice is about 1377 kg/ha, which is far less than the total rice productivity in Manipur. However, maximum acreage (29,110 ha) of *jhum* rice is in the district of Churachandpur followed by Tamenglong (26,330 ha) which is about 13.98 and 12.75 per cent of total rice acreage in the state. On the other hand, the productivity of *jhum* rice is higher in the Tamenglong district followed by Churachandpur and Chandel district of Manipur.

The chemical fertilser consumption in the hill districts of Manipur where *jhum* rice covers major portion is meager or very low with only 8 per cent of the total consumption in the state. Among the hill districts, Senapati consumed comparatively higher fertilsers (Table 19.2).

Constraints

The major constraints faced in enhancing the rice productivity in *jhum* areas of the state are:

Table 19.1: District Wise Area, Production and Productivity of *Jhum* Rice in Manipur

A = 000'ha, P = 000'Mt, Y = kg/ha

District/State	*2003-04*			*2004-05*			*2005-06*			*2006-07*			*2007-08*			*Average*			*Per cent of Jhum Rice to Total Rice*		
	A	*P*	*Y*	*A*	*P*	*Y*	*A*	*P*	*Y*	*A*	*P*	*Y*	*A*	*P*	*Y*	*A*	*P*	*Y*	*A*	*P*	*Y*
Chandel	9.47	7.42	784	9.47	6.27	662	9.32	6.21	666	9.65	4.52	468	8.45	5.59	662	9.27	6.00	648	4.45	01.39	31.18
Chura-chandpur	29 33	22.72	775	29.33	22.44	765	29.02	22.10	762	29.52	18.49	626	28.34	18.35	647	29.11	20.82	715	13.98	04.81	34.41
Tameng-long	27.01	24.71	915	27.01	21.6	791	26.20	21.06	804	26.53	16.11	607	26.00	16.26	625	26.55	19.90	748	12.75	04.60	36.00
Senapati	9.45	5.62	595	9.45	6.49	687	9.20	6.90	750	9.42	6.10	648	8.46	4.16	492	9.20	5.85	634	4.42	01.35	30.51
Ukhrul	8.46	3.67	434	8.46	6.56	775	8.26	6.33	766	8.60	5.06	588	7.47	4.36	584	6.56	5.20	629	3.15	01.20	30.27
Total state *Jhum* rice	83.72	64.14	766	83.72	63.10	754	82.0	62.60	763	83.37	50.28	603	78.72	48.72	619	65.56	57.77	701	31.49	13.35	33.73
Total state rice	211.53	473.38	2238	209.0	456.86	2186	200.0	393.35	1967	209.85	418.36	1994	210.66	421.86	2003	208.21	432.75	2078	–	–	–

Source: State Department of Agriculture, Government of Manipur, 2008.

Table 19.2: Percentage of Chemical Fertilizer Consumption to Total Fertilizer cConsumption in the Hill Districts of Manipur

Districts	*2000-01*	*2001-02*	*2002-03*	*2003-04*	*Mean*
Chandel	2.09	2.34	2.28	2.35	2.27
Churachandpur	0.87	1.28	1.25	1.55	1.24
Tamenglong	1.28	1.66	1.79	1.74	1.62
Senapati	0.98	1.43	1.15	1.10	1.17
Ukhrul	1.21	1.43	1.24	1.60	1.37
Hill Total	6.43	8.14	7.71	8.40	7.67

Source: Statistical Abstract of Manipur (2005), Directorate of Economics and Statistics, Govt. of Manipur.

- ✰ Low temperature during reproductive phase, high rainfall and humidity and low sunshine hours are the characteristics feature of the growing season
- ✰ Lack of suitable HYVs for *jhum* rice
- ✰ Lack of improved crop management practices for *jhum* rice
- ✰ Lack or no plant nutrient supplementation for higher yield of *jhum* rice
- ✰ Remoteness of *jhum* rice areas
- ✰ Lack of infrastructure facilities
- ✰ High rainfall causes very high humidity during growth stages of rice leading to increased incidence of diseases and pests.
- ✰ Land tenure system is one of the social constraint which discourages actual growers to invest on farm improvement
- ✰ Actual cultivators are generally tenants rather than owners. Thus, in many cases even the choice of varieties of a crop determined by the land owners.

Status and Strategies to Enhance and Sustain the Productivity of *Jhum* Rice

Varieties

The productivity of *jhum* rice is very low in the state due to indigenous method of cultivation, use of local tall varieties, moisture stress, poor soil fertility as well as heavy infestation of weeds and insects-pests. The peoples of hill district of Manipur are growing 256 local cultivars due to lack of availability of seeds of HYVs timely and adequately in hills. Following are the varieties that developed and recommended for cultivation in *jhum* lands for stability of production in hill districts. The local genotypes recommended for *jhum* lands are Bali, Balung, Chanapaddy, Teke etc. Besides, other improved varieties recommended for *jhum* land cultivation are RC Maniphou 6, Bhalum 1, Bhalum 2, IRAT 109, IR 50, Pusa 743 etc in Manipur. But the productivity of these varieties still very much dependent on the good weather conditions, therefore, there is a need for developing and using high yielding, short duration, drought resistant *jhum rice* varieties with tolerance to low temperature in high altitudes.

Nutrient Management

Jhum cultivation is based on the concept that cultivating the field after burning the plants and trees and leaving the land when the fertility is exhausted. But as the farmers are taking crops two or more seasons without application of nutrients, it leads to depletion of soil fertility. Therefore, the following measures can be taken up for maintaining the soil fertility and productivity of *jhum* rice fields:

- ✰ The nitrogen fixing plants such as subabul (*Leucaena leucocephala*), pigeon pea (*Cajanus cajan*), alder (*Alnus nepalensis*), Baga medola (*Tephrosia candia*) etc are planted at closer spacing to establish hedge row across the slope. The *jhum* rice can be grown in between the hedgerows by incorporating these trees to enrich the nutrient content in the soil. Growing of nitrogen fixing hedge species on the field bunds helps in fixing of atmospheric nitrogen and reduces the leaching losses of mineral nitrogen. Their vigorous root system helps in mobilization of phosphorus, potassium and other trace elements. Decomposition of organic matter liberates CO_2 and organic acids and solubilises P, K, Ca and other trace elements. Accumulations of organic matter through leaf litter of hedge species improve the water holding capacity of soils and other physical properties.
- ✰ Growing of seasonal leguminous crops such as pea, cowpea, arhar, beans, soybean, rice bean etc as intercrop to improve the soil fertility of *jhum* rice field is very essential
- ✰ Supply of nutrients in adequate amount and balanced proportion by adopting integrated plant nutrition systems (IPNS) *i.e.* on - farm composting of straw and locally available materials, incorporation of fresh chopped tree leaves, grasses, weeds etc. and inorganic chemicals to *jhum* rice should be promoted for higher yield for the *jhum* plots. A small composting pit should be dug out in each *jhum* area and filled up with these materials before sowing of crops for decomposition and applied to *jhum* rice to enhance the soil fertility and productivity. Some African worms should be released into the compost pit for fast decomposition of waste materials.
- ✰ Strip intercropping of rice with legumes like groundnut, soybean, black gram, and red gram along the contour helps in resisting the soil erosion and improves and maintains soil fertility and productivity.
- ✰ After harvesting of *jhum* rice, growing of pulses like rice bean can also increase the soil fertility and reduce the weed population besides conservation of the soil moisture
- ✰ The soils of *jhum* rice areas are acidic in nature and highly deficient in nutrients and phosphorus particular; therefore, the application of lime and phosphatic fertilizers like basic slag, rock phosphate, bone meal etc is necessary to improve soil physical condition and P availability.

Water Management

Jhum rice is normally grown as rainfed without any artificial water application causing low productivity as even in small dry spell in monsoon affects the growth and development of the crop. Therefore, the following options should be considered in order to meet out the above adverse effect of weather.

- ✰ *Jalkunds* should be constructed at different places for life saving irrigation to *jhum* rice during dry spells.
- ✰ Proper use and diversion of water streams to *jhum* rice fields wherever available in a zigzag way by making small earthen channels or through plastic pipes or bamboos should be explored for improving the productivity of *jhum* rice.
- ✰ Water harvesting tank should be made up in all suitable locations with the use of earthen dams across the slope at 2 m vertical intervals.
- ✰ Small trenches in between rice strips should be made to improve the availability of moisture for a longer period to the rice crop.

Insect Pests and Diseases

Rice blast (leaf and neck) is endemic and major diseases of *jhum* rice in Manipur. Sheath blight has emerged as an important disease and is well distributed in all hill districts of Manipur. Besides, Bacterial Leaf Blight is also reported to be affecting *jhum rice.* Among the insects-pests, the most important are stem borer, gundhi bug, leaf hopper etc in *jhum* rice in Manipur. The farmers of *jhum* areas normally do not take up any control measures for the insect pest and diseases. Lack of awareness as well as non-availability of plant protection chemicals in time is the main bottleneck to control the insect pest and diseases in hill areas which leads to lower yield of *jhum* rice. Infact, farmers adopt a number of indigenous technical khowhows (ITK) practices like application of leaves and twigs of some herbs, goat excreta, cow urine, peel of pummelo, suspension of rotten crabs etc to control insect pests in the field. Therefore, the development of suitable insect pest and diseases management system based on locally available indigenous materials because traditional approaches of the farmers already included natural biological control and cultural control components as well as ITKs. The above techniques needs standardization for better results in controlling insect pest and diseases and realizing yield potential of *jhum* rice in the state. Further, spraying of Quinolphos 25 EC @ 2 ml/litre of water or Endosulphan 35 EC @ 1.5ml/lit of water to control insect-pests and Hinosan 50 EC @ 1ml/litre of water should be integrated in pest and diseases management strategy of *jhum* rice incase the pest and diseases could not be controlled with the traditional methods. Application of Neem, Kochila and Karanj seed powder at 100 kg/ha is effective for stem borer control. On the other hand, leaf extract of water pepper@10-12 g/lit, Neem or Karanj oil at 3 per cent and Neem seed extract at 5 per cent is effective for BPH, WBPH and leaf folder in rice.

Weed Management

Weed is a greater problem in *jhum* rice field which compete with main crop for moisture, nutrients, light, and space. This may be one of the reasons why the farmers prefer to grow tall local cultivars, which show fast initial growth. The intensity of both narrow and broad leaved weeds is very high because of high intensity of rainfall and optimum temperature for weed growth that prevails during monsoon season. The most important weeds in *jhum* rice fields are *Eupatorium odoratum, Eupatorium adenphorum, Imperata cylindica, Cynadon dactylon, Micania micrantha, Cyperus rotundus, Ageratum conyzoides* etc (Singh and Sharma, 2007). In *jhum* land manual weeding is done for at least three times during the crop season, which results in higher cost of production. Further weeds could not be controlled timely in *jhum* areas due to shortage of labour during peak period leading to delay in weed management. But the highest weed competition occurs during the first 25-30 days after sowing, therefore, care should be taken to control the weeds by hand weeding or spraying of herbicides.

Socio-cultural Aspect

The productivity of *jhum* rice in hills is linked to the socio - economic status of the people, lack of suitable technology, inadequate infrastructure, poor input-out put, marketing, insufficient extension network are some of the problems which need to be resolved to achieve higher and sustained *jhum* rice productivity in the state. The strong political will and Government support system is the key for improving the livelihood of the *Jhumias* in the hill district of Manipur. The traditional system should be augmented with new appropriate agro-techniques to enhance productivity of *jhum* rice without affecting the socio-cultural milieu and the environment. The introduction of modern method of terraces cultivation in *jhum* rice areas in lower slopes of hills by practicing line sowing/planting and use of small implements can reduce the pressure in *jhum* areas and *jhum* cycle can be maintained, if not increased.

References

Anonymous (2007). *Status paper* submitted by Department of Agriculture, Govt. of Manipur for National Workshop on Rice Portal held at CRRI, Cuttak, Orissa August 17-18, 2007.

Anonymous (2005). *Statistical Abstract of Manipur*. Directorate of Economics and Statistics, Govt. of Manipur pp. 144-145.

Pattanayak, A., Bujarbaruah, K.M., Sharma, Y,P, Ngachan, S.V., Dhiman, R,K., Munda, G.C., Thakur, N.S.A., Satapathy, K.K., Datta, K.K., Prakash, N., Devi, Pramila and Viswakarma, A.K. (2006). Steps towards optimizing rice production in NE India. *Research Bulletin No 21*. ICAR Research Complex for NEH Region, Umiam pp 25-42

Sharma, T.C (1976). The prehistoric background of shifting cultivation. Shifting cultivation in NE India. *North East India Council for Social Science Research*, Shillong.

Singh, N.L. and Sharma, P.T. (2007). Weeds of rice and their Scientific Control, *Technical bulletin No. 1*, CAU, Imphal pp 2-3

Viraktamath, B.C., Ahmed, M.I., Subba, L.V. and Mishra, B. (2004). Boro Rice and strategies for enhancing rice production and productivity in the Northeast. *Proceedings of Seminar* Nov. 23-24, 2004 held at SIPARD, Agartala, Tripura, sponsored by DRR, Rajendranagar Hyderabad pp206

Chapter 20

Prospects of Cashew Industry in Meghalaya

R. Ao

Research Scholar,
Department of Rural Development and Agricultural Production,
North-Eastern Hill University, Tura Campus, Tura – 794 002, Meghalaya

ABSTRACT

Cashew, often referred as 'wonder nut' is an export oriented crop grown for its nuts. The cashew industry plays a vital role in the Indian economy providing employment to more than 5 lakh people both directly and indirectly. In India cashew cultivation is confined mainly to the peninsular India. Major cashew producing states are Karnataka, Kerala, Maharashtra, Orissa, Andhra Pradesh, Tamil Nadu and West Bengal. It is also grown to a limited extent in Goa, Andaman and Nicobar Islands, Madhya Pradesh, Manipur, Meghalaya and Tripura. In the year 2007, India contributed 17.3 per cent of the total world cashew production thus showing its production potential. Its importance in Meghalaya can be visualized from the manifold increase in the area and production of this crop during the last decade. The area and production of cashew in Meghalaya was 6188 hectares and 6000 metric tonnes during 1995-96 and it increased to 6823 hectares with 11300 metric tonnes in 2006-07. The productivity has been steadily increasing from 970 kg/ha. in 1995-96 to 1656 kg/ha.(2006-07). Under the propitious agro-climatic conditions of this region, the crop grows successfully without intensive care. The total production of Cashewnut in Meghalaya in the year 2004-05 was 8846 metric tonnes and the total raw material consumption capacity of the processing units was 245 metric tonnes. There was huge surplus of raw material (8601 metric tonnes) which was sent outside the state as raw nuts. The cashew crop apart from having prospect for setting up of processing unit provides ample scope for creation of self-employment opportunities from its by-products offering opportunities for development of local industry which in the long run can change the entire economic situation of the people of the state.

Keywords: Cashew, Area and production, Processing unit, Meghalaya.

Introduction

Cashew (*Anacardium occidentale*) often referred as 'wonder nut' is an export oriented crop grown for its nuts. Cashew is a native of Brazil which was spread by the Portuguese to different parts of the world primarily for soil conservation, afforestation and wasteland development. It is one of the most important crop in India as it stands in a premier position in international cashew trade for the last one-century and has the largest area harvested under raw cashewnuts in the world. As per FAO statistics, 21.6 per cent of the world's harvested area under cashewnut was in India and the country accounted for 17.3 per cent of the total world cashew production during the year 2007. During 2007-08, cashew covered an area of 8.68 lakh hectares in the country with a production of 6.65 lakh metric tonnes (DCCD, Cochin). In India, cashew cultivation is confined mainly to the peninsular India. Major cashew producing states are Karnataka, Kerala, Maharashtra, Orissa, Andhra Pradesh, Tamil Nadu and West Bengal. It is also grown to a limited extent in Goa, Andaman and Nicobar Islands, Madhya Pradesh, Manipur, Meghalaya and Tripura.

The Indian Cashew industry provides employment to more than 5 lakh people both directly and indirectly, particularly in the rural areas and it thus plays a very vital role in the economy. Today nuts constitute an important part of diet in several countries of the world. Umesh *et al.* (2002), indicated that the strong and established research and development network, the availability of good number of cashew varieties suitable for varied situations across the country along with scientific production management practices offered good opportunity for the development of cashew industry in the country. There has always been a stable price in the international market for cashew when compared to any other edible nuts. They found that cashew represents a diversification option for inferior/degraded lands which are less suitable for commercial cultivation of other food crops. Its importance in Meghalaya can also be visualized from the manifold increase in the area and production of this crop during the last decade. The area and production of cashew in Meghalaya was 6188 hectares and 6000 metric tonnes during 1995-96 and it increased to 6823 hectares with 11300 metric tonnes in 2006-07. The productivity has been steadily increasing from 970 kg/ha. in 1995-96 to 1656 kg/ha. in 2006-07 (Department of Agriculture, Meghalaya, 2006). Under the propitious agro-climatic conditions of this region, the crop grows successfully without intensive care.

Its nutritional composition compares well with almond, hazelnut, walnut etc. The trees are usually grown for their kernels which when roasted have a very pleasant taste. Rehm and Espig (1991), Deckers *et al.* (2001) and CTA (2007), reported that raw nuts which are much in demand in industrialized countries are processed into kernels that constitute a valuable export product for confectionery. They are used as an ingredient for making fruit paste, candied fruit, canned fruit, cashew apple, resins, jams and jellies, chutney, fruit juice, alcohol and vinegar. It is a zero-cholesterol nut with high unsaturated fatty acids, proteins and carbohydrates with a right combination of amino acids, vitamins and minerals. Wine and pulp are produced from cashew apple. Apple is eaten as a raw fruit or is fermented to produce a delicious alcoholic drink. The pulp is the fibrous part obtained after extracting juice from the apple and could be used as animal feed or dried and processed into diet fiber biscuit.

Another product of cashew is the Cashewnut Shell Liquid (CNSL) which is greatly valued in the international market as a raw material for brake and clutch linings, paints, and varnishes. It is also used in lacquers, agglutinants, insecticides and fungicides (Rehm and Espig, 1991 and Deckers, *et al.*, 2001). Cashew tree is also found useful in making live fence, shade trees, firewood and charcoal. Rehm and Espig (1991) added that in India, a gum is obtained from the trunk, which can be used as a substitute for gum Arabic.

There is ample scope for creation of self-employment opportunities in this sector which has been performing well and is coming up with great promise which in the long run can change the entire economic situation of the people of Meghalaya. With this background, a study was conducted to examine the production of cashew, assessment of its marketable surplus and its potential for industrial development.

Methodology

The present study is undertaken in Garo hills of Meghalaya as cashew is extensively grown in Garo Hills. The Garo hills comprise mainly East Garo hills district, West Garo hills district and South Garo hills district. The present study is mainly based on secondary data. The secondary data on area, production and productivity of cashew was collected from the Department of Agriculture, Govt. of Meghalaya for the period 1995-96 to 2006-07. The status of cashewnut processing units in Meghalaya was collected from the District Horticulture Office, West Garo Hills, Tura for the year 2004-05 as the processing units are located only in this district. The various tools and techniques used in the present study are discussed below:

Analytical Framework

In the present study Compound Growth Rates (CGR) and CV were computed for area, production and productivity. The compound growth rate was computed as follows:

Quadratic function $Y = a + bx$

Where, x is the time variable, Y is the variable for which growth rate is calculated and b is the regression co-efficient of Y on x. The Compound Growth Rate percentage is as follows:

(CGR per cent) = (antilog 'b'-1) x 100

The significant of growth rates was tested by applying student't' test where $t = g/SE\ (r)$, with (N-2) d. f. where r is the growth and N is the total number of years considered under study.

$$SE\ (r) = 100\ b/0.4329 \sqrt{[\ (\Sigma \log r^2) - (\Sigma \log Y)^2/N - (\log b)^2 \Sigma x^2]/(N\text{-}2)\ \Sigma \times 2}$$

To measure the magnitude of variability in area, production and productivity for the total period, the co-efficient of variation (per cent) was computed. The CV was calculated by using the following formula:

CV (per cent) = (Standard deviation/mean) x 100

Results and Discussion

Cashew is extensively grown in Garo Hills. However, the bulk of the produce is sold outside the state as raw nuts. The increase in area under this plantation crop is due to the effort for reclamation of jhuming/shifting lands. The computed growth rate (Table 20.1) for the period 1995-96 to 2006-07 indicates an increased growth rate of 1.02 per cent, 5.40 per cent and 4.32 per cent in area, production and productivity respectively. Thus it is clear that the area, production and productivity cf cashewnut in the state as well as in the Garo hills is encouraging and thereby enhancing the industrial potential.

Cashewnut Processing Industry in Meghalaya

There were only 6 processing units in Meghalaya located in West Garo hills district for the year 2004-05 (District Horticulture Office, West Garo Hills, Tura). The capacity of processing plant depends upon the duration of operation per year, drying capacity of processed nuts, number of labour employed and quantity of raw materials purchased. Generally the processing units are engaged in processing of Cashew for 8 to 9 months in a year starting from the month of May.

It is evident from Table 19.2 that 245 metric tonnes of raw material were consumed per year by the processing units of West Garo hills and the total production of processed nuts was 120.25 metric tonnes. The production capacity of the units varied from 4–10 Quintals/day. The processed nuts were sold in the local market and also sent to Guwahati, from where it is distributed to various parts of the N.E. States and the rest of the country.

Since the secondary data available on processing unit is relating to 2004-05, thus taking the total production of Cashew in Meghalaya in the year 2004-05 as 8846 metric tonnes (Table 19.1), the total raw material consumption capacity of the processing units was 245 metric tonnes. There was huge surplus of raw material (8601 metric tonnes) which was sent outside the state as raw nuts.

The production of Cashewnut in Garo hills is more than its processing capacity and offers immense scope for establishing processing unit in the state. Trained and cheap labours are also easily available from the adjoining state of Assam. Moreover, the cashew crop apart from having prospect for setting up of processing unit provides ample scope for creation of self-employment opportunities from its by-products offering opportunities for development of local industry which in the long run can change the entire economic situation of the people of the state.

Some of the opportunities are mentioned as below

1. The cashew tree is multi-functional: apart from exportable cashewnuts, it supplies timber, firewood, medicines, fodder, etc.
2. By-products (cashew balm, cashew apple and shell) offer opportunities for development of local industry.
3. The world market for cashew is dynamic.
4. The nuts produced in the region are particularly appreciated for their taste.

Table 20.1: Area, Production and Yield of Cashewnut in Meghalaya

Area = in hectares
Production = in metric tones
Productivity = in kgs/ha

Year	*Cashewnut*		
	Area	*Production*	*Productivity*
1995-96	6188	6000	970
1996-97	6212	6284	1015
1997-98	6238	6313	1013
1998-99	6248	6335	1014
1999-00	6401	6419	1003
2000-01	6320	6294	996
2001-02	6347	6418	1011
2002-03	6649	6730	1012
2003-04	6708	6787	1012
2004-05	6765	8846	1308
2005-06	6785	11207	1652
2006-07	6823	11300	1656
CGR per cent	1.02	5.40*	4.32*
CV per cent	3.87	26.14	5.99

*significant at 5 per cent probability level

Source: Meghalaya Agriculture Profile 2006 (third edition), Department of Agriculture, Govt. of Meghalaya.

Table 20.2: Status and Processing Capacity of Cashewnut Processing Industry in Meghalaya (2004-05)

Type of Grade Being Produced	*Raw Materials Consumption/Yr*	*Production Capacity/Day*	*Production/ Year*	*Marketable Surplus Outside*
6–J	245 M.T	43.5 Qtls.	120.25 M.T	64 M.T
4–J				
2–J				
2–Piece				
4–Piece				
Powder/Broken				

Source: District Horticulture Office, West Garo Hills, Tura (2004-05).

However, in spite of having immense potential, the cashew industry is not up to the mark in the state due to various problems as discussed below:

1. Most farmers are lacking the knowledge in handling the crop for maximum yield.

2. In many cashew farms, a lot of the fleshy apple and some nuts waste away despite their potential economic uses; because many farmers do not have the capacity to adequately turn the produce into valuable products for local and international consumption and marketing.
3. Most of the farms are located far from the village where transport and communication is not adequate.
4. There is lack of organized market due to which producers are forced to distressed sale and ultimately getting low price.

References

Anonymous (2006). District Horticulture Office, West Garo Hills, Tura.

Anonymous (2006). Meghalaya Agriculture Profile 2006 (third edition), Department of Agriculture, Meghalaya.

Anonymous (2007). FAO statistical database.

Anonymous (2011). Directorate of Cashew Development, Cochin.

CTA (2007). CTA spore food sovereignty the right to food (128). 4 April. Published by CTA, The Netherlands.

Deckers, J., Cundali, S.H., Shomari., Ngatunga, A. and Bassi, G. (2001). Cashew Crop Production in Tropical Africa. Romain H. Reamaekers (Ed.) Published by Directorate General for International Cooperation (DGIC), Brussels, Belgium. pp. 236-238.

Rehm, S. and Espig, G. (1991). The cultivated plants of the Tropics and subtropics published by verlag Josef Margraf scientific books-CTA, Weikersheim, West Germany. pp. 236-238.

Umesh, K.B., Vishnuvardhana and Thirmualaraju, G.T. (2002). Cashew Industry. *Agro India*, 6(9&10): 16-18.

Chapter 21

Scientific Seed Production Technology for Sustainable Paddy Cultivation: A Study of Bongaigaon District of Assam

Jaynal Ud-din Ahmed, Sahnaj Rahmatulla Laskar** and Bhriguraj Bordoloi****

Department of Management, North Eastern Hill University, Tura Campus, Tura – 794 002

ABSTRACT

Assam is predominantly an agrarian state. The yield rate of rice in Assam still needs to be improved in comparison with the average yield rate in rest of India. In Assam, cultivation is still carried on with traditional techniques, and modern inputs like fertilizers, HYV seeds, irrigation and pesticides, etc. are yet to be adopted fully. Seed is a basic and crucial input for attaining sustained growth in agricultural production. High Yielding Variety (HYV) seeds play a vital role in the progress of agriculture. High Yielding Variety Seeds produce more yield, takes less crop maturity time, multiple crops can be grown easily, seeds are mostly flood and drought resistant, more prone to pests and diseases as compared to the use of traditional seeds. They are considered as 'miracle seeds'. But in Northeastern states, the spectacular improvement in rice production due to adoption of high-yielding varieties (HYVs), utilization of irrigation facilities, application of balance use of fertilizer, plant protection measures and use of improved farm implements has not been yet realized.The present paper has tried to make an attempt to focus on the positive impact of seed technology programme in *Bongaigaon* district of Assam initiated by *Krishi Vigyan Kendra (KVK), Chirang*.

E-mail: *jaynaluahmed@yahoo.co.in; **sahnaj077@gmail.com; ***bhriguraj_bordoloi@yahoo.com

The paper has also highlighted one of the major issues like shortage of supply of HYVs and attempted to suggest some of the solutions to those problems.

Keywords: ***Scientific seed production technology, Paddy cultivation, Sustainable agriculture.***

Introduction

With the advent of agriculture, it was evident that a change has occurred, or to put it more appropriately, it started the growth of civilizations. Agriculture has helped the human race to settle down. People started growing food and began to live in fixed settlements. Suddenly these nomadic food gatherers were transformed into settled food growers. Agriculture brought with itself a change, a change that would transform the ways of the world and make it a better place to live in. As time passed, advancements in agriculture were inevitable and completely desirable as well as justified. But there was a price to be paid for this indiscriminate human interference in the form of ecological imbalance. For some time, it seemed as if there was no getting out of this ever-circling and gradually engulfing whirlpool. But the time was apt for something new and something completely out of the box that could reverse the effects of excessive human interference with nature and could attain harmony among these two ever so contrasting components.

Then came the savior in the form of a completely new concept, it was the concept of '*Sustainable Agriculture*'. It is basically a practice in agriculture that takes into consideration the principles of ecology; the relationship between the living entities and their environments and reveals their interdependence. Is merely seeks to strike a balance between these two elements and this is what makes it so special. This concept turned out to be a '*Messiah*' in the given context and was able to take over the minds and the hearts of agriculturists all over the world and is now one of the most widely practiced form of agriculture. The basic idea behind Sustainable Agriculture is to satisfy the needs of the humans for their food and fiber. To be able to sustain their livelihoods and also enhance the environmental quality and also the natural resources which tends to determine the agricultural produce. It also earnestly advocates the efficient use of the scarce resources and helps maintain balance between the humans and the environment. In a nutshell, Sustainable Agricultural practices help sustain the viability of livelihood and also preserve the environment.

Seed technology plays a very important role in Sustainable Agriculture. Unavailability of quality seeds has spelt doom for the farmers trying to practice sustainable agriculture. This problem is more acute in developing countries like India, particularly in the less developed regions of the country like the North-East. In order to meet this demand for quality seeds, mere laboratory-field production is not adequate. The situation demands shifting of these production beds to the farmer's field in order to achieve bulk increments in quantities. Then only, the industrial or economic production capacities can be attained and the forces of economies of scale comes into play in order to minimize the cost involved which will further the cause of sustaining research and development in this field for the developers. It will help increase the cropping intensity and improving the efficiency of the farms. It will ensure high quality seed material supply to the farmers which will result in unbelievably high production compared to the local variety. It will be certain of the timely supply and the quick spread of new varieties developed by the plant breeders.

Assam is mainly an agricultural state with an estimated 39.44 lakh hectares under cultivation. It lies on the *Brahmaputra* flood plain valley which is a rich fertile alluvial tract of land with heavy annual rainfall and high humidity that provides optimum conditions for good agricultural produce. It experiences a Tropical Monsoon Rainforest type climate. Agriculture accounts for more than a third of Assam's income and employ s69 per cent of the workforce. It has an organic nature of cultivation. Its proximity with the other South Asian countries favors it over other destinations for business, topped with cheap labor costs.

Recognizing this feasible opportunity, coupled with the above benefits, the Assam Agricultural University, *Jorhat*, has initiated a Technology Showcasing programme in the farmer's field in the early part of the year 2010-11 in order to promote scientific technologies in the field of Seed production. During *Kharif*, 2010, *Krishi Vigyan Kendra, Chirang* (the organization under study), conducted Technology Showcasing programme in an area comprising 100 hectares of cultivable lands under the *Boitamari* revenue circle of *Bongaigaon* district (the place under study) for the production of quality seeds of the winter rice variety *Ranjit*. It was an overwhelming success story in the making with the active participation of 300 practicing farmers of the *Kayethpara* Part 1 and Part 2 villages. Keeping this in mind, several training were conducted on different crop management aspects of rice along with other related crops like pulses and oilseeds for the knowledge Upgradation of the beneficiaries, under the programme. In continuation with the programme, after the harvest of the winter rice variety *Ranjit*, Technology Showcasing in seed production of summer rice variety *Jaymoti and Kanaklata* and experimenting with certain varieties of pulses and oilseeds were carried out at the same site during the *Rabi* season 2010-11. With the ongoing harvesting of this phase of summer rice, the whole biennial cycle is nearing completion and the programme is to complete its first successful year of its ambitious geometric progression technique of seed production. The harvest id partly bought by the *Krishi Vigyan Kendra, Chirang* and subsequently sold to the state government for further distribution.

Importance of Scientific Seed Production Technology for Sustainable Paddy Cultivation in Assam

Quality seeds play a very important role in increasing the crop yield; however, seed replacement rate in the district is very low which may be attributed to ignorance of farmers on seed production technology. The scientific technologies have a high importance for increasing the cropping intensity in Assam. There is high demand for quality seeds but as compare to the demand supply is very less as these are largely produced in few labs having limited capacities. This certified seed technology, if can be adopted with proper knowledge, the farmers do not have to depend on other supply sources and can produce quality seed materials in a huge quantity in their large fields itself. It will completely transform their lives by increasing their income levels and improving general standards of life. They will be able to double their returns by merely adopting this technology with a bare minimum increase in costs. Moreover, it will be very helpful to increase efficiency of their farms which was so far lacking. From the state's point of view, it will save considerable amount of currency which was formerly being used to import grains from other states and go a long way in solving the food shortage problems of the state by making it self-sufficient.

Types of Seeds

Four classes of seeds are generally recognized in seed certification namely breeder seed, foundation registered and certified. These classes are recognized by AOSCA *i.e.* Association of official seed certifying agencies.

1. *Breeder's seed*: Is a seed or vegetative propagating material which is directly controlled by sponsoring breeder of institution and which provides increases of foundation seeds.
2. *Foundation seed*: is a seed stock so as to maintain specific genetic identity and purity and may be designated or disturbed by agriculture experiment station. Production must be carefully supervised by representatives of the station. Foundation seed is the source of all other certified seed classes, either directly or through registered seed.
3. *Registered seed*: Registered seed is the progenies of foundation and it is handled so as to maintain genetic identity and purity and that has been approved by and certified by certifying agencies.
4. *Certified seed*: Is the progeny of foundation, registered or certified seed, that is handled to maintain genetic identity and purity and that has been approved by and certified by certifying agencies.

Objectives of the Study

1. To identify the need of seed technology for sustainable paddy cultivation in Assam
2. To assess the impact of seed technology in the productivity and profitability of the farmers of *Bongaigaon* district
3. To identify the problems faced by the farmers in adopting and adapting to this technology.
4. To suggests some of the measurable solutions in order to popularize this practice.

Methodology

The study is based on 4 villages from different blocks of *Bongaigaon* district of Assam *viz. Pub Enkorbari* Village from *Sidli* Block, *Pub Khamarpara* village from *Borobazar* block, *Alengmari* village from *Manikpur* block, *Bashbari* village from *Boitamari* block.The study is based on both primary and secondary data. Total sample size:105.Out of which total no. of farmers: 100. Total no. of KVK,Chirang staffs :5.The secondary datas were collected from KVK, Chirang annual reports, various research papers,journals, different official websites.Pie charts has been used for analyzing the various datas.

Bongaigaon District Profile

The following tables show some of the statistics relating to the status of agriculture and other related matters of the district.

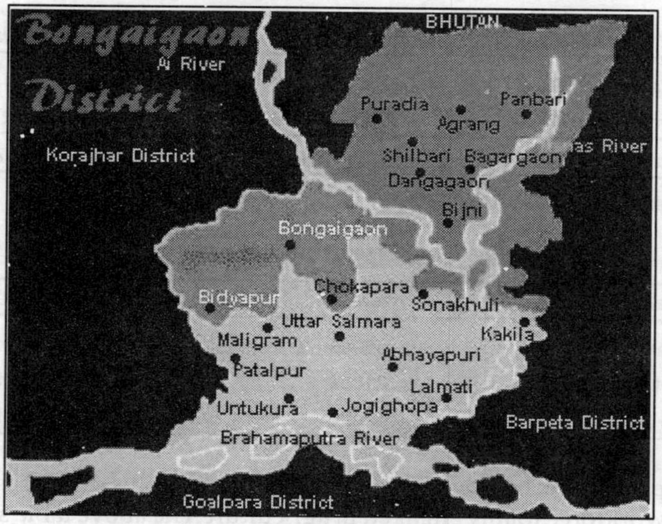

Figure 21.1
(Source: http://bongaigaon.gov.in/hist.htm)

It is located in the North Eastern Himalayan sub-region of India with a Longitude range of 89° East to 90°96' East and a Latitude range of 26°28' North to 26°54' North. It has an average rainfall of 250 cm to 350 cm with temperatures ranging from a Min: 13°C to a Max: 32°C. It has a net cropped area of 93,875 hectares.

Table 21.1: Profile of *Bongaigaon* District

District	*Area in Sq. Km. (2001 Census)*	*Population*	*Rural Population*	*Urban Population*	*Population Density per Sq. Km.*
Bongaigaon	1725	621136	523922	97214	360

Source: Economic Survey Assam 2010-11.

Table 21.2: Towns, Sub Divisions, C.D. Blocks, and Villages of *Bongaigaon* District

District	*Towns (2001 Census)*	*Sub-Division (2007)*	*C.D. Block (March, 2007)*	*No of Villages (2001 Census)*	
				Total	*Inhabited*
*Bongaigaon**	4	2	5	626	593

* Provisional figure for Area, Population, Village and Towns.

Source: Statistical Handbook Assam 2010.

Table 21.3: Number of Main Workers, Marginal Workers and Non-workers in Assam, as per 2001 Census

District		Total Workers	Main Workers	Marginal Workers	Non-Workers	Cultivators	Agricultural Laborers Workers	House-hold Industries	Other Workers
Bongaigaon	**Male**	225300	200644	24656	239940	86876	34221	4113	100090
	Female	67943	32443	35500	371652	26915	17751	4827	18450
	Total	273925	209928	63997	548110	101107	50105	12879	109834

Source: Economic Survey Assam 2010-11.

Table 21.4: Size of Land Holdings

Size	Number	Percentage
Less than 1 Ha	66065	66.1 per cent
Between 1 to 2 Ha	21802	22 per cent
Above 2 Ha	12390	12 per cent

The prospects of Technology Showcasing are good in this district as they have a considerable workforce totally engaged in agriculture The above table suggests that there is a considerable amount of fragmented land holdings which do not produce much revenue. Increasing Efficiency of the farms is the only solution.

Table 21.5: Description of Major Agro Ecological Situations (Based on Soil and Topography)

Sl.No.	Agro Ecological Situation	Characteristics
1.	Foot hill old mountain valley alluvial plain	The northern part of the district comprising this situation contains old mountain valley alluvial soils (Alfisol and Ultisol). It is buildup of alluvial materials washed down from the hill slops. The surface soil is light yellow to pale brown, compact, sticky and plastic. Generally, medium to heavy in soil texture. The elevation is higher towards foot hills which gradually slop towards south.
2.	Flood prone recent riverine alluvial plain	Recent riverine alluvial (Entisol), sandy to sandy loam in soil texture. This situation is represented by an almost flat topography which often experiences flood hazard. Apart from some natural depressions, some riverine islands are also in existence.
3.	Flood free riverine alluvial middle plain	Old riverine alluvial type (Inceptisol). The texture of the surface soils ranges from sandy loam to loam, silty clay loam, silty clay and clay. The topography is almost plain.
4.	Char land	New alluvial plains, neutral in reaction, sandy-silty-clayey, and sandy-silty and sandy in soil texture (Entisol). Chronically flood affected areas except the stable chars.
5.	Hill and Hillock	Old alluvial type (Alfisol), sandy to sandy loam in texture and acidic in nature. The topography is undulating.
6.	*Beels*	Entisols, usually peaty in nature and texturally these are silty and clay. Low lying waste land areas.

Source: Annual Report, KVK, Chirang 2010-11.

Major Problems Identified in the Four Villages of Bongaigaon District Under the Study

The study is based on 4 villages from different blocks of *Bongaigaon* district of Assam *viz. Pub Enkorbari* Village from *Sidli* Block, *Pub Khamarpara* village from *Borobazar* block, *Alengmari* village from *Manikpur* block, *Bashbari* village from *Boitamari* block.

The yield rate of rice in Assam is still pretty dismal and needs to be improved as it is very low in comparison with the average yield rate of the country. In Assam, cultivation is still carried on with traditional techniques, depending on the rain fed farming and modern inputs like fertilizers, HYV seeds, irrigation and pesticides etc. are yet to be adopted fully. Summer rice has been cultivated in limited areas of the district that too, with some unknown, intruded varieties without following proper methods of cultivation. Besides due to other factors like frequent floods, draught, erosion of soil etc. the yield rates fluctuates widely in Assam. Beside this, Soil acidity, Yield gap in paddy, Low rate of seed replacement and poor adoption of HYVs, Poor fertility management, Un-organized marketing system, Injudicious use of chemical fertilizers, Excessive use of chemical pesticides are some of the major problems that are threatening in the district. Moreover post-harvest activities like storage and marketing facilities are lacking and if present, in a sorry state.

The Agriculture Status of Assam

Assam is predominantly an agrarian state. About 99 per cent area of total land mass of the State is rural. The net cultivated area of the State is 28.10 lakhs hectares (2008-09). More than 83 per cent of the farmer family is small and marginal farmers (Agricultural Census, 2005-06). The agriculture in the state mostly depends on rainfall. One of the most serious problems adversely affecting the state's agriculture is the recurrence of devastating floods almost every year.

(*Source*: NEDFI Data Bank).

Table 21.6: Trends of Production of Food Grains in Assam (Figure in 1000 tonnes)

Item	*2006-07*	*2007-08*	*2008-09*	*2009-10*
Total Rice	2916	3319	4008	4408

Source: Source: Economic Survey Assam 2010-11.

Table 21.7: Production of Rice in 2010–11 (Figure in 1000 tonnes)

Crop	*Area (in lakh hectare)*	*Production (in lakh MT)*	*Average Yield (in Kg./hectare)*
Autumn Rice	3.65	4.11	1126
Winter Rice	18.00	30.60	1700
Summer Rice	5.00	11.60	2325
Total Rice	26.65	46.31	1738

Source: Economic Survey Assam 2010-11.

Table 21.8: Production of Principal Crops in Assam (Thousand Tonnes)

Crops	*1960-61*	*1980-81*	*1990-91*	*1993-94*	*1994-95*
1	*2*	*3*	*4*	*5*	*6*
Autumn Rice	267	502	522	587	619
Winter Rice	1,367	1,978	2,565	2,556	2,447
Summer Rice	7	43	183	219	213
Rice (Total)	1,641	2,523	3,270	3,362	3,279

Source: Mirror of Assam – Vegetation, Retrieved from http://www.vedanti.com/Assam_Mirror/Vegetation.htm on 05/12.11.

Role of KVK, Chirang in Popularizing the Technology in the Region

From last few years *KVK, Chirang* has been working hard to popularize scientific seed production technology of 'Jaymoti' and 'Kanaklata' in the farmer's field through training programme, front line Front Line Demonstrations programme, advisory services etc.

Table 21.9: Frontline Demonstrations

Sl.No.	*Crop/ Enterprise*	*Thematic Area*	*Technology Demonstrated*	*Details of Popularization Methods Suggested to the Extension System*
1	Rice	Crop management	Seed production in rice (Var. *Ranjit*)	– Organizing training programme – Interaction with farmers – Providing information about the source of seed – Farmer's field visit- Advisory services

Source: Annual Report, *KVK, Chirang* 2010-11.

Technology Showcasing Programme Initiated By KVK, Chirang

The *'Technology Showcasing Programme'* is a foundation and certified seed production programme in the farmer's field is a positive step initiated by *KVK, Chirang* in various parts of Assam and *Bongaigaon* district is one of them. In this programme several progressive farmers were promoted as seed growers in both *Bongaigaon* and *Chirang* districts. Altogether six farmers are now engaged in certified seed production programme covering an area of 10.0 ha under rice (Var. *Ranjit and Kanaklata*) and 2.0 ha under toria (Variety. TS-36).Considering the importance of scientific technologies for increasing the cropping intensity in the state as well as to produce quality seed material in the farmer's field, Assam Agricultural University has launched technology showcasing programme in the farmer's field during 2010-2011 to ensure adequate supply of quality seed at the farmer's level.

During *Kharif*, 2010, *KVK, Chirang* conducted technology showcasing programme in 100.0 hectare land area with the active participation of 300 practicing farmers of *Kayeth Para* Part 1 and Part 2 villages under *Boitamari* Revenue Circle of *Bongaigaon* district for production of quality seed of winter rice (Var. *Ranjit*). In connection with the programme, several trainings on different crop management aspects of rice were organized for knowledge up gradation of the beneficiaries. Crop cutting results reported production of 8.1, 6.1, 7.8, and 5.7 tonnes per hectare in four different locations. Similarly, seed production of summer rice (Var. *Jaymoti and Kanaklata*) has been going on at the same site during *Rabi* from 2010.

(*Source*: *KVK, Chirang* Newsletter, Vol. 1, April, 2010-March 2011)

Some of the Impact of the Programme in the District

1. A steady increase in area (Approx. 120.0 ha) under these two HYVs of summer rice (*'Jaymoti' and 'Kanaklata'*) was achieved and also an increase in crop yield (60.0 q/ha).
2. About 120.0 ha area was brought under seed production programme of *Kharif* rice (var. *Ranjit*) and which produced approximately 3000.0 q quality certified seed during kharif, 2010.
3. In the same locality, farmers started seed production programme with summer rice (var. *Kanaklata and Joymoti*) in 100.0 ha area during summer, 2010-11.
4. Because of the sincere efforts, farmers have started adopting improved production technology of *Sali* rice especially in respect of quality seed, fertility management and pest management.
5. At present HYV of *Kharif* rice is cultivated more than 40 per cent of rice growing areas of the district.
6. Considering the high yield potential of HYVs of *Sali* rice, it is expected that more farmers will come forward to adopt these varieties in near future. With the consistent effort and intervention of *KVK, Chirang* in 2010, value of HYVs of summer rice has been popularizing among the farmers.
7. Further, because of the continuous effort made by *KVK, Chirang* to popularize system of rice intensification (SRI) technology in summer rice, about 60.0 ha in *Kokila* village and 12.0 ha in *Kayethpara village* under *Bongaigaon* district have been put under summer rice cultivation with system of rice intensification.

Table 21.10: Cropping Details

Sl.No.	*Crop*	*Technology Demonstrated*	*Variety*	*No. of Farmers*	*Area (Ha.)*
1	Rice	Seed production technique in winter rice	*Ranjit*	300	100.0
2	Rice	Seed production technique insummer rice	*Joymoti and Kanaklata*	300	100.0

Source: Annual Report, *KVK, Chirang* 2010-11.

Interpretation: The initial turnout was good with 300 farmers coming up and king the initiative with 100 hectares of cultivable lands. It was a good beginning for neprogramme.

Table 21.11: A Comparative Analysis: Economic Impact

Average Cost of Cultivation (Rs./ha) Demon-stration	*Average Cost of Cultivation (Rs./ha)* Local Check	*Increase in Cost (per cent)*	*Average Gross Return (Rs./ha)* Demon-stration	*Average Gross Return (Rs./ha)* Local Check	*Increase in Gross Return (per cent)*	*Average Return (Profit) (Rs./ha)* Demon-stration	*Average Return (Profit) (Rs./ha)* Local Check	*Net Increase in Net Return (per cent)*	*Benefit-Cost Ratio (Gross Return/ Gross Cost)*
16,220	15,000	8.13	76,500	36,000	112.5	60,280	21,000	187	4.72
20,500	19,000	7.89	NA	NA	NA	NA	NA	NA	NA

Source: Annual Report, *KVK, Chirang* 2010-11.

Interpretation: This table shows the relative advantage of the technology over traditional practices. It compares relative cost, the relative gross returns and the relative average net returns and their respective increase. This gives us a clear picture of the potential benefits of the technology and gives us more than enough reasons to adopt the technology.

Table 21.12: Impact of the Seed Technology

Name of Specific Technology/ Skill Transferred	*No. of Participants*	*Per cent of adoption*	*Change in Income (Rs.)* Before (Rs./Unit)	*Change in Income (Rs.)* After (Rs./Unit)
Seed production technique in *Kharif* rice (Variety: *Ranjit*)	300	50	28,000.00/ha	76,000.00/ha

Source: Annual Report, *KVK, Chirang* 2010-11.

Interpretation: This table shows the response of the initial phase and it also depicts the huge turnaround that has been achieved in the very first phase of the programme.

Problems Faced by the Farmers in Adopting and Adapting to this Technology

Although since last few years *KVK, Chirang* has been taken initiative to popularize scientific seed production technology for improvement of paddy cultivation by using HYV of 'Jaymoti' and 'Kanaklata' in the farmer's field through training programme, front line Front Line Demonstrations programme, advisory services etc. there are several number of problems has been identified among the farmers in adopting and adapting to this technology. Some of problems are mentioned below:

- ☆ *Availability and conservation HYV: (both Govt. and Local market supply problems in HYV)*: As per interaction with some farmers it has been found that though KVK bring some HYV breed but supply for the breed were found to be limit. Though some farmers are showing their keen interest for those HYV breeds but supply of these seeds are not enough from government or from other sources. Thus problem of non availability in local market is one of the major issues which we need to rethink again.

- ☆ *Belief on traditional practice*: Most of the farmers were found that they still practicing their age old traditional way of practicing the paddy farming.
- ☆ *Knowledge of Scientific Cultivation*: Knowledge of Scientific Cultivation among the farmers in north eastern part especially some of the interior part of Assam are still lagging behind from the mainstream.They are still deprived from the modern facilities, adequate knowledge about updated and upcoming technologies, market in formations.
- ☆ *Use of pesticides and chemical fertilizer*: Most of farmers of the district have lack of knowledge of proper utilization of fertilizer and pesticides. To increase productivity they are highly using chemical fertilizers and loosing the productivity of soil day by day.
- ☆ *Training and extension services*: Although it has been taken initiative by the KVK, Chirang through several training programme, demonstration advisory services so on but in the true sense number of participants is limited as compare to total numbers of farmers. Some of the farmers are not aware of those training and advisory services.
- ☆ *Soil suitability for HYV breeds*: It has been found that all type of soil is not suitable HYV breed. Most of the farmers have confusion on HYV breeds whether these are suitable for their soil or not. Moreover there are less number of farmers interested for their soil testing for lack of knowledge soil test centre and cost for the same.
- ☆ *Financial problem and financial support from institution*: Most of farmers of the district are not economically viable. Due to poor economic condition they are unable to bear the cost of maintaining a sustainable agricultural practice. The farmers are also not getting proper support from different financial institution.
- ☆ *Irrigation facility*: Large number of the farmers of the district still depends on natural source of irrigation. They don't have proper irrigation facility for their cultivation and are based on rain fed farming. Most of villages of district are deprived from the supply of electricity in adopting cheaper irrigation facility. If we can compare the status of farmers of Punjub, Haryana, the farmers of this area are still deprived of various facilities.

Findings and Discussions

1. With the harvest results of the Winter/*Kharif Ranjit* variety rice under Technology Showcasing in 2010-11, there has been a considerable improvement in the yield per hectare with an average yield amounting to 51.0 Qtl/Ha over the previous yield of the local check average of 36.0 Qtl/Ha, which is an increase of 41.7 per cent.
2. The average cost of cultivation has gone up marginally with the introduction of technology showcasing, which currently is Rs. 16,200 per hectare over the previous spending on the local variety of Rs. 15,000 per hectare, which amounts to a negligible increase of mere 8.13 per cent. But

those who seem to be discouraged by this, needs to understand that good things come with a price, which in this case, is quite negligible.

3. The Gross Returns of the total harvest shows a very encouraging stat amounting to Rs. 76,500 per hectare over the previous local check of Rs. 36,000 per hectare which is a more than double increase of 112.5 per cent.
4. But the real income or the Average Net Return went over the roof when the stats have shown a record breaking increase in returns of 187 per cent. The Average Net Return of the total programme shows a whopping figure of Rs. 60,280 per hectare over its local check of Rs. 21,000 per hectare.
5. Trying to imitate the overwhelming success of its *Kharif* harvest, the *Rabi* season 2010-11 also followed the similar lines by the sowing of summer rice variety of *Jaymoti* and *Kanaklata* in around 60 hectares and experimenting with 35 hectares of pulses and 5 hectares of oilseeds.
6. The farmers sell their seeds back to the *Krishi Vigyan Kendra* or to the market which gives them the seed price of Rs. 15-20/Kg for foundation seeds and Rs. 25-30/Kg on the certified seeds, whereas previously they got the grain price of Rs. 4-6/Kg as they sold it mostly in the form of grains.
7. By analyzing the data obtained, we can conclude that this initiative will be a basic tool for secured food supply as it leads to remarkable increase in production. It will result in fewer imports from other states like *Punjab, Haryana* and the southern states and the excess yield can be considered for export as well. It will help feed the ever increasing population.
8. One problem that occurred was the reluctance of some farmers to adopt the new technology due to sheer ignorance, lack of information and illiteracy, due to which they were unable to judge the incredible results of the technology.
9. The marketing aspect of the programme is not very clear and the arrangements made seem to be blurred. It's not decided yet as to how much will be bought by *Krishi Vigyan Kendra* or the government and how much will be sold in the market.
10. Handling new crops can be tricky for the farmers as they behave completely different from their indigenous counterparts. They require different set of conditions which should be understood by the local farmers, or else this technology might even back-fire.
11. The farmers enrolled under the Certified seed programme has to deal with one major drawback of buying fresh seeds after every two generations from the *KVK*, as after two generations of seeds of seed production, there are chances of the strains getting contaminated or mixed with the local varieties, thereby decreasing yield on subsequent farming. Hence, pure variety seeds need to be purchased after every two generations.
12. Although this strategy has done well for its first year, but this humble beginning should be converted to reach more people and proper plans to engulf and engage more farmers should be made in the years to follow.

13. Technology Showcasing has been a suitable means of the quickest possible spread of new varieties developed (released) by the plant breeders. It has been exemplary in this regard. This also keeps the farmers well informed of the latest developments in the area of seed technology.
14. This programme will be carried out throughout the state so that everyone may benefit from its application. Each region will be on its own to carry out the process after completion of the first year of the programme.
15. Following this technology over and over again will lead to the production of quality seeds in a geometric progression fashion that will allow the economies of scale to set in and as a result seeds will be available for everyone at reasonable prices.

Suggestions

1. More field demonstration techniques should be organized.
2. Adequate steps should be taken to educate the farmers regarding the benefits of new technologies.
3. Proper and periodic training should be given in order to properly manage the new crops in order to reap maximum yield.
4. Seeds should be made available in the markets for everyone to adopt the technology.
5. The marketing aspect of the programme should be taken seriously in order to avoid loss. *KVK* and the state government should step in to remove the intermediaries and also other steps so that the maximum profits remain with the farmers eventually.
6. Certified seeds should be made available by the government at reasonable prices as and when required. Supply should be constant and adequate.
7. Continuous experiments with newer varieties should be done in order to be self-sufficient in all crop categories.
8. The span of the programme should be increased with each passing year so that everyone can have the benefits from the programme. Without increasing the reach and span over the coming years, the very purpose of the programme will be defeated.
9. Proper survey should be done to obtain feedback from the farmers as to know their status and compatibility with the technology.
10. Even though the process will be left in the hands of the farmers for its continuation in the future, but periodic monitoring and proper attention should be paid to address the grievances or any other problem occurring in the future.

Conclusion

With the development of irrigation facility and with the availability of other facilities, many farmers have come forward to cultivate summer rice in some new areas also. Farmers are very much impressed with this new technology. They are getting much more profits than before. Front Line Demonstrations on seed production technology in rice helped the farmers in understanding of seed production procedure and would ensure adequate supply of quality seeds at the farmer's level. As most of the farmers utilize intruded varieties of summer rice, therefore, HYVs of summer rice *'Kanaklata' and 'Jaymoti'* may be considered as replacement of these intruded varieties because of their higher yield potential and availability of quality seeds in time. Adoption of improved production technology of *Kharif* rice in the farmers' field is not yet satisfactory. HYV seeds yield more under irrigated conditions. The HYVs bred could be grown with assured nutrient and water supply and cannot sustain in local adverse growing conditions of the region but also demands controlled irrigation. These crops should be irrigated at right time as per the schedule and in right quantity. Farmers should be made aware of the new rice varieties, technologies and adequate knowledge for practicing sustainable agriculture. It is still early days for Technology showcasing and there is always room for improvement in the future. With official reports yet to be drafted, till then its efficacy remains a bit blurred, although all efforts have been made in this report to throw as much light as possible on this promising technology. Every project suffers from some initial hiccups, but efforts are being made to flawlessly execute the technology. This programme might experience some bad obstacles in its path but plans have been made in such a way so that these can be removed with ease, like the consent of the government officials is very important in this regard. Now, the programme will be taking a different turn and will have to face new challenges. It will no longer be compared with its incompetent local counterpart that has helped this technology look so good with results which goes over the roof and very hard to believe and people tend to get carried away. But now, it has to compete with its own statistics of previous years which will not only be hard to beat but also will stabilize the overtly optimistic, upward shooting graphs of the first year. It has yet to prove its mettle and stand the tides of the more demanding time to come.

References

Annual Report, *KVK Chirang*, 2010-11.

Assam Small Farmers' Agri-Business Consortium, Uniqueness of Assam and Business Opportunities in the Sector: Agriculture. http://assamagribusiness.nic.in/agri4.htm (retrieved on 09/12/12)

Assam, http://en.wikipedia.org/wiki/Assam (retrieved on 09/12/12)

Bongaigaon District Profile, http://bongaigaon.gov.in/dist_profile.htm (retrieved on 09/12/12), http://bongaigaon.gov.in/hist.htm (retrieved on 04/12/12)

Singh, B.N. *et al.*, Towards Rice Self-sufficiency in Northeastern India, Central Rice Research Institute, Cuttack, Orissa.

http://www.ncap.res.in/upload_files/workshop/wsp10/html/chapter11.htm (retrieved on 05/12/12)

Introduction to sustainability, what is sustainable agriculture, http://www.sustainabletable.org/intro/whatis/(retrieved on 06/12/12)

KVK Chirang, Newsletter, Volume 1, April, 2010-March, 2011 http://aau.ac.in/dee/kvkbongaigaon/index.php (retrieved on 05/12/12)

My agricultural information bank, Role and Goal of Seed Technology in Crop Production, http://www.agriinfo.in/?page=topic&superid=3&topicid=79, Principles of Quality Seed Production, http://www.agriinfo.in/?page=topic&superid=3&topicid=82, Genetic Purity during Seed Production, http://www.agriinfo.in/?page=topic&superid=3&topicid=83, Seed Certification, http://www.agriinfo.in/?page=topic&superid=3&topicid=85 (retrieved on 06/12/12).

Sustainable Agriculture, http://en.wikipedia.org/wiki/Sustainable_agriculture, (retrieved on 06/12/12).

Chapter 22

Rural Entrepreneurship Development in Northeast India: Potential of Beekeeping in West Garo Hills District of Meghalaya

Tarun Kumar Das[1] *and Amit Das*[2]

[1]*Subject Matter Specialist (Agri. Extension),*
[2]*Research Associate (NAIP) KVK, ICAR, Tura*

ABSTRACT

The Northeastern region of India is very rich with diverse natural resources. However, in the face population boom the equilibrium of nature often threatened by indiscriminate exploitation of these resources. At this juncture, alternative agricultural practices such as beekeeping are emerging as a promising and sustainable venture for livelihood and income generation in northeastern region in general, and Garo Hills of Meghalaya in particular. Honey is a product of beekeeping that has numerous uses, and beekeeping refers only to rearing of domesticated honeybee species and their management. Out of 4 species commonly found in Garo Hills, *Apis cerena indica* or *Bija khol* is most extensively reared for commercial honey production. Beekeeping, at present, attracts small to marginal farmers for subsidiary income generation owing to its minimum input requirement. Given scientific practices coupled with community participation, beekeeping has the potential of wide socio-economic upliftment as exemplified by some of the societies in this region. Destruction of natural habitat of honeybees and unstable market pose greatest threat and challenge for beekeeping from becoming an ideal rural entrepreneurial activity, however, realizing it's potential various government and non government organization are putting efforts to safeguard the interest of farmers as well as to protect the environment. Moreover, responsible use of natural resources would in the long term decide whether beekeeping would be widely accepted as a profitable entrepreneurial activity which at present seems to be a very good alternative.

Keywords: ***Beekeeping, Alternative agriculture, Rural entrepreneurship, Natural habitat, Garo Hills.***

Introduction

The unique terrains of northeast India are covered in green vegetation and inhabited by a mixed population comprising of different ethnic groups of people. The northeastern region of India is referred to as one of the biodiversity 'hotspots' of the world. However, this part of the world also has its problem due to its geographic remote location coupled with numerous socio economic issues. Many of the components of development do not reach to the inmates of rural villages in this region resulting in poor living conditions for many. For their livelihood a large number of people depend on harvesting the raw natural resources and cultivation methods that are shifting (*Jhum*) in nature. These practices are observed across many hilly places among many cultures in this region. Hence, a large number of people in this region still live in economically deprived condition despite having rich and diversified natural resources. Moreover, the uncontrolled exploitation of these natural resources is leading to an ecological imbalance and may not sustain forever since the natural resources are finite and the population is ever increasing. In the face increase in population and depletion of natural resources, the only way to address the major concern of food and nutritional security of the people of this region as well as to protect the environment is to adopt sustainable strategies of food production and economic development which are in harmony with nature. Meghalaya, one of the states of Northeast India has been blessed with unique floral and faunal varieties spread out in diversified agro climatic areas *viz.* tropical and sub-tropical to temperate or near temperate conditions. West Garo Hills district in particular, experience a warm climatic condition, which harbours numerous life forms in nature and has great potential for domesticating and rearing in captivity. This has opened up ways for the rural population in this place to earn their livelihood with sustainable use of natural resources. Beekeeping is one such activity that seems to hold immense potential for entrepreneurship development in the rural parts of Northeast region in general, and West Garo Hills district of Meghalaya in particular keeping in view that it plays a pivotal role in agricultural diversification besides generating additional farm income. Although there is vast similarity in climatic conditions and occurrence and distribution of floras and faunas in all three districts of Garo Hills in Meghalaya, this paper focuses on field reports from emerging practices in sustainable use of natural resources and entrepreneurship development in West Garo Hills district in Meghalaya.

Honey, Honeybees and Beekeeping

Honey is a nutritious consumable product, which also has extensive medicinal use. Natural honey is obtained from honeybees and beekeeping has been reported as profitable subsidiary activity or even a whole time occupation for small and marginal farmers in many places. Honeybee is a common name for any of several species of highly social bees known for their honey-hoarding behavior and their use as a domesticated species. *Beekeeping* refers only to rearing of domesticated honeybee species and their management. The practice of rearing bees for honey production is also known as *Apiculture.* Honeybees can be easily reared, easily adaptable to different climates and to laboratory or captive conditions, and have a complex social life. Honeybees do not only produce honey but also plays a very important role in pollinating many of the important cross-pollinated crops. Particularly horticultural crops such as orange, pear, guava, litchi, mango etc. that are abundantly found in West Garo Hills are greatly benefitted by honey bee pollination and thereby result in increase of yield. Besides, beekeeping generates certain value added products such as beeswax, pollen, propolis, bee venom, royal jelly and bee brood etc. Although Beekeeping is not a traditional industry in India, However, collection of honey from wild bee colonies has been practical since time immemorial. In a world of deteriorating crop yields where traditional farming methods are no longer sufficient, beekeeping has emerged over the years as a sustainable option that has increased crop fertility and provided a new product to the farmers for income generation.

Species of Honeybees in West Garo Hills

Among the common species of honey bees that are found in India 4 species have been observed in West Garo Hills of Meghalaya *viz.* Rock bee (*Apis dorsata,* locally known as *Bija ama*), Little bee (*Apis florae,* locally known as *Mengkhcri*), Indian bee (*Apis cerena indica,* locally known as *Bija khol*), and Stingless bee (*Trigona iridepennis,* locally known as *khol muk*). There are scattered reports of occurrence of another species, which resembles to Rock bee but it could also be European bee or Italian bee (*Apis mellifera*) as reported by some local beekeepers. Of all the species found locally in West Garo Hills, Indian bee *Apis cerena indica* or *Bija khol* is most widely reared in captivity for commercial production of honey.

Honeybee Rearing: A Tryst of Art and Science

In nature honeybees are social insects and exhibit colonial living behaviour. The members making the colony typically consist of the *queen* (reproductive female), the *worker* (non reproductive female) and the *drone* (male). The number of bees in a single colony may range from 30,000 to 50,000, which may include one queen, 200-300 drones and rest workers. *Swarming* is the natural instinct of honeybees to reproduce its colonies. By this method the large colonies are divided naturally when the colony population reaches maximum. Swarming tendency of honeybees may be increased by natural or artificial factors like sudden failure of queen to lay eggs, imbalanced number of worker bees of different ages and heat, lack of ventilation, lack of space for laying eggs and storage of honey etc. all of which create congestion in the hives.

Practices of Beekeeping in West Garo Hills

For beekeeping it is essential to possess certain knowledge about the honeybees, their characteristics and behavioural pattern. In rural West Garo Hills the tribal people have learnt about the occurrence and behaviour of the local bee species over the years from their ancestors. For domestication, the beekeepers of this region collect the mass of bees (colonies) from undisturbed branches of trees or crevices found naturally in forests (Figure 22.1) and introduce the whole hive into the wooden hive boxes for the purpose of rearing in captivity. Hiving of wild bee colony is usually carried out during evening hours by slightly shaking or smoking the colony in a cut out process. The queen is first captured by blowing smoke and then introduced into the box, side by side the combs are physically cut out from the natural habitat (cavities in trees etc.) one by one and tied to the brood frames by some plant-based fibre to create similar environment inside the bee box and then collection of bees take place in total of 3-4 hours.

Bee-keeping activity is ideally carried out in places with minimum disturbances from human beings and other domesticated animals. Forest areas with flowering plants preferably with sufficient source of nectar, pollen and water thus provide optimum conditions for rearing honeybees. Movable wooden frames with boxes (hive boxes) are placed at such locations after cutting out the combs from natural habitat. The resident bees leave fresh honey sucked from flowers in the cells of honeycomb provided in the boxes to eat bee feed. When these cells are full of honey they are hermetically sealed by capping with wax. Then honey is extracted manually by scrapping off wax caps with knives and then putting them through honey extractor (Figure 22.2). In West Garo Hills extraction of honey is carried out periodically from November to May excepting the months between June to October because of rainwater causing dilution and thus deterioration in quality of honey. Hence, these 5-6 rainy months are observed as resting period with regard to honey production in West Garo Hills. Thus the extraction of honey is carried out about three to four times in a year depending on volume of honey deposition. Although the productivity varies depending on numerous factors, about 2.5-3 kgs of honey is obtained per hive box per extraction in West Garo Hills district.

Entrepreneurial Prospects and Importance in Livelihood

The potential beneficiaries of beekeeping are largely landless labourers, small and marginal farmers, ex-serviceman and retired persons, house wives/farm women, unemployed youth and Student etc. apart from secondary activity of traditional farmers. At present in West Garo Hills beekeeping is carried out in scattered places, however emerging slowly as more organized business ventures by individuals and co-operatives. One inspirational example of rural entrepreneurship development through beekeeping is seen in Chandigre locality in West Garo Hills district where the community participation has led to formation of Nokrek Beekeeping Society. Through the community movement the society is able to produce honey for local and external markets and the members are able to earn subsidiary income from beekeeping. At present the honey is sold @ 300 INR/litre in the markets of West Garo Hills with minimal initial investment which indicates great potential for entrepreneurship development through beekeeping in this region. Realizing its potential, there have been efforts from government to encourage beekeeping in rural areas. Khadi and Village Industries Board has been extending many incentives as well as marketing support to rural people since many years and their initiatives have helped the bee keepers in West Garo Hills as well. Khadi and Village Industries Commission (KVIC),

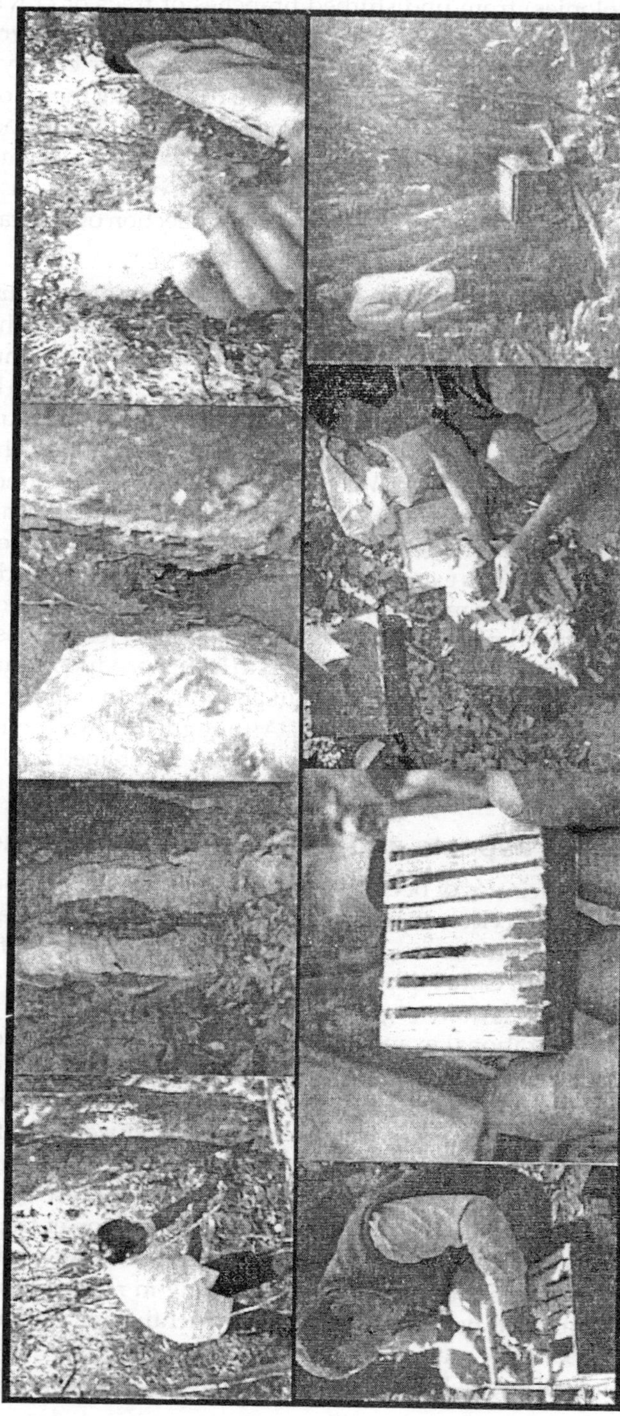

Figure 22.1: Collection of Bee Colonies from the Wild

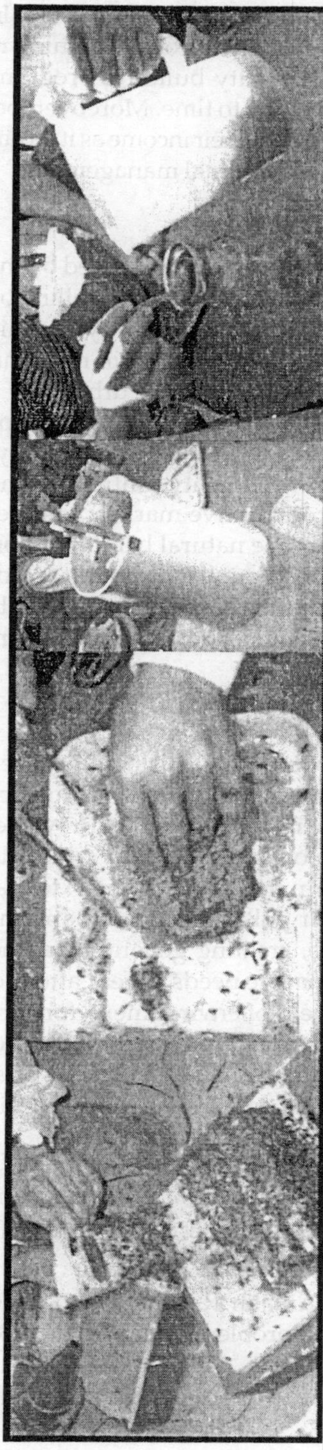

Figure 22.2: Processes in Honey Extraction in West Garo Hills

Krishi Vigyan Kendra (KVK) under ICAR and District Industries and Commerce Centre (DICC) have been playing direct role in training rural people of West Garo Hills in beekeeping through capacity building programmes and also providing technical and financial support time to time. Moreover, people are showing interest in beekeeping as alternative source of their income as it requires low initial expenditure, no elaborated infrastructure and minimal management.

Challenges and Threats

The largest challenge in bee production as faced by the beekeepers of Garo Hills at present is non-availability of stable market for selling of honey as the demand for honey keeps fluctuating. The most common way of selling the honey is to supply them in government enterprises and retailing in rural markets and occasionally in few established business outlets in relatively urbanized markets. Thus, there is an urgent need for establishment of a supply chain for their product. The future success of beekeeping as a profitable entrepreneurial activity will largely depend on addressing these issues along with quality control of honey in a increasingly health conscious consumer driven competitive market. Another important aspect in this regard to consider is to protect the natural habitat of honeybees. According to the beekeepers in West Garo Hills there has been a large decrease in the number of natural bee colonies in wild, which is likely to be caused by increase in construction of brick chimney/burning of charcoal in the vicinity, deforestation and rapid climatic changes.

Conclusion

Whether a subsidiary or whole time business, beekeeping can play a vital role in agricultural diversification and additional product development creating employment opportunities, thereby livelihood improvement especially in rural areas. Besides, as honey has many food and medicinal values, it also addresses the issue of nutritional security of the farm families to some extent as well as their well-being. Since human disturbances, air pollution and climatic changes adversely affect the natural habitat of honeybees, creating general awareness among the people for necessity of conservation of nature needs urgent attention and related individuals and organizations from the field of environment, forestry and agriculture can play direct role in this regard. Thus, there exists a great potential in beekeeping for entrepreneurship development in northeastern hill areas particularly in West Garo Hills, however, sound marketing policies and the responsible use of natural resources are likely to determine the future path in beekeeping.

References

Chaturvedi, P.L., Hussain, S.R. and Nath, S. (1969). A tip for migratory beekeepers of Kumaon region. *Indian Bee J.*, 31: 48–51.

Gatoria, G.S., Chhuneja, P.K., Aulakh, R.K. and Singh, J. (2001). Migratory beekeeping in India: its prospects and problems. *Indian Bee J.*, 63(1-2): 23–34.

Goyal, N.P. and Rana, B.S. (1992). Economics of migratory beekeeping between hills and plains of north-west India. *Indian Bee J.*, 54(1-4): 19–25.

Nigam, P.M. and Kumar, A. (2001). *A Textbook of Agricultural Entomology*. Emkay Publications, Delhi, pp. 238–274.

Sarkar, B.B. *et al.* (2011). Training Manual–Scientific Apiculture and Honey Production for the Farmers of Meghalaya, pp. 18–23.

Shahid, M. (1992). Beekeeping in the North West Frontier Province of Pakistan. In: Honeybees in Mountain Agriculture (Ed. Verma, L.R.). Proc. Int. Expert Meeting on Beekeep. Dev. In the Hindu Kush - Himalayan Region, Kathmandu, Nepal, June 21 - 23, 1989. Oxford and IBH Pub. Co., New Delhi, pp. 193–209.

Singh, T.S.M.S., Subba Rao, K. and Mohana Rao, G. (1998), Migratory routes for honey production and colony multiplication in Bihar, India. *Indian Bee J.*, 60(4): 207–209.

Suryanarayana, M.C. and Rao, K.S. (1998). *Apis cerana* F. for Indian apiculture and its management technology. In: *Perspectives in Indian Apiculture*, (Ed.) R.C. Mishra and R. Garg. Agro Botanica, Bikaner, India. pp. 66–108.

Chapter 23

Lesser Known Medicinal Plants of Uttarakhand and Hills: Their Uses and Conservation Strategies

Shailesh Tripathi[1]**, *V.K. Sah*[2]*** *and Rahul Dev*[3]****

[1]*Assistant Professor, Horticulture (MAP);* [2]*Professor (Ecology)*
College of Forestry and Hill Agriculture,
G.B. Pant University of Agriculture and Technology, Hill Campus,
Ranichauri (Tehri Garhwal), Uttarakhand – 249 199
[3]*Research Scholar, IARI, New Delhi*

ABSTRACT

In literal sense species are called underutilized whose potential has not been properly realized. A particular plant may be underutilized in a specific region but may be fully exploited in other parts of the world. Similarly, a plant may be used as a food or fibre crop because of its economic value but its medicinal properties may be relatively less or unknown and yet to be exploited. However, there are some plants which have specific economic value with respect to a particular target group or which can be of use in future but could not be exploited widely so far, may be termed as underexploited and underutilized. Therefore, the term underutilized should be used with caution.Uttarakhand, a treasure of medicinal plants, has been worst affected by the over extraction of this precious green gold, thereby threatening the very existence of these important plants in the state. Despite a rich heritage and enormous herbal wealth existing in the state, it is disheartening to point out that the knowledge acquired and disseminated by the great *saints* and scholars is facing extinction on the very land of its origin and there is no consolidated strategy to conserve medicinal plants for their use. Therefore, efficient and safe conservation of these resources have to be achieved through appropriate combination of various plant *viz.*, conservation of eco-system, *ex-situ*

E-mail: *shaileshgbpuat@gmail.com; **sahvk_111@yahoo.co.in; ***rahul2iari@gmail.com conservation methods (field genebanks, botanical garden, herbal garden, seed genebanks, cryobanks etc.), *in-situ* conservation approach, legal and administrative measures etc.The agro-climatic conditions in the country provide an ideal habitat for the natural growth of variety of medicinal plants but cultivators are still not fully aware of their potential. The cultivation, processing and production technologies are not available at grass root level. The commercial cultivation of medicinal plants is also often found uneconomical mainly because of poor and unorganized marketing infrastructure. Therefore, our research efforts have to be relevant need based, demand and market orientated effective and visible and in order to achieve these targets some of the priority areas warrant immediate attention such as: exploration, identification, characterization, documentation and conservation of natural diversity have to be undertaken on a nationwide basis; establishment of adequate number of nurseries for developing good quality planting material and develop better varieties for commercial cultivators of many more medicinal plants that are required in bulk; marketing facilities and price support policy for growers are needed with a tie up or buy back arrangements; many medicinal plants like Kuth, Satavar, Vankakri etc. need to be promoted for cultivation at household level which would provide primary health care; the linkages with industries, sister organization in the country as well as abroad need to be developed and strengthened.

Keywords: Underutilized, Medicinal plants, Biodiversity, Conservation, Traditional uses.

Introduction

In the Indian context no plant is considered useless although the potential of the plant may not be known to the users. In ancient Indian literature, this wisdom has been transmitted to the succeeding generation about the real worth of the plant and its multifarious uses. At the time of settled agriculture when mankind started the domestication of the plant for economic use, the list of plants referred to by our ancestors could have been endless. But slowly over the generations and with the passage of time, urbanization and ecosystem degradation due to industrialization and other factors have taken a heavy toll of this great boon of the nature. As per the estimate of the F.A.O. about 75 per cent of the biodiversity has been lost for ever.

If we take a look at the major food crops being used by man, it becomes clear that how narrow our food basket has become as only three crops *viz.*, rice, wheat, maize provide 60 per cent of our food requirement. Same is the case with medicinal plants as well. While our ancestors had deep knowledge about most of the plants growing in their vicinity and for all kinds of ailments, there was a specific plant for use; we hardly have knowledge about a few of them. Since *Ayurvedic* knowledge is considered to be sacred and was hardly used for any pecuniary gain, *Vaids* and *Hakeems* used to pass on the knowledge only to the deserving ones who were capable of using them wisely for the benefit of the users alone. This process of transmission of the knowledge from the *Guru* to the deserving *Shishya* alone has caused the loss of the knowledge in the process due to its secretive nature. Moreover, written documents were not possible during those ancient times and it was through the ears that the knowledge was percolated. Scientific documentation of the plants and their uses are only a few century old while, plants were being used from the dawn of the civilization. Fortunately, we have ancient treaties like the *Vriksha Ayurveda*, *Charak Samhita*, *Sushruta Samhita* etc.

which record faithfully the value of individual plants. Many of the plants having wide economic importance have been extinct, some are endangered and others still unknown. The widely used medicinal plants are few which are being cultivated for the extensive use and export potential. However, there are some plants which have specific economic value with respect to a particular target group or which can be of use in future but could not be exploited widely so far over a reasonable cultivable area, may be termed as under exploited and underutilized. The lack of scientific knowledge and paucity of the resources may have prevented large scale exploitation of these medicinal plants (Tripathi and Sah, 2010).

In literal sense species are called underutilized whose potential has not been properly realized. Their cultivation is restricted to specialized geographical pockets in different agro-ecological regions mainly by the poor farming communities who have little access to modern agro inputs and well organized marketing and communication infrastructure. The term under utilization has specific context *viz.*, area, time and user group. A particular plant may be underutilized in a specific region but may be fully exploited in other parts of the world. Similarly, a plant which might have been of great use at a particular point of time in our civilization might have lost its value as a major crop subsequently. Again a plant may be used as a food or fibre crop because of its economic value but its medicinal properties may be relatively less or unknown and yet to be exploited. Therefore, the term underutilized should be used with caution. The lesser known medicinal plants in the Himalayan region are grown and consumed by local people from very ancient time. The Himalayan region is one of hot spot for genetic variability for different characters in underutilized medicinal plants. Keeping this in view, we have categorized a group of medicinal plants as underutilized in the Himalayan context and hence, are purely tentative. The list of underutilized medicinal plants along with their traditional uses has been given in the Table 23.1.

Indian Himalayan Region possesses many endemic medicinal plants, which have versatile potential to generate a variety of unique natural products. Currently, only few of them are being utilized commonly for limited uses due to lack of R&D work on these plants, *e.g.*, the oil of *Saussurea costus* (Kuth) is utilized in producing world class perfume and has a very high price in world market, but the production of this oil in India is negligible. Likewise many other plants, such as *Nardostachys jatamansi, Acorus calamus, Origanum vulgare, Inula racemosa* etc. are used in both, medicinal and aromatic purposes, but the production of these plants in India is very limited. But these genetic resources are threatened-a process switched on by the exploding human and livestock population, unsustainable resource extraction and unfriendly waste dumping. Many species have already disappeared or are on the verge of extinction, designated as endangered species. Such disappearance of genetic resources is related to the extinction of innumerable species with which it is intrinsically linked through food chain and food web. This has also led to serious ecological consequences with an eventual bearing on productivity, which is affecting the basic life support system and livelihood of millions of organisms living on the earth. Therefore, many lesser-known, rare and endangered medicinal plants also need to be studies and exploited for industrial uses in order to break the current

Table 23.1: Lesser Known Medicinal Plants of Uttarakhand Hills with Traditional Uses

Sl.No.	Botanical Name	Vernacular Name(s)	Family	Altitude (m)	Habit	Plant Part Used	Major Chemical Constituents	Traditional Uses
1.	*Achyranthes aspera* L.	Latjira	Amaranthaceae	Upto 2000	Herb	Whole plant	Triterpenoids, Saponins	Rheumatism, Cancer
2.	*Aesculus indica* Hook.	Pangar, Horse chestnut	Hippocastanaceae	1500-3000	Deciduous tree	Seeds	Saponins, Aesculine	Hemorrhoids, Thrombosis
3.	*Aesandra butyracea* Baehni (Roxb.)	Indian Butter Tree, Chiura	Sapotaceae	Upto 1700	Medium size tree	Seed, Bark, Fruit pulp, Leaves	α-spinasterol, β Sitosterol, α-spinasterone	Emollient, Antirheumatic, Skin affections
4.	*Agave americana* L.	Century Plant, American Aloe, Rakas Pallah	Agavaceae	Upto 1800	Stout shrubby rhizomatous plants	Roots, Leaves, Gum, Flowers	Chlorogenin, Kaempferol-3-glucoside, Kaempferol-3-rutinoside	Diaphoretic, Diuretic, Anti-syphilitic, Anti-scorbutic, In boils
5.	*Ainsliaea aptera* DC.	Karvi-Buti, Kauru, Khad-Jari	Asteraceae	2500-3500	Perennial herb	Roots	–NA–	Roots are effective in acute gastric pain, Vermicide and in dirrhoea
6.	*Ajuga bracteosa* Wall ex. Benth.	Nilkanthi	Lamiaceae	1000-2700	Perennial herb	Whole plant	β and γ Sitosterol	Diuretic, Malarial fever
7.	*Alangium salvifolium* Wang.	Ankola, Akola	Alangiaceae	Upto 1000	Deciduous, rambling shrub or tree	Root, Root-bark, Seeds Leaves, Fruits	Alangine A&B, Emetin, Marckine, Psychotrine	Single drug for the treatment of rabies, increases peristaltic movement of intestine, useful in fever and skin diseases
8.	*Anagallis arvensis* L.	Jonkmari	Primulaceae	Upto 2000	Annual herb	Aerial parts	Saponins, Tannins, Cucurbitacins	Epilepsy, Mental problem
9.	*Androsace rotundifolia* Hardw.	Nirodhak-buti	Primulaceae	1500-3400	Perennial herb	Leaves	Primulagenin	Birth control

Contd...

Table 23.1–Contd...

Sl.No.	Botanical Name	Vernacular Name(s)	Family	Altitude (m)	Habit	Plant Part Used	Major Chemical Constituents	Traditional Uses
10.	*Angelica glauca* Edgew	Gandrayan	Apiaceae	3000-4000	Perennial herb	Dried roots	Algelicin, Bergaptan	Carminative, typhoid, diuretic, stomach pain
11.	*Apium graveolens* L.	Ajmoda	Apiaceae	Upto 2400	Biennial or annual herb	Fruits	Choline, Myristic acid	Carminative, Stomachic, Nerve tonic
12.	*Arisaema tortuosum* (Wallich) Schott.	Bag-mungri	Araceae	1500-3000	Stem-less perennial herb	Tubers	Saponins	Anticonvulsant, Psycho medicine
13.	*Arnebia benthamii* Johnston	Laljari	Boragina-ceae	3000-4000	Perennial herb	Roots	Shikonine, Shikalkin	Cardiac disorder, Colouring matter in hair oils
14.	*Artemisia maritima* L.	Worm Seed, Patti, Chhuari	Asteraceae	2100-2800	Dedidous shrub	Whole plant	Santonin, Devanone, Thujone	Anthelmintic, laxative, stomachic, appetizer, cure *"tridosha"*
15.	*Artemisia vulgaris* L.	Nagdona	Asteraceae	Upto 2200	Perennial herb	Whole plant	Cineol, α-Thujone	Heamagogic, Antispasmodic, Leucorrhoea
16.	*Asphodelus tenuifolius* Cav.	Piazi	Liliaceae	Upto 1000	Herb	Seeds	Lupeol	Diuretic, Antiseptic, applied in ulcers
17.	*Astragalus candolle-anus* Bunge.	Milk vetch, Rudravanthi	Fabaceae	2700-4000	Perennial herb	Roots	Isoflavonoids, Saponins	Anti-hypertensive, Anti-ulcer
18.	*Baliospermum montanum* (Willd.) Muell-Arg.in DC	Jangli jamaghota	Euphor-biaceae	Upto 1000	Undershrub	Seed	Phorbol esters, Montanin	Purgative, substitute of jamal-ghota, seed oil applied in rheutamic pain, leaf and root extract given in jaundice and bronchial diseases
19.	*Berberis aristata* DC.	Kilmoda, Barberry, Daru-haldi	Berberida-ceae	1800-2500	Shrub	Root bark	Berbarine, Berbamine	Eye diseases, Jaundice, Malaria, Anti-cancer

Contd...

Table 23.1–Contd...

Sl.No.	Botanical Name	Vernacular Name(s)	Family	Altitude (m)	Habit	Plant Part Used	Major Chemical Constituents	Traditional Uses
20.	*Berginia ligulata* (How) Stern.	Pashanbhed, Velvet Leaf	Saxifraga-ceae	2000-3200	Perennial herb	Rhizomes	Bergenin and its glucosides	Anti-lithic, diuretic, earache, antiscorbutic, astringent, abortifacient, antidiarrhoeal, febrifuge
21.	*Betula utilis* D. Don	Bhojpatra, Paper birch	Betulaceae	3000-4000	Medium size tree	Bark	Betulin, Betulic acid	Astringent, Anticonvulsant, Epilepsy
22.	*Bidens pilosa* L.	Kumra	Asteraceae	Upto 2500	Annual herb	Aerial parts	Flavonoids, Xanthophylles	Peptic ulcer, Ulcerative colitis, Leucoderma
23.	*Boenninghausenia albiflora* Reichb	Pissumar, Upania Jhar	Rutaceae	1350-2000	Deciduous under shrub	Roots, Plants	Xanthyletin, Acridone	Roots applied to old wounds
24.	*Caltha palustris* L.	Marsh merigold	Ranuncula-ceae	3000-4000	Perennial herb	Whole plant	Helliborin, Veratrin	Diabetes, Bronchitis, Uterine cancer
25.	*Cannabis sativa* L.	Indian hemp, Bhang, Bhangi	Cannaba-ceae	Upto 2500	Annual	Whole plant	Essential oil, Cannabinoids, Flavonoides glycosides	Intoxicant, Stomachic, Analgesic, Stimulent, Aphrodisiac, Sedative
26.	*Cassiope fastigiata* (Wallich) D. Don	Hieunsheli	Ericaceae	2800-4000	Branched shrub	Twigs	–NA–	Fire burns, Cooling effect, Wound healing
27.	*Celtis australis* L. Kharik	Nettle tree,	Ulmaceae	Upto 2000	Tree Fruits	Leaves, Mucilage	Tannins,	Reduces uterine bleeding, Peptic ulcer
28.	*Cirsium arvense* (L.) Scop.	Japanese thistle, Bhungchee	Asteraceae	2500-3000	Perennial herb	Whole plant	Flavonoids	Pneumonia, Cough, Cooling effect
29.	*Cleome viscosa* L.	Jakhya	Cappara-ceae	Upto 2000	Annual herb	Seed	Macrocyclic diterpene, Cleomeolide	Rheumatic arthritis, Worm infections

Contd...

Table 23.1–Contd...

Sl.No.	Botanical Name	Vernacular Name(s)	Family	Altitude (m)	Habit	Plant Part Used	Major Chemical Constituents	Traditional Uses
30.	*Coleus barbatus* Willd.	Pathurchur, Kabaduthuir	Lamiaceae	500-2000	Perennial herb	Fasciculated Roots, Leaves	Forskohlin, Barbatusin	Leaves used in stomach disorders, bronchodilator
31.	*Corydalis govaniana* Wallich	Indrajata, Chingti	Papavera-ceae	2400-4000	Perennial herb	Tubers	Choline, Corulumine	Uterine bleeding, Abdominal pains, Chronic fever, Liver complaints
32.	*Cynoglossum glochidiatum* Wallich ex Benth.	Lichkura, Kumbro	Boragina-ceae	Upto 2100	Annual herb	Roots	Cynaustraline, Cynaustine	Dyspepsia, Digestive disorder
33.	*Datura stramonium* L.	Jimson weed, Datura	Solana-ceae	Upto 2700	Annual herb	Leaves, Flowering tops	Hyoscyamine, Hyoscine Atropine	Used in asthama, nose troubles, boils, in Parkinson's disease, as pre anesthetic in surgery and child birth, Anodyne
34.	*Delphinium denudatum* Wall	Nirvishi, Himalayan Delphinium	Ranuncula-ceae	2200-4000	Perennial herb	Roots and Rhizome	Delphisine, Delphinodine	Analgesic, aphrodisiac, anti-toxic, toothache
35.	*Desmodium elegans* DC.	Chamra, Chamkat	Fabaceae	1200-3500	Deciduous shrub	Roots	Tyramine, Hordenine	Carminative, useful in discharges, piles, oedema, scorpion sting
36.	*Dicliptera bupleuroides* Nees	Kuthhi, Saamp ki buti	Acantha-ceae	Upto 2000	Herb	Aerial parts Leaves	–NA–	Check bleeding, Cough, Gastroenteritis, Snake bite
37.	*Didymocarpus pedicellatus* R. Br.	Pathar-phori	Gesneria-ceae	1000-2800	Herb	Leaves	Pashanone, Pedicellin, Didymocarpin	Cures kidney stones
38.	*Dioscorea bulbifera* L.	Genthi, Potato yam	Dioscorea-ceae	1500-2100	Herb	Tuber, Leaves	Diosgenin, Cryptoxanthin	Anti dysenteric, useful in dyspepsia, urinary discharges, bronchitis, "*Vata*", piles

Contd...

Table 23.1–Contd...

Sl.No.	Botanical Name	Vernacular Name(s)	Family	Altitude (m)	Habit	Plant Part Used	Major Chemical Constituents	Traditional Uses
39.	*Doronicum roylei* DC.	Kauni, Plague nashak buti	Asteraceae	3000-3600	Shrub	Roots	Pardalianchol	Prevent giddiness, Nervous depression
40.	*Ephedra gerardiana* Wallich ex Stapf.	Somlata	Ephedera-ceae	3000-4000	Under shrub	Stem	Ephedrine	Cough, Asthma, Heart stimulant
41.	*Echinops cornigerus* DC.	Kantela, Globle thistle	Asteraceae	1000-2000	Herb	Root	DIPPA-amyrin, Lupeol	Roots extract for teething of infants, root juice given in fever and urinary troubles
42.	*Eupatoriun adenophorum* Sprengel	Kala Bansa	Aesteraceae	Upto 2200	Shrub	Leaves	Volatile oil, Ayapanen	Crushed leaves applied on wounds
43.	*Fagopyrum esculentum* Moench	Kutu, Buck-wheat	Polygona-ceae	2000-2700	Herb	Leaves, Flowers	Flavonoids (Rutin, Quercetin)	Treat the harmful effects of X -rays, useful in hemorrhagic, capillary fragility, hypertension, bleeding from the kidneys
44.	*Fumaria indica* (Haussknecht) Pugsley	Khairua	Fumaria-ceae	Upto 2000	Annual herb	Flowers	Fumaric acid, Fumarine	Stimulant to liver and gall bladder, Diuretic
45.	*Gaultheria fragrantissima* Wallich	Checker berry	Ericaceae	1800-3000	Small shrub	Leaves	Gaultherilene, Gaultherine	Rheumatism, Acute gastritis
46.	*Gentiana kurroo* Royle	Neelkanth, Karu, Kadu, Indian Gentian	Gentiana-ceae	1300-3500	Perennial herb	Roots and rhizome	Gentiopicroside Gentiopicrinin	Blood purifier, carminative, diaphoretic and digestive, hepatic stimulant, used in urinary troubles
47.	*Geranium wallichianum* D. Don ex Sweet	Ratanjot	Gerania-ceae	2400-3600	Herb	Roots	Gallic acid, Succenic acid	Stomach ulcers, Uterus cancer

Contd...

Table 23.1–Contd...

Sl.No.	Botanical Name	Vernacular Name(s)	Family	Altitude (m)	Habit	Plant Part Used	Major Chemical Constituents	Traditional Uses
48.	*Geum elatum* Wallich	Masreen	Rosaceae	2700-4300	Perennial herb	Roots	Phenolic glycosides, Tannins	Astringent, Deep cuts, Wounds, Pharynx infection, Gastrointestinal
49.	*Girardinia diversifolia* (Link) Friss.	Himalayan neetle, Bara bichhua	Urticaceae	1200-2600	Herb	Leaves	Formic acid	Leaf juice is given in gonorrhea, Swollen joints, Headache
50.	*Gnaphalium affine* D. Don	Marsh Cudweed	Asteraceae	1200-3000	Annual herb	Aerial parts	Volatile oil, Tannins	Tonsillitis, Reduces high blood pressure
51.	*Habenaria intermedia* D.Don	Ridhi-Bidhi	Orchidaceae	Upto 2500	Herb	Tubers	–NA–	Used as tonic
52.	*Hedychium spicatum* Buch.	Kapoor-kachri, Banhaldu, Spiked Gingerlily	Zingiberaceae	Upto 2200	Herb	Rhizome	Essential Oil, Eucalyptol	Diarrhoea, cough, asthma, dysentery, dropsy, hair falling, liver complaints and rheumatism
53.	*Heracleum candicans* Wall. ex DC.	Patrala, Kashmiri-Kaindal	Apiaceae	3000-3500	Perennial herb	Fruits, Root, Seed	Xanthotoxin, Xanthotoxol	Aphrodisiac, nervine
54.	*Hippophae salicifolia* D. Don	Seabuckthorn, Amesh, Aamil, Tserma	Elaegnaceae	2000-3500	Thorny deciduous shrub/tree	Fruits, Leaves	Flavonoids, Organic acids, Vitamin C, A, B_1, B_2, E, K&P	Improves resistance to several diseases, Skin diseases, Boils, Blisters, Asthma, Jaundice, Gastric diordore
55.	*Holarrhena pubescens* (Buch.-Ham.) Wallich	Kutaja, Kurchi, Dola-kuda	Apocynaceae	Upto 1000	Deciduous laticiferous shrub or small tree	Bark, Seeds, Leaves	Conessine, Conesimine, Kurchine, Gum resin and tannins	Astringent, tonic, used in chronic dysentery, colic pain, diarrhoea

Contd...

Table 23.1–Contd...

Sl.No.	Botanical Name	Vernacular Name(s)	Family	Altitude (m)	Habit	Plant Part Used	Major Chemical Constituents	Traditional Uses
56.	*Hypericum perforatum* L.	Choli-phulya	Hyperica-ceae	1000-3000	Herb	Flowers	Hypericin, Volatile oil	Antispasmodic, antiviral, stimulates bile flow
57.	*Inula cappa* (Buch-Ham. Ex D.Don) DC	Athhu, Tamagari	Asteraceae	2000-2500	Shrub	Whole plant	Inulin, Inunolide, Alantolactone	Roots are given in suppressed urination
58.	*Jurinea dolomiaea* Boiss	Dhoop	Asteraceae	3000-4300	Prostrate Perennial herb	Roots	–NA–	Given in fever after child birth, Stimulant
59.	*Justicia adhatoda* L.	Vasaka, Adusa, Malabar Nut	Acantha-deae	500-1600	Dense Shrub	Leaves, Root, Bark, Flowers	Vasicinolone, Vasicine, Vasicinine	Household remedy, Bronchial asthma, Wonderful cough syrup, Tuberculosis
60.	*Leucas lanata* Benth.	Bis-Kapra, Dhurlughas	Lamiaceae	700-2000	Small herb	Leaves, Flowers	Alkaloides, Glucoside	Anti-inflammatory, Whooping cough, Antipyretic
61.	*Litsea chinensis* Lam.	Singrau	Lauraceae	1500-2500	Evergreen tree	Bark	Limonene, Citronellal, Citral	Bark applied on fractured bones
62.	*Megacarpea polyandra* Benth.	Badmula, Rokki	Brassica-ceae	3000-4300	Perennial herb	Roots	–NA–	Diuretic, Antipyretic, Abdominal diseases
63.	*Micromeria biflora* Benth.	Van-ajwain	Lamiaceae	1000-2500	Small herb	Aerial parts	–NA–	Cold, Gastroenteritis, Sinusitis
64.	*Mucuna pruriens* (L.) DC.	Kaunch, Kiwanch	Fabaceae	Upto 1000	Climbing annual	Roots, Leaves, Seeds	Prurieninine, Mucunine	Improves Digestion, Stomach problems, Aphrodisiac
65.	*Murraya koenigi* (Linn.) Spreng.	Curry-leaf tree, Kadi-patta, Kaidarya	Rutaceae	Upto 1000	A small tree	Leaves, Roots	Girinimbin, Mahanimbine, Koeingin	Tonic, Stomachic, Purgative, Promote appetite and digestion, used to cure eruptions

Contd...

Table 23.1–Contd...

Sl.No.	Botanical Name	Vernacular Name(s)	Family	Altitude (m)	Habit	Plant Part Used	Major Chemical Constituents	Traditional Uses
66.	*Myrica esculenta* Buch.-Ham ex D. Don	Kaphal, Bay Berry	Myrica-ceae	1000-2300	Tree	Root bark	Triterpenes, Myrictrin	Cold, Sore throat
67.	*Oenothera lamarckiana* Linn.	Dheriani, Evening Primrose	Onagra-ceae	Upto 2000	Annual herb	Seed	–NA–	Cardiovascular disorder, metabolic disorders, derma-tological conditions, diabetes, premenstrual syndromes
68.	*Onosma bracteatum* Wall	Ratanjot, Cerinthe	Boragina-ceae	3000-4000	Perennial herb	Roots	Onosone A and B	Rheumatic arthritis, wound, blood purifier, good for heart
69.	*Criganum vulgare* L.	Ban Tulsi, Jakham-buti	Lamiaceae	1500-3600	Branched perennial herb	Aerial parts, Essential oil	Volatile oil, Carvacrol, Thymol	Strong antiseptic, Bronchitis, Asthma, Healing of wounds
70.	*Oroxylum indicum* Vent.	Tentu, Arlu	Bignonia-ceae	Upto 1200	A small to medium sized deciduous tree	Root bark, Stem bark, Fruit, Seed	Oroxylin, Baicalein	Bitter, Diuretic, Carminative, improves appetite, Expec-torant
71.	*Oxalis corniculata* L.	Tipatia, Chilmora	Oxialida-ceae	Upto 2400	Creeping perennial herb	Whole plant	Oxalic acid, Vitexin	Convulsions in infants, Anti-diarrhoel
72.	*Paeonia emodi* Wallich ex Royle	Peony, Chandryan	Paeonia-ceae	1800-2500	Perennial shrub	Roots	Paeonin, Peonol	Irregular menstruation, Anemia
73.	*Paris polyphylla* Smith	Satwa	Liliaceae	2000-3000	Herb	Whole plant	Strychinen, Glycosides	Aches and soreness of head, eyes, Bronchitis
74.	*Perilla frutescens* (L.) Britton	Bhang Jira, Jhutela, Bhangra	Lamiaceae	Upto 3500	Bushy annual	Whole plant	Volatile oil, Perillene, Phellandrene	Sedative, Stomachic, Anti-spasmodic, Diaphoretic
75.	*Phytolacca acinosa* Roxb.	Jarak	Phytolacca-ceae	1500-2600	Perennial herb	Roots	Phytolaccatoxin, Picro toxin	Oedema, Malignant boils

Contd...

Table 23.1–Contd...

Sl.No.	Botanical Name	Vernacular Name(s)	Family	Altitude (m)	Habit	Plant Part Used	Major Chemical Constituents	Traditional Uses
76.	*Plantago lanceolata* L.	Jangli Isabghol	Plantaginaceae	1500-2500	Herb	Leaves, Seeds	Polysaccharides, Flavonoids	Cuts, Wounds, Piles, Blood dysentery
77.	*Plumbago zeylanica* L.	Chittrak	Plumbaginaceae	Upto 2500	Perennial shrub	Leaves, Roots	Plumbagic acid, Plumbagin	Abortifacient, Skin diseases, Wounds, Rheumatism
78.	*Polygonatum verticillatum* Allioni	Maha maida Indian Polygonatum	Liliaceae	2000-2800	Herb	Roots and Rhizome	Lysine, Serine, Threonine, Aspartic acid	Diuretic, appetizer and tonic, emollient, galactagogue, aphrodisiac, fever, weakness
79.	*Potentilla fulgens* Wallich ex Hook.	Vajradanti	Rosaceae	1500-3000	Herb	Roots, Stem	Catechins, Tannins	Toothache, Stomatitis, Wounds
80.	*Prinsepia utilis* Royle	Bhainkal	Rosaceae	1500-2700	Shrub	Root bark, Seeds	–NA–	Rheumatic pain, Diarrhoea, Tonic
81.	*Prunella vulgaris* L.	Common Self Heal, Kalaveuth	Lamiaceae	1500-3000	Perennial herb	Whole plant	Essential oil	Gastritis, Breathing problems
82.	*Pyracantha crenulata* (D. Don) M. Roemar	Ghingaru	Rosaceae	1800-3000	Shrub	Berries	Cynogenetic glycosides, Crategin	Cardiac diseases, High blood pressure, Reduces cholesterol
83.	*Rhododendron anthopogon* D. Don	Dhoop, Burans	Ericaceae	3000-4800	Shrub	Leaves	Triterpenoids, Flavonoids	Stimulant properties
84.	*Roscoea alpina* Royle	Jilsua, Kakoli	Zingiberaceae	1800-2400	Herb	Leaves	–NA–	Dried powder of leaves used in wounds and cuts of cattle, Plant extract used as a tonic
85.	*Rubia cordifolia* L.	Manjistha	Rubiaceae	1200-2700	Climbing perennial herb	Roots, Stem	Munjistin, Perpurin	Tonic, astringent, Bacillary dysentery

Contd...

Table 23.1–Contd...

Sl.No.	Botanical Name	Vernacular Name(s)	Family	Altitude (m)	Habit	Plant Part Used	Major Chemical Constituents	Traditional Uses
86.	*Rubus ellipticus* Sm	Pila Hissalu, Yellow Raspberry	Rosaceae	1200-2400	Tall shrub	Roots, Young stem	–NA–	In colic pain
87.	*Ruta graveolens* L.	Rue, Salep	Rutaceae	Upto 1500	Herb	Aerial parts	Rutin, Glucosides	Antispasmodic, abortifacient, reduces blood pressure
88.	*Rumex nepalensis* Sprenges	Pahari Palak Khourasho	Polygona-ceae	1200-4300	Perennial herb	Roots, Leaves	Anthraquinones, Oxalates	Dysmenorrheal, Constipa-tion, Stomachache
89.	*Satyrium nepalense* swartz.	Salang mishri	Orchida-ceae	2000-3000	Herb	Rhizome.	–NA–	Aphrodisiac, used in diarr-hoea, tubers edible
90.	*Selinum tenuifolium* Wallich ex C. B. Clarke	Bhutkeshi	Apiaceae	2700-4000	Perennial herb	Roots	Coumarins, Selinone, Selinidin	Asthma, Incense of roots are beneficial in hysteria
91.	*Siegesbeckia orientalis* L.	Hiskura, Gobariya	Asteraceae	Upto 2000	Herb	Whole plant	Darutigenol, Darutoside	Decoction of plants with rice water taken in diarrhoea and bowl complaints
92.	*Skimmia laureola* DC.	Nairpatti, Kasturipatra	Rutaceae	1500-3000	Evergreen shrub	Leaves	Skimmin, Lina-lool Scopoletin, β-phelandrene	Cooling effect, Tonic, Fever
93.	*Smilax aspera* L.	Chhob-chini Kukurdara	Smilaca-ceae	Upto 2000	Climber	Roots	Sarsasopogeni Asperagenin	Diuretic, Diaphoretic, Rheu-matic-arthritis
94.	*Solidago virguarea* L.	Pinja-phool, Golden Rod	Asteraceae	2000-3000	Herb	Whole plant	Quercitrin, Rutin, Kaempferol, Astragalin	Used in kidney troubles, decoction of whole plant used for the treatment of asthama, rheumatism and wounds
95.	*Sonchus oleraceus* L.	Dudiya, Milk thistle	Asteraceae	Upto 2200	Herb	Whole plant	Apigenin, Luteolin, Stigmasterol	Plant used as a liver tonic

Contd...

Table 23.1–Contd...

Sl.No.	Botanical Name	Vernacular Name(s)	Family	Altitude (m)	Habit	Plant Part Used	Major Chemical Constituents	Traditional Uses
96.	*Stephania glabra* Roxb.	Gindaru, Gangaru	Menisperma-ceae	Upto 2200	Climber	Tubers	Stepharine, Stepharotine	Psycho medicine, Asthma, Tuberculosis
97.	*Strobilanthes auriculatus* Nees	Kapurminjri, Pondai	Acanthaceae	Upto 1000	Shrub	Leaves	Aromatic oil	Febrifuge, Expectorant, Tonic
98.	*Swertia chirayita* Buch	Chirayta, Chiretta	Gentiana-ceae	2000-3200	Perennial herb	Whole plant, Roots	Gentianine, Swerchirin, Gentiopicroside	Anti-inflammatory, piles, ulcers, skin diseases, sto-machic, anthelmintic
99.	*Tagetus minuta* L.	Jangli Gainda	Asteraceae	Upto 2000	Herb	Whole plant	Tagetone, Ocimens	Otalgia, Stomachic, Aperient
100.	*Tanacetum longifolium* Wallich ex DC.	Guggal, Amritdhara-ghas	Asteraceae	1000-4000	Perennial herb	Aerial parts	Volatile oil, Sesquiter-penes	Nasal drop, Headache, Bodyache, Antispasmodic
101.	*Taraxacum officinale* Weber.	Kanphuliya, Dudhla	Asteraceae	800-4000	Small herb	Whole plant	Taraxacerine, Taraxacin	Migraines, Hepatitis, Head-ache
102.	*Taxus baccata* L.	Yew, Thuner, Talispatra	Taxaceae	1800-3300	Small or medium size evergreen tree	Twigs, Bark Leaves, Berries	Taxines A&B, Taxicatin, Ephedrine, Millosine	Extraction of important anti-cancer drug taxol, used in treatment of headache, giddiness, rheumatism liver and urinary tract affections
103.	*Thalictrum foliolosum* DC.	Mamira	Ranuncula-ceae	1300-3500	Perennial herb	Roots	Jatrorrhizine, Berberine	Improves eye sight, Diarrhoea, Piles
104.	*Trichosanthes tricuspi-data* Lour.	Indrayan	Cucurbita-ceae	600-2100	Herbaceous climber	Roots, Fruits	Cucurbitacins, Punicic acid	Bronchitis, Asthma, Diabetes
105.	*Urgenia indica* (Roxb) Kunth	Jangli Piaj	Liliaceae	Upto 1400	Glabrous herb	Digitoxin, Digitalis	Dried bulbs	Heart ailments, Epilepsy, Diabetes

Contd...

Table 23.1–Contd...

Sl.No.	Botanical Name	Vernacular Name(s)	Family	Altitude (m)	Habit	Plant Part Used	Major Chemical Constituents	Traditional Uses
106.	*Urtica dioica* L.	Bichhu Ghas, Stinging Nettle	Urtiacaceae	Upto 3000	Perennial shrub	Roots	Flavonoids, Amines, Formic acid	Sciatica, Rheumatism, Skin diseases, Hidden muscular pains/hurt
107.	*Verbascum thapsus* L.	Ban Tambaku	Scrophularia-ceae	Upto 2500	Perennial herb	Leaves, Flowers	Flavonoids, Volatile oil, Corcetin	Bronchitis, Asthma
108.	*Viola serpens* L.	Vanphsha, Sweet ciolet	Violaceae	1000-2700	Perennial herb	Flower, Aerial parts	Violin, Violoquernitin	Asthma, bleeding piles, cancer of throat, cough fever, headache and skin diseases
109.	*Vitex negundo* L.	Nirgundi	Verbenaceae	Upto 1500	Small size tree	Leaves	Nishidine, Hydrocotylene	Cold, Headache, Swelling in joints, Flatulence, Improving eye sight
110.	*Woodfordia fruticosa* (L.) Kurz	Dhaula	Lythraceae	Upto 1600	Shrub	Flowers	Polystachoside, Ellagic acid	Febrifuge, Hemorrhoids
111.	*Zanthoxylum armatum* DC.	Timroo	Rutaceae	Upto 1800	Shrub	Bark, Fruits, Leaves	Pipevine, Linalool	Liver tonic, Diuretic, Anti-septic

Sources: Polunin and Stainton, 1985; Rastogi and Mehrotra, 1990; Rastogi and Mehrotra, 1991; Badoni, 1995; Kaul, 1997; Gaur, 1999; Joshi and Kumar, 2000; Prajapati *et al.*, 2003; Joshi *et al.*, 2004; Jain, 2004; Maheshwari, 2005; Tripathi and Sah, 2010.

stagnation in the production and marketing sectors. R&D work on lesser-known plants and subsequent product development and their commercialization will certainly open new markets in India, which will put India on number one positions in the world in case of many plants and plant products. Further large scale cultivation of these plants may also contribute towards the socio-economic upliftment of hill farmers (Raghav, 2005).

While declaring the 'Health for All' by 2000 A.D. the World Health Organization (W.H.O.) has emphasized the need for revitalization and expansion of traditional system of medicine, and has listed 21,000 plants species that have been reported for medicinal use all over the world (Singh and Singh, 2004). The global interest in plant-based pharmaceuticals has grown rapidly in recent years leading to remarkable increase in demand for medicinal plants. Naturally growing wild population have remained the main source of these raw-materials in the form of 'Sacred-grooves' and establishment of several herbal gardens.

The agro-climatic conditions in the country provide an ideal habitat for the natural growth of variety of plants which are suitable for beneficial uses, particularly from health point of view. India has the richest bio-diversity, wherein more than 9,500 plant species are identified and documented, which have importance in the pharmaceutical industries. In this context, it is heartening to note that about 80 per cent population in developing countries rely on the traditional medicines, mostly the plant drugs for their healthcare as per the assessment of World Health Organization. Besides, these herbs and plants are the raw material for more than 25 per cent of drug of modern medicine. This is because of the fact that the medicines derived from the medicinal plants are non-narcotic having no side-effects. With rising awareness globally for health, there is increasing consciousness for the use of herbs for the healthcare. In India, more than 25,000 effective plant based formulations are available and more than half a million practitioners of Indian System of Medicine use medicinal plants in preventive and curative applications. Internationally, medicinal plants are traded for more than Rs. 6,000 million per year and the growth in this trade is estimated to be more than 7 per cent annually. The Indian share in the global trade is negligible compared to those of many countries including China, although potential exists. Thus, it is imperative to harness the potential and capitalize on opportunities of increasing demand for medicinal plants globally.

Uttarakhand: Treasure Trove of Medicinal Plants

The newly created beautiful state "Uttarakhand", having a total geographical area of about 53,484 sq. Km, out of which about 63 per cent is the forests and about 12.50 per cent, is cultivated land. The floristic diversity is manifested through different phyto-climatic and topographic regimes. The temperate climate of the state is a rich source of biodiversity, which has abundant medicinal and aromatic plants wealth that has tremendous commercial national and global potential. But this state has been worst affected by the over extraction of this precious green gold, thereby threatening the very existence of these important plants in the state. However, the land has tremendous potential for cultivation of medicinal. The cultivation of these plants on commercial scale would provide self-employment and improve the economic standard of the people because these are cash crops.

Uttarakhand is committed for its commercial exploitation, through research and development to achieve, the vision of herbal state. Herbal Research and Development Institute, Gopeshwar, an autonomous Institute of Government of Uttarakhand and a nodal agency for Medicinal Plant Board is dedicated for sustainable development through effective coordination. G.B. Pant University of Agriculture and Technology is committed for the research on medicinal plants. Through research, potential cultivars/species, agro-technologies and processing techniques for large number of medicinal plants have been developed. There is an agri export zone for promotions of medicinal plants for export. Expectedly, with existing infrastructure the state is poised to take lead (Sah and Tripathi, 2009).

Despite a rich heritage and enormous herbal-wealth existing in the state, it is disheartening to point out that the knowledge acquired and disseminated by the great *saints* and scholars is facing extinction on the very land of its origin. Owing to ignorance and lack of proper documentation, the practice of herbal-cure, which mostly, often literally, depended on the phenomenally good memory of the practitioner, had to suffer a lot during the course of time. The person, who knew the art of healing, usually did not reveal his knowledge to anyone for fear of being misused. As a result, much of the valuable knowledge of the practitioners used to be lost along with the death of such a person.

Biodiversity: A One-Time Endowment

Biodiversity is the variation of life forms within a given ecosystem, biome or for the entire Earth. Biodiversity is often used as a measure of the health of biological system. The biodiversity found on Earth today consists of many millions of distinct biological species, which is the product of nearly 3.50 billion year of evolution.

It has increasingly clear in recent years that human activities have contributed significantly towards accelerating the rate of 'extinction' of species and rapidly degrading and destroying their natural habitats. Franatic search for target species and ruthless over-exploitation over decades have alarmingly depleted the vast region. The pressure on drastically shrunk natural populations of some threatened species of medicinal plants is so great that they are hardly able to regenerate themselves, particularly wherever roots (containing desired chemicals) are greedily dug-out exposing the soil to forces for erosion (Singh, 2005a).

Threat Assessment and Priority Setting

The potential value of India's medicinal plant wealth is enormous. Over exploitation of more than 35 species has resulted in their endangered/vulnerable status. Over one-third of the plants listed in "Red Data Book of India" are of medicinal value. Unlike other crops, the systematic collection and characterization of germplasm of medicinal plants is limited. Considering the richness of genetic resources, their regional distribution, need to be surveyed and priorities have to be worked out, keeping in view the relative degree of threat of extinction (Gautam, 2005).

Of the many classes of threats to wild medicinal plant species, trade plays a considerable part (25 per cent). Besides that, human interference, fragmentation, over-exploitation and indiscriminate harvest pose the severe danger. IUCN, IDRC, WWF and others (1998) have assessed the alarming stage of medicinal plant in the Himalayan region of India and red listed them. The Government of India has already imposed ban on export of those plants species.

India has made exceptional efforts to document its medicinal plant flora and to identify threatened species. *A Red Book of Threatened Medicinal Plants* has been produced. Between 1990 and 1996, 6 of 9 reports of 'Red List' category assignments in the sources reviewed for the MPSG Bibliography were from India.

Areas of Biodiversity

The vegetation and flora both show great variety due to significant altitudinal variations from sea-level to the highest mountain ranges of the Himalayas, the longest coastline, hot deserted conditions in the west, cold deserted conditions in the north and humid tropical conditions in some parts of the country. The plant wealth of the country provided us with a large variety of economic products, as also a variety of medicinal plants, drugs-tannins, fibers, gum etc. From the view point of general vegetation and floristic elements, India can be divided into the eight regions (Tyagi *et al.*, 2005).

General Causes of Threat to Plant Species

Both natural as well as man-made causes are posing a great threat to the survival of many medicinal plant species. Certain natural disasters leading to destruction of habitats can not be blamed for, but certain factors can be controlled or their effects can be diluted to a great extent. The major causes leading to loss in biodiversity are listed below (Chauhan, 2005):

Abiotic Factors

- ☆ Floods droughts, earthquakes, landslides etc.
- ☆ High rate of endemism
- ☆ Critically low population (less than 100 individuals plants)
- ☆ Pathogens/Disease
- ☆ Absence of pollinators
- ☆ Invasion by exotics or other aggressive species
- ☆ Land, water and air pollution

Biotic Factors

- ☆ Destruction and modification of habitats
- ☆ Over exploitation for commercial, scientific and educational purposes
- ☆ Over grazing by domestic animals
- ☆ Regeneration of obnoxious weeds
- ☆ Change/expansion in farming activities

- ☆ Disturbing permanent grasslands and pastures
- ☆ Afforestation and deforestation
- ☆ Industrialization, urbanization, building, townships, roads, dams
- ☆ Tourism development
- ☆ Pressure from exotic introductions

Conservation Strategies

Traditionally medicinal plant communities have depended on plants to meet their basic needs over centuries. The pharmaceutical industry views plant wealth as a source of income. The modern medicine world perceives plants as a panacea for many ills. Despite this, there is no consolidated strategy to conserve medicinal plants for their use (Purohit and Vyas, 2005). Medicinal plants are a very diverse group of species with very diverse conservation needs. Most of the species occurs in wild and very few under cultivation. Therefore, efficient and safe conservation of these resources can best be achieved through an appropriate combination of different methods.

Conservation of Ecosystem

This can be achieved by careful conservation of selected areas of suitable size. All elements of plant life in such areas should be allowed to regenerate in its own fashion and no interference, encroachment or disturbance of any type is allowed in these. In these selective sites, suitable management practices can be applied so that all elements of ecosystems *i.e.*, plants, mammals, birds, insects etc. are preserved.

Ex-situ Conservation Approach

It requires maintenance of germ-plasm outside their original habitats. The diversity is being conserved in field genebanks, seed gene bank, *in-vitro* banks and cryobanks. The *ex- situ* conservation, however, lacks the natural evolution process that is always operative in nature.

Field Genebanks

This involves maintenance of collections in fields, herbal gardens and botanical gardens. The clones are maintained in vivo in green houses or as field planting. Although field genebank assures ready supply of planting materials, it requires considerable space and high costs.

Botanical Gardens

The conservation of living plants on botanical gardens involves substantial investments in terms of facilities and maintenance. Therefore, the scope has to be limited accordingly.

There are about 1,500 botanical gardens and *arboreta* in the world (Heywood and Heywood, 1991) Needless to say that such gardens can contribute immensely by generating information on seed physiology, reproductive biology and on the techniques of propagation, maintenance and protection.

The main limitation in conserving threatened species in such gardens is that the species suffer from inadequate population/genepool, as they have limited individuals to represent genetic diversity.

Herbal Gardens

These are more specific and specialized mainly on cultivation and maintenance of medicinal plants.

A number of herbal gardens are being maintained at various Universities, States and the private sector. Gupta (1993) has reported that the Ministry of Agriculture (Horticulture Division), Govt. of India has sanctioned funds to set up 16 medicinal plants gardens in State Agriculture Universities (SAUs) and Research Institutes all over the country. These gardens will collect and maintain diversity in 100 economically important plants and 35 endangered species in their respective regions. Such herbal gardens not only help in conservation of important species but also create public awareness by holding short-term training programmes for students, farmers and growers. The National Bureau of Plant Genetic Resources (NBPGR), New Delhi under the Indian Council of Agricultural Research (ICAR) and its regional stations at Bhowali, Shimla, Shillong and Thrissur are actively engaged in collection and conservation of medicinal plants.

Seed Genebanks

A major global emphasis on conservation of germplasm diversity has been to conserve ex situ in long-term repositories called genebanks. The genebank facilities at NBPGR, New Delhi, include 12 long term modules at -20°C and one medium - term storage module at 10°C. Further, medium-term storage facilities have also been provided in the network at some of its regional stations and at National Active Germplasm Sites (NAGS).

This enhanced capacity of the National Genebank can accommodate 10 lakh seed samples. The workshop on "National Concern for Management, Conservation and Use of Agro biodiversity" held at Shimla (October 15-16, 1997), recommended that the National Genebank be declared as National Heritage. All the participants in the workshop were requested to deposit germ-plasm of different crops (including medicinal plants) in the National Genebank at NBPGR.

In order to further strengthen conservation activities of medicinal plants, a proposal on "Establishment of G-15 Genebank" for Medicinal and aromatic Plants (GEBMAP) was approved for implementation in the 'Summit at Caracas' by the group on South-South Consultation and Co-operation (G15 countries) in November, 1991. As part of this programme, the Department of Biotechnology, Government of India, has identified NBPGR as one of the three centers recognized to carry out survey, collection and conservation of medicinal and aromatic plants in north and north-west Himalayas. Rajasthan and Gujarat and adjoining northern plains are also to be covered by its network system.

Two more gene banks setup under G-15 (GEBMAP) is located at the Tropical Botanic Garden and Research Institute (TBGRI), Thiruvanathapuram and the Central

Institute of Medicinal and Aromatic Plants (CIMAP), Lucknow. The G-15 genebank at NBPGR, New Delhi, became operational in 1993.

'In-vitro' Conservation

The *in-vitro* technology has witnessed tremendous advancement during the past four decades. Every plant part (root, stem, leaf, hypocotyls, axillary buds, embryo, single cell and naked protoplast) can be grown successfully in culture under completely aseptic conditions. With this technology, tropical temperate plant species can be maintained on normal or modified culture media, in normal culture room (25 to 28°C) for a number of years.

At NBPGR, New Delhi, *in-vitro* genebank conserves 'threatened species' of medicinal and aromatic plants in nutrient media. Concerted efforts have been made towards evolving simple, efficient and economical protocols that ensure rapid multiplication and genetic stability of germ-plasm.

Cryo-banks and Conservation of Germplasm

The long- term conservation of vegetaively propagated species, orthodox and recalcitrant seeds can be achieved by using cyropreservation. The reproductive material such as tissues, pollen, embryo, embryonic axis and seed material can be preserved in liquid nitrogen (-196°C) and vapour phase (-150°C) NBPGR has created a facility of this kind where 2.5 lakhs of germplasm lines can be stored through this technique.

DNA Conservation

DNA can be used for conservation of threatened species. Preservation can be in the form of extracted DNA or as genomic DNA libraries. Additionally, DNA can be extracted from dead tissues, *i.e.* herbarium specimens, providing information that would be otherwise lost if living tissues are not available.

In-situ Conservation Approach

The *in-situ* conservation, in contrast to the *ex-situ* approach, allows for continued, dynamic adaptation of plants to the environment. The distribution pattern of the wild PGR in different phyto-geographical regions and the areas of their concentration where rich diversity of wild species still continues to perpetuate are of special significance for undertaking programs on their *in- situ* conservation.

In- situ conservation of wild flora through protected habitats and ecosystems is being implemented by the Ministry of Environment and Forests, Govt. of India. Fourteen Biosphere Reserves have been identified on the basis of survey of data. Seven of them are already being made operational under UNESCO's Man and Biosphere (MAB) project. The NBPGR, New Delhi, has active collaboration with the Ministry of Environment and Forests in establishing the first Biosphere Reserves or *Citrus genetic* resources in Meghalaya.

In-situ approaches have to be different in the case of cultivated and non-cultivated genetic resources. Wild relatives of crop plants require an altogether different approach. Inventories are required to be prepared for each target area using public information/literature, herbarium collections and satellite imageries.

To strengthen the *in-situ* conservation of the medicinal plants resources, Chandel *et al.* (1996) cited three critical factors to be incorporated in all conservation oriented programmes: (*i*) prevention of destruction of populations and their habitats; (*ii*) maintenance and/or enhancement of the population level and variability; (*iii*) prevention of collection and excessive commercial exploitation.

Legal and Administrative Measures

These measures should invariably include the following elements if conservation of biodiversity is to succeed:

1. Strict control on habitat
2. Regulated extraction for trade and industry including bonafide local use
3. Ban on collection of endangered species (under act of law)
4. Maintenance of germplasm/gene pool
5. Encourage large scale cultivation by providing market support

Various central and state acts have been promulgated from time to time which includes:

- ☆ Forest Conservation Act 1980
- ☆ Land Reclamation Act, 1937
- ☆ Environment Protection Act. 1986
- ☆ National Forest Policy, 1988
- ☆ National Conservation Strategy, 1992

Inculcating Awareness among the People for Conservation

It is a proven fact that no policy of the government can succeed without active involvement and participation of the people. Until and unless people are educated, motivated and persuaded to adopt the policies, no tangible results can be achieved. For the success of conservation policies, we need to create awareness among the people through:

- ☆ Educational mechanism
- ☆ Creating awareness about local bio-resources and their importance
- ☆ Training's to provide scientific skills
- ☆ Standardizing of packages of practices
- ☆ Providing knowledge about value additions
- ☆ Keeping consideration of the social changes and economic considerations
- ☆ Inviting people's active participation through Co-operatives, Voluntary Organizations, Mahila Mandals, Youth Clubs etc.

Techniques for Production of Medicinal Plants

The medicinal plants are non-conventional in nature. The cultivators are still not fully aware of their potential. The cultivation, processing and production technology are not available at grass root level. Availability of genuine, high value planting material and facilities for quality evaluation is lacking the other features of the products resembles the typical characteristics of agricultural commodities. The active principles of the plant are generally, secondary metabolites and their biosynthesis, though controlled genetically, is strongly affected by environmental and cultural factors. It is, therefore, advised not to use chemical fertilizers, insecticides and pesticides in cultivation of medicinal plants. While planning for cultivation of medicinal plants, the factors like geographical and edapho-climatic conditions like availability of sunlight (quantity and duration), shade etc. during recent years, and several recommendations have emerged for undertaking cultivation of medicinal plants on marginal/problematic soils like saline, alkaline or wastelands. The cultivation on these soils requires special agronomic management and amelioration before starting the cultivation. The plants have varying degree of tolerance to such soils and the cropping systems shall be adopted accordingly. The agronomic management factors like water, fertilizer, weed and pest management require specific attention for cultivation of medicinal plants. The use of chemical fertilizers should strictly be avoided as this may reduce the value of the herbs. Use of synthetic chemicals for the control of insects and diseases should be avoided due to their great hazard to humans, lower forms of the animal life and also to achieve principle of the medicinal herb. Harvesting of herbs requires careful planning so as to retain their active ingredients (Tewari, 2004)

Bottlenecks in Production of Medicinal Plants

The commercial cultivation of medicinal plants is often found uneconomical mainly because of poor marketing infrastructure. The price of produce is totally controlled by the retail market and is quality based. In absence of testing and standardization facilities, there is wide fluctuation in the price. Small farmers on marginal lands are generally cultivating medicinal plants with low input resources. Besides this, information on several aspects of their agricultural productivity is also not easily available to the farmers, which make their venture a failure. The produce from the medicinal trees require a long gestation period, thereby making the venture unsuccessful. The recommendations regarding agro-technology for medicinal plants are available only in piece meals, for individual crops, totally neglecting the important aspect of cropping system. Simultaneously, the recommendations about the possibility of growing medicinal plants as inter or mixed crop with other medicinal plants or traditional agricultural crops are also not available. Such incomplete recommendations targeting to increase the production of individual plants have limited use only as they are difficult and uneconomical to be followed by the farmers. It is, therefore, necessary to carry out research work with an object to develop the most efficient cropping system, involving medicinal plants, agricultural crops, cash crops and trees for different agro-climatic regions (Tewari, 2004).

Market Scenario of Medicinal Plants

Medicinal plants market in the country is today unorganized due to several problems. Medicinal plants are a living resource, exhaustible if overused and sustainable if used with care and wisdom. Current practices of harvesting are unsustainable and many studies have highlighted depletion of resource base. Many studies have confirmed that pharmaceuticals companies are also responsible for inefficient, imperfect, informal and opportunistic marketing of medicinal plant. There is a vast, secretive and largely unregulated trade in medicinal plants, mainly from the wild, which continues to grow dramatically in the absence of serious policy attention materials where the origin of a particular drug is assigned to more than one plants, due to which, adulteration is common in such cases. All these affect the market both directly and indirectly.

The medicinal and aromatic plants provide the raw material for the production of medicines, flavors, herbal cosmetics, perfumery etc. There are over 1200 aromatic plant species found in the wild, out of which only 40 species are traded. Demand for these herbs are increasing progressively with increase in number of star hotels and multinationals establishing and consumer oriented cosmetics, biscuit and pharmaceutical units. Currently, most of these herbs are grown in large quantities and marketed by France, U.K., Canada. Turkey and U.S.A. It is estimated that Indian consumption alone of these herbs is approximately 200 tonnes per annum, and only about 60 tonnes are produced indigenously. Bulk of these herbs (188 tonnes) is used for culinary purpose and about 12 tonnes are consumed for medicinal and cosmetic preparations. The annual exports of the derivatives from these plants are to the tune of Rs. 600-700 million. The growth and demand of the natural perfumery and cosmetics and flavouring materials in the world is thus one of the fastest growing sectors in the global herbal market. It is estimated that the total world market of essential oils, aroma chemicals and fragrance chemicals in 1978 was to the tune of 3600 million US $ which rose to 7 billion dollars in 1987. It is estimated to cross to at least 20 billion US$ till 2005.

The potential for earning foreign exchange by India from the exports of medicinal plants is estimated to be over US$ 3000 million per annum. The export of these plants and their products has a tremendous potential particularly to advanced countries in Europe, USA and Japan. However, India's present share in the global market is not very significant. Inadequate market support, unregulated exploitation from forests, poor quality consciousness, inadequate availability of planting material are some of the hindrances which prevented rapid development of the sector. However, realizing the vast potential, a number of steps have been taken by the Government for promoting this sector (Singh, 2005b).

According to the report of the World Health Organization (WHO), a large population of the world requires traditional system of medicines, largely plants based to meet their primary health care needs. The EXIM bank of the India, in its report has reported the value of medicinal plants related trade in the India of the order of 5.5 billion US dollar and is growing rapidly. According to WHO, the international market of herbal production is estimated to be US $ 62 billion which is poised to grow to US $ 5 trillion by the year 2050. India's share in the global export market of medicinal plants related trade is less than 0.5 per cent. The Chinese export based on the plant including raw drugs, therapeutics and other is estimated to be around Rs. 18,000- Rs. 22,000 cores. In view of the innate Indian strengths which interalia include diverse eco-systems, technical and farming capacity and a strong manufacturing sector, the medicinal plants area can become a huge export opportunity after fulfilling domestic needs.

Future Thrust

If past achievements are analyzed, it gives a satisfaction that India has done fairly well in a few areas. But in future, it has to face stiff competition in the global arena. The GATT, Intellectual Property Rights (IPRs) and Trade Related Intellectual Property Rights (TRIPs) demand liberalization of economy and globalization of market. In the changing scenario, India has to enhance its research and developmental preparedness in the field of medicinal plants to a level to meet not only our current needs but it has also to emerge as front runner at International level. Our research efforts have to be relevant, need based, demand and market oriented, effective, efficient and visible. In order to achieve these targets, some of the priority areas warrant immediate attention as outlined by Ghosh, (2000).

- ✩ Exploration, identification, characterization, documentation and conservation of natural diversity have to be undertaken on a Nation wide basis.
- ✩ Based on experience gained in the development of medicinal plants, it would be necessary to pay attention to some of the critical issues. Since, a number of organizations are involved in the development of this sector, it would be necessary to have strong networking and coordination among these organizations for effective implementation of the programmes. The market propelled demand of medicinal plants in the country is currently met from natural ecosystems. Out of the approximately 4000 species of plants that have been identified to have medicinal value only very few (20-30) are cultivated. There is urgent need to domesticate and develop better varieties for commercial cultivation of many more medicinal plants that are required in bulk.
- ✩ Major constraints in the development of this sector experienced by the farmers are non-availability of quality planting material of improved varieties. Establishment of adequate number of nurseries for developing good quality planting material of medicinal and aromatic plants would be needed. The mechanism for ensuring quality planting material is essential to promote the production.
- ✩ Marketing is a major problem, particularly in the medicinal plants sector. Realistic assessment of the requirement of raw material by the industry would be necessary before embarking on large-scale cultivation of these crops. Lack of development and extension support in the cultivation and processing, and unorganized marketing. Marketing facilities and price support policy for growers are needed with a tie up or buy back arrangements.
- ✩ Contract farming or buy back arrangements should be encouraged to ensure remunerative price to the farmers. This is more needed for the medicinal plants, which have only pharmaceutical uses.

- ✰ A majority of the medicinal plants used in the Indian systems of medicine come from the forest areas. Since the forest resources are dwindling due to overexploitation, urgent steps need to be taken for *in-situ* conservation of the medicinal plants. This would require reform in forest legislation so that cultivation could be encouraged.
- ✰ Medicinal plants have got little extension support. State Agriculture Universities and Research Institutions may take up this part to train the growers. The linkages with industries, sister organizations in the country as well as abroad need to be developed and strengthened.
- ✰ There should be a drive to locate prospective molecules possessing anti-arthritic, anti-diabetic, anticancer and antiviral properties in our diverse flora. Extensive work has to be taken up to develop drugs, elucidate its chemical and pharmacological action, and prove its effectiveness and safety and also to patent them in favour of our country's economy.
- ✰ There are many medicinal plants like Kuth, Satawar, Van kakri etc. that need to be promoted for cultivation at household level which would provide primary health care. Besides, there are plants like aonla, bael which are horticultural crops having medicinal value, need to be promoted on a large scale. Network of analytical laboratories and other infrastructure like plant health clinics, disease forecasting units etc. need to be established for providing support to the sector.
- ✰ Post harvest processing of medicinal plants with emphasis on quality enhancement and value addition should ascribe paramount importance. Chemists should develop fast, less cumbersome and inexpensive analytical procedures to determine the quality of medicinal plants, crude drugs and medicinal formulations.
- ✰ In order to achieve the target of export and to promote cultivation of medicinal plants convergence and synergy among various programmes have to be established having vertical and horizontal integration in mission mode with end to end approach.

Conclusion

As a result of unscientific exploitation of plant wealth, particularly from the natural resources like the forest areas, the available bio-diversity is under stress and many of the plant species are on the verge of extinction. Efforts have been made by different organizations like the Ministry of Environment and Forests, Department of Indian Systems of Medicine and Homoeopathy, besides the Department of Agriculture and Cooperation, which need integration. The role of National Medicinal Plants Board becomes very important which could strive to bridge the gaps for enhancing the growth of the sector. Marketing and quality control are key areas needing attention. Research efforts through the ICAR-SAU System have helped in evolving agro-techniques and propagation of new varieties. However, commercial cultivation of these plants would require reduced cost of cultivation and proper market tie-up. The Working Group on Horticulture for the Formulation of Plan have projected a high growth in the MAP sector and have recommended for plan investments not only for continuing the ongoing programme but also on new activities like establishment of disease forecasting, leaf analysis labs use of IT etc. to be implemented in Mission Mode approach. Now it is time to harness the potentiality, to achieve visualized growth.

References

Badoni, A.K. (1995). *Conservation of Biological Diversity in Garhwal Himalaya for Herbal Industry* (In Hindi), Society for Himalayan Environmental Research (SHER), Dehradun, India. 92p.

Chandel, K.P.S., Shukla, G. and Sharma, Neelum (1996). *Biodiversity in Medicinal and Aromatic Plants in India: Conservation and Utilization*, NBPGR, New Delhi, pp 1-239.

Chauhan, N.S. (2005). Conservation of endangered species of medicinal and aromatic plants of Western Himalaya. (*in*) *Winter School on Medicinal, Aromatic and Under-utilized Plants Research*, held during 29 September-19 October 2005 at CCSHAU, Hisar, pp 12-20.

Gaur, R.D. (1999). *Flora of the District Garhwal North West Himalaya (With Ethnobotanical Notes)*, Trans Media, Srinagar (Garhwal), India. 811p.

Gautam, P.L. (2005). Biodiversity in medicinal and aromatic plants: Strategies for their conservation. (*in*) *Souvenir on National Consultative Workshop on Medicinal and Aromatic Plants*, held during 25-27 June 2005 at GBPUA&T, Pantnagar, pp 1-8.

Ghosh, S. P. (2000). Medicinal and aromatic plants: Present status and future perspective. *Indian J. Arecanut, Spices Med. Pl.* **2**(4): 145-147.

Gupta, R. (1993). Conservation and utilization of Indian medicinal plants. *Indian J. Pl. Genetic Resources*, 6, 131-138.

Heywood, C. and Heywood, V.H. (1991). *International Directory of Botanical Gardens*, V. Koeltz Scientific Book, Koenigstein, Germany.

Jain, S.K. (2004). *Medicinal Plants*, NBT, New Delhi, India. 216p.

Joshi, N. and Kumar, N. (2000). *Aromatic and Medicinal Plants in Central Himalayas*, DRDO, Pithoragarh, India. 124 p.

Joshi, P., Negi, K.S. and Joshi, V.K. (2004). *Devbhoomi Ki Divya Oshadhiyan*, first edn. Gyanoday, Nainital, India. 198p.

Kaul, M.K. (1997). Medicinal Plants of Kashmir and Ladakh: Temperate and Cold Arid Himalaya, Indus Publishing Company, New Delhi. 173 p.

Maheshwari, R.C. (2005). *Cultivation, Post-Harvesting and Quality Assurance of Medicinal Plants in Uttaranchal*, Centre for Rural Development, IIT, New Delhi, India. 454p.

Polunin, O and Stainton, A. (1985). *Flowers of the Himalaya*, Oxford University Press, Delhi, India.

Prajapati, N.D., Purohit, S.S., Sharma, A.K and Kumar, T. (2003). *A Handbook of Medicinal Plants: A Complete Source Book*, Agrobios, India. 554p.

Purohit, S.S and Vyas, S.P. (2005). Organic farming and medicinal plant cultivation. (*in*) *Souvenir on National Consultative Workshop on Medicinal and Aromatic Plants*, held during 25-27 June 2005 at GBPUA&T, Pantnagar, pp 58-69.

Rastogi, Ram P. and Mehrotra, B.N. (1990). *Compendium of Indian Medicinal Plants* Vol. 1 (1960-1969), Central Drug Research Institute, Lucknow and Publication and Information Directorate, New Delhi. 497 p.

Rastogi, Ram P. and Mehrotra, B.N. (1991). *Compendium of Indian Medicinal Plants* Vol. 2 (1970-1979), Central Drug Research Institute, Lucknow and Publication and Information Directorate, New Delhi. 833 p.

Sah, V.K. and Tripathi, Shailesh (2009). Medicinal plants in North-West Himalaya: Perspectives and issues. (*in*) Bisht, B.S. and Singh, D.P. eds. Agriculture in Uttarakhand Hills. Directorate of Experiment Station, G.B.P.U.A.&T., Pantnagar. pp. 174-183.

Singh, B.M. and Singh, P. (2004). Collection, characterization and conservation of medicinal and aromatic plants. (*in*) *Proceedings of National Seminar*, held during 27-29 February 2004 at CCSHAU, Hisar, p 133.

Singh, B.M. (2005a). Collection and conservation of endangered medicinal and aromatic plants, (*in*) *Winter School on Medicinal, Aromatic and Under-utilized Plants Research*, held during 29 September-19 October 2005 at CCSHAU, Hisar, pp 1-7

Singh, H.P. (2005b). Promotion of medicinal and aromatic plants sector in Uttaranchal: Need of hour. (*in*) *Souvenir on National Consultative Workshop on Medicinal and Aromatic Plants*, held during 25-27 June 2005 at GBPUA&T, Pantnagar, pp 9-15

Tewari, S.K. (2004). Prospects and techniques of production of medicinal plants, (*in*) *Organic Farming in Horticulture, Proceedings of National Symposium*, held during 29-30 August 2003 at CISH, Lucknow, pp 115-6. Pathak R K, Ram Kishun, Khan R M and Ram R A (Eds). CISH, Rehmankhera, Lucknow, India

Tripathi, Shailesh and Sah, V.K. (2010),Underutilized medicinal plants of Uttarakhand hills, (*in*) Peter K.V. ed. Underutilized and Underexploited Horticultural crops, Vol.5, new India Publishing Agency, New Delhi, pp. 425-448

Tyagi, C.S., Yadava, O.P and Verma, P.K. (2005), Conservation of biodiversity of medicinal and aromatic plants, (*in*) *Winter School on Medicinal, Aromatic and Under-utilized Plants Research*, held during 29 September-19 October 2005 at CCSHAU, Hisar, pp 8-11

Chapter 24

Scope of Backyard Poultry Production with Improved Germplasm in N.E. Hill Region

Sunil Doley

Senior Scientist
Divions of Poultry Science,
ICAR Research Complex for NEH Region,
Umroi Road, Umiam – 793 103, Meghalaya

ABSTRACT

The northeastern region of India possesses 36.46 millions poultry, out of which 84.50 per cent is located in rural areas. The total annual egg production of the region is 907.9 millions and contributing 7.46 and 2.25 per cent towards national poultry population and egg production respectively. Out of the total poultry population in the region, more than 60 per cent is indigenous fowls, mostly located in the hands of rural farmers. The region lagged behind in terms of poultry production and productivity mostly because of extensive type of production system with low producing indigenous stocks. As a result, to meet the requirement, the region imports eggs and broilers from other states of the country.The indigenous peoples of the region are mostly non-vegetarian for which meat and eggs play an important role in their food habits. Majorities of the population in the region live in rural areas and depend for their livelihood on agriculture and allied enterprises and poultry farming can play an important role in the economic development of these farming communities.

Introduction

Backyard poultry farming is an age-old farming practice in the rural areas of the region, where few numbers of indigenous poultry birds are reared in the backyard for their self consumption as well as for sell in the hardship. But the production potentials of these indigenous stocks are very low, compared to the improved chicken varieties like Vanaraja, Gramapriya, Giriraja, Girirani, Krishna – J, etc., which are exclusively developed for backyard poultry production under low input systems. These chicken varieties have all the desirable qualities of indigenous chicken combined with better production potential, both for meat and egg. They are hardy, relatively resistant to various poultry diseases; thrive well on locally available feed resources, good scavengers, attractive plumage color and long legs to escape from predators. The performance of Vanaraja and Gramapriya varieties under backyard system of rearing has been studied in northeastern region and found to be better performing bird compared to the desi chickens under backyard farming. Therefore, there is a growing demand for these improved chicken varieties, which closely resembles local chicken in look and taste, and hence consumers are willing to pay higher prices for meat and eggs of improved chicken varieties compared to commercial broilers and layers. It is also low cost, less intensive and self-sustaining farming which is least dependent on external inputs.

Lost Cost Poultry Housing

Low cost poultry house can be constructed with locally available materials like bamboo, wooden planks, thatch, etc. Houses are preferably constructed on elevated areas to keep the floor above the ground level in order to avoid seepage of moisture from surrounding. Orientation of the houses is usually in east-west direction to receive maximum sunlight. It is preferable to construct the floor of poultry houses with pucca, as it is durable, rat proof, easy to clean and disinfect. The height of the poultry house should be 2.1 to 2.4 m at side wall and 3.0 to 3.3 m at the centre to provide sufficient slope on either side. The locally available roofing materials like thatches or tins, asbestos sheets may be used for construction. The overhang of roof should be sufficient (1 to 1.5 m) to provide full protection from entering rain water. The $2/3^{rd}$ portion of the side wall should be fitted with solid material from floor level and remaining $1/3^{rd}$ portion may be fitted with crossed split bamboo or wire netting for open air ventilation.

Brooding Management

The proper care of baby chicks with provision of supplemental heat to maintain their body temperature from 0-6 weeks of age is very important and is known as brooding management. There are two types brooding management:

Artificial Brooding

In artificial brooding, an artificial source of heat is provided to warm the baby chicks to maintain their body temperature. Hover type of brooder can easily be fabricated locally with wooden planks fitted with bulbs as provision of heat and light sources. The standard temperature requirement in the brooder during first week of brooding period is 35°C, which has to be reduced @2.8 °C per week, till the end of 6th week up to 21 °C. Moreover, where the provision of electricity is not available in the remote areas, a Bukhari or Charcoal Angathi, with the provision of kerosene light during night hours may be used as the source of artificial heat.

Brooder house and equipments like brooders, waterers, feeders, etc. should be thoroughly cleaned, disinfected, and fumigated. Overnight fumigation of house should be done before 24 hours and should be ventilated in at least for 3-4 hours to remove traces of poisonous gas before placement of chicks in the house. Spread litter material like dried sawdust, paddy husk, wood shavings etc. on the floor uniformly up to 5-10 cm thickness.

Brooder guard is provided to prevent straying away of baby chicks from the sources of heat, feed and water and also to protect from air drafts. A cardboard or metal sheet of 0.45 m in height should be placed in circular manner at a distance of 0.85 to 0.90 m from the edge of brooder. For the first 7-10 days, spread old news papers over the litter material in order to prevent the chicks from eating litter material. Plenty of clean drinking water with supplementation of electrolyte and vitamin B-complex is recommended for first 3-4 days and also ground maize or starter mash should be spread on the paper or egg trays. The floor space requirement under the brooder is 50-66 cm^2 floor space/chick.

Natural Hatching and Brooding

The improved chicken varieties like Vanaraja, Gramapriya, Giriraja, Girirani, Krishna – J, etc. are lacking broodiness character for which they do not sit on eggs to hatch their chicks. On the other hand, desi hens are very good sitter for hatching of their chicks. Therefore, the broody desi hen can be utilized for hatching of fertile eggs of improved chicken varieties for multiplication chicks at farmers' field. A broody hen can comfortably take care of hatching and brooding of 12-15 chicks. The broody hen and day old chick should be provided brooding nest and protection in the night, although they may be allowed to take out newly hatched chicks in daytime for foraging of food with proper care to protect them from predators.

Grower and Layer Management

The ranges of recommended floor, feeder and water space per bird in the poultry house are given below which can be monitored accordingly during the growing and laying periods.

Table 24.1

Sl.No.	*Age in Weeks*	*Space*		
		Floor (cm^2)	*Feeder (cm)*	*Water (cm)*
1	Growing period (6-18 weeks)	900-1900	7-10	1.5-2.5
2	Laying period(>18 weeks)	1800-2200	12-15	2.5

Lighting

There is no need of providing artificial light during growing period and during laying period they should be provided 15-16 hours of light per day. The extra light required may be provided either in the evening or early morning in the poultry house which will encourage egg production.

Feeding Management

Feed alone accounts for 70 per cent of the total expenditure in commercial poultry production system. Therefore, low cost formulation of poultry ration by incorporating locally available low cost feed ingredients is very important to minimize the cost on feeding in the backyard poultry production system. Household waste, crop by products, broken grains, etc. may be offered. Moreover, locally available feed ingredients like Job's tear grain (corticated and decorticated), broken rice, rice bean fodder and ground nut straw can be included in the ration or may be supplemented depending upon their nutrient requirement in addition to allow for scavenging in the day time, as some of their nutrient requirement can be met from grazing on tender grasses, seeds of weeds, insects, snails, etc in the backyard. During egg laying period, pieces of shell grit or limestone may be supplemented to meet extra calcium requirement for egg production.

Nutrient Requirements in the Ration of Improved Poultry Germplasms

Table 24.2

Sl.No.	*Types of Feed*	*Growth Stage (Weeks)*	*Crude Protein (per cent)*	*M.E. (Kcal/Kg)*
1	Chick starter	0 to 8	20	2800
2	Grower feed	9 to 18	16	2600
3	Layer feed	19 to 72	18	2650

Health Management

The following general managerial practices should be followed to prevent the occurrence of diseases in the backyard poultry production systems.

1. Procure disease free, healthy chicks from reliable hatchery.
2. Poultry houses and equipments like feeders, waterers and hovers should be thoroughly cleaned and disinfected.

Table 24.3: Recommended Vaccination Schedule for Backyard Poultry

Diseases	*Name of the vaccine*	*Age*	*Dose*	*Route*
Marek's disease	HVT MD vaccine	Day old	0.2 ml	S/C or I/M
Ranikhet disease	F-1/Lasota/	4-7 days	1 drop	Eye and nostril
Gambaro disease	Intermediate plus	14-18 days	–	Drinking water
Ranikhet disease	F-1/Lasota/	35 days	1 drop	Eye and nostril
Fowl pox	Chicken embryo adapted	6-7 weeks	0.5 ml	Wing stab method
Ranikhet disease	R2B strain killed vaccine	8-10 weeks	0.5 ml	S/C or I/M

Note: Vaccination schedule varies from place to place depending on out break of particular diseaseas.

3. Optimum floor space should be provided and overcrowding should be avoided.
4. Litter should be kept dry and houses should be well ventilated.
5. Provide plenty of fresh and clean drinking water.
6. Diseased and dead birds should be immediately removed.
7. Birds should be vaccinated and de-wormed at regular intervals as per the direction of veterinarian.

References

Banerjee, G.C. *Poultry*. Oxford and IBH Publishing Company, New Delhi.

Gueye, E.F. and Van'T Hooft, K. (2002). Networking for family poultry development. *LEISA Magazine*, 18(1): 36.

Huque, Q.M.E. (1996). Improving skills of the small farmers in poultry management. In: *Proceedings 20thWorld's Poultry Congress*, New Delhi, India, Vol. 1, pp. 47-60.

Information dissemination for family poultry: *E.F. GuèyeWorld's Poultry Science Journal*, Vol. 65, March 2009 123.

World's Poultry Science Association (2009). *World's Poultry Science Journal*, Vol. 65, March 2009.

Chapter 25

Sustainable Development through Modern Agricultural Techniques

Teena Rani[1] and Manish Kumar[2]*

[1]Department of Molecular Biology and Biotechnology, CCS HAU, Hisar, Haryana

[2]Veterinary Surgeon, Department of Animal Husbandry and Dairying, Haryana

ABSTRACT

Until about four decades ago, crop yields in agricultural systems depended on internal resources, recycling of organic matter, built-in biological control mechanisms and rainfall patterns. Agricultural yields were modest. Most of the labor was done by the family with occasional hired help and no specialized equipment or services were purchased from off-farm sources. The vital importance of food to physical, economic and cultural development, together with the importance of efficient, sustainable production makes modern techniques crucial–in fact, there is strong evidence that only such approaches have any significant chance of meeting the world's basic food needs in the next few decades. The aim of using modern agriculture techniques is to produce abundant food without depleting the earth's resources or polluting its environment. Conventional 20th-Century agriculture took industrial production as its model and verti-cally-integrated agri-business was the result. The industrial approach, coupled with sub-stantial government subsidies, made food abundant and cheap. Producers and researchers are annually increasing the pace of improve-ments in agriculture production with improved techniques and machinery. In modern agricultural systems, farmers believe that they have much more central roles and are eager to apply technology and information to control most components of the system. In contrast to the isolation inherent in traditional arrangements, modern agriculture tends to see its success as dependant on linkages–access to resources, technology, management, investment, markets and supportive

* *Corresponding Author:* E-mail: drtinamehta@gmail.com

government policies. Because each farm is differ-ent, there's no single formula for sustainable success, but modern agricultural principles are good places to begin learning what it will take. Some of the key consider-ations for sustainable development are knowledge about markets, adding value to products, building soil structure and fertility, selection of cropping pattern as per the need of crop and the land, sowing disease resistant crop varieties, scientific use of growth regulators to promote plant growth, using modern irrigation techniques in water scarcity areas, protecting water quality on and beyond the farm, manage pests ecologically, scientific use of pesticides, maximizing biodiversity on the farm, harvesting with minimum yield loss and proper storage of the produce.

Keywords: Agriculture, Crop, Fertilizer, Irrigation, Technology, Yield.

Introduction

The word *agriculture* is the English adaptation of Latin *agricultūra*, from *ager*, "a field", and *cultûra*, "cultivation" in the strict sense of "tillage of the soil". Thus, a literal reading of the word yields "tillage of a field / of fields". Agriculture (also called farming or husbandry) is the cultivation of animals, plants, fungi and other life forms for food, fiber, and other products used to sustain life. Agriculture was the key implement in the rise of sedentary human civilization, whereby farming of domesticated species created food surpluses that nurtured the development of civilization. Technology can be termed as the application of knowledge for practical purposes. In this context, technology allows people to become more efficient or to do things that were not possible before.

India is primarily an agricultural country dating back to more than ten thousand years. Today, India ranks the second largest in agricultural output worldwide. Agriculture contributes approximately 33 per cent of the Gross Domestic Product and round about 70 per cent of Indian population is engaged in agriculture sector. Most of the agricultural outputs including wheat, tea, coffee, cotton etc. are exported to foreign countries contributing about 8.56 per cent of India's exports. About 43 per cent of geographical land is used for agricultural activity. India accounts world's number one country in sugarcane and stood second in rice output.

The history of agriculture dates back thousands of years and its development has been driven and defined by greatly different climates, cultures, and technologies. However, all farming generally relies on techniques to expand and maintain the lands suitable for raising domesticated species. For plants, this usually requires some form of irrigation, although there are methods of dry-land farming; pastoral herding on rangeland is still the most common means of raising livestock. In the developed world, industrial agriculture based on large-scale monoculture has become the dominant system of modern farming, although there is growing support for sustainable agriculture (*e.g.* permaculture or organic agriculture).

Until the Industrial Revolution, the vast majority of the human population labored in agriculture (Altieri, 1992). The type of agriculture developed was typically subsistence agriculture in which farmers raised most of their crops for consumption on farm and there was only a small portion left over for the payment of taxes, dues or trade. In subsistence agriculture, cropping decisions are made with an eye to what the family needs for food and to make clothing and not the world market place. Development of agricultural techniques has steadily increased agricultural productivity and the widespread diffusion of these techniques during a time period is often called an agricultural revolution. A remarkable shift in agricultural practices has occurred over the past century in response to new technologies and the development of world markets. This also led to technological improvements in agricultural techniques such as the Haber-Bosch method for synthesizing ammonium nitrate which made the traditional practice of recycling nutrients with crop rotation and animal manure less necessary.

Synthetic nitrogen along with mined rock phosphate, pesticides and mechanization has greatly increased crop yields in the early 20th century. Increased supply of grains has led to cheaper livestock as well. Further, global yield increases were experienced later in the 20th century when high-yielding varieties of common staple grains such as rice, wheat and corn (maize) were introduced as a part of the Green Revolution. The Green Revolution exported the technologies (including pesticides and synthetic nitrogen) of the developed world to the developing world. In 1798, English philosopher Thomas Malthus famously predicted that the Earth would not be able to support its growing population but technologies such as the Green Revolution have allowed the world to produce a surplus of food.

The cereals rice, corn and wheat provide 60 per cent of human food supply (Matson *et al.,* 1997). Between 1700 and 1980, the total area of cultivated land worldwide increased 466 per cent and yields increased dramatically, particularly because of selectively bred high-yielding varieties, fertilizers, pesticides, irrigation and machinery (Matson *et al.,* 1997). For example, irrigation increased corn yield in eastern Colorado by 400 to 500 per cent from 1940 to 1997 (Matson *et al.,* 1997). Breeding programs have developed highly specialized animal, plant and poultry varieties thus increasing production efficiency. The development of genetic engineering has given rise to genetically modified transgenic crops and to a lesser degree livestock that possess a gene from an unrelated species that confers a desired quality. Such modification allows livestock to be used as "factories" for the production of growth hormone and other substances.

The last century has seen the intensification, concentration and specialization of agriculture, relying upon new technologies of agricultural chemicals (fertilizers and pesticides), mechanization and plant breeding (hybrids and genetically modified crops). With the rapid rise of mechanization in the late 19th century and the 20th century, particularly in the form of the tractor, farming tasks could be done with a speed and on a scale previously impossible. These advances have led to efficiencies enabling certain modern farms in the United States, Argentina, Israel, United Kingdom, Germany and a few other nations to output volumes of high-quality produce per land unit at what may be the practical limit. In the past few decades, a move towards sustainability in agriculture has also developed, integrating ideas of socio-economic justice and conservation of resources and the environment within a farming system. This has led to the development of many responses to the conventional agriculture approach, including organic agriculture, urban agriculture, community supported agriculture, ecological or biological agriculture, integrated farming and holistic management as well as an increased trend towards agricultural diversification.

Modern science has also revolutionized food processing; refrigeration, for example, has made possible the large meatpacking plants and shipment and packaging of perishable foods. Urbanization has fostered the specialties of market gardening and truck farming. Harvesting operations have been mechanized for almost every plant product grown. In the developed nations, the family farm is disappearing, as industrialized farms, which are organized according to industrial management techniques can more efficiently and economically adapt to new and ever-improving technology, specialization of crops and the volatility of farm prices in a global economy. In Third World countries, where small farms using rudimentary techniques, still predominate.

However, concerns have been raised over the sustainability of intensive agriculture. Intensive agriculture has become associated with decreased soil quality in India and Asia and there has been increased concern over the effects of fertilizers and pesticides on the environment, particularly as population increases and food demand expands. Further, industrialization leads to the use of monocultures, when one cultivar is planted on a large acreage. Because of the low biodiversity, nutrient use is uniform and pests tend to build up, necessitating the greater use of pesticides and fertilizers (Acquaah, 2002). Integrated pest management (IPM), which has been promoted for decades and has had some notable successes has not significantly affected the use of pesticides because policies encourage the use of pesticides and IPM is knowledge-intensive (Matson *et al.*, 1997). So there is need for scientific and judicious use of modern agricultural techniques for sustainable agricultural development.

To increase our food production modern agricultural techniques can be used like sowing good quality seeds, improve the methods of sowing making the soil rich in nutrients and use better techniques for harvesting the crops. Some of the agricultural practices which scientists have developed are explained here.

Preparation of Soil

This is an important practice which helps to enrich the soil and make it more fertile and aerated. It involves addition of manure followed by turning, loosening and leveling of the soil, using agricultural implements like spade, plough or mechanical farm implements. Prior to planting, the soil needs to be prepared, usually by some form of tillage or chemical burn-down to kill the weeds in the seedbed that would crowd out the crop or compete with it for water and nutrients. Tillage methods can be divided into three major categories, depending on the amount of crop residue they leave on the surface. Residue slows the flow of runoff that can displace and carry away soil particles.

- ✩ *Conventional Tillage*: Until the last decade or so the standard tillage practice for corn was use of the moldboard plow for primary tillage followed by several secondary tillages and mechanical cultivation after the crop was up. Now about two-thirds of row crops are planted without use of the moldboard plow (Allmaras *et al.*, 2000) and mechanical cultivation is often limited to one or no operations.

- ☆ *Reduced Tillage* is usually done with a chisel plow and leaves 15 per cent to 30 per cent residue coverage on the soil.
- ☆ *Conservation Tillage* leaves at least 30 per cent residue coverage on the soil. Conservation tillage methods include no-till, where no tillage is done at all and seeds are placed directly into the previous season's crop residue; strip-till, in which only the narrow strip of land needed for the crop row is tilled; ridge till; and mulch till.

Tillage is a labour-intensive activity in low-resource agriculture practiced by small land-holders and a capital and energy-intensive activity in large-scale mechanized farming (Lal 1991). For any given location, the choice of a tillage practice will depend on one or more of the following factors (Lal 1980; Unger 1984) (Table 25.1).

Table 25.1. Factors Affecting Choice of Tillage Practice

A. Soil factors	**B. Climatic factors**
1. Relief (slope)	1. Rainfall amount and distribution
2. Erodibility	2. Water balance
3. Erosivity	3. Length of growing season
4. Rooting depth	4. Temperature (ambient and soil)
5. Texture and structure	5. Length of rainless period
6. Organic-matter content	
7. Mineralogy	
C. Crop factors	**D. Socio-economic factors**
1. Growing duration	1. Farm size
2. Rooting characteristics	2. Availability of a power source
3. Water requirements	3. Family structure and composition
4. Seed	4. Labour situation
	5. Access to cash and credit facilities
E. Other Government policies	**F. Objectives and priorities**

The effect of tillage systems on crop yield is not uniform with all crop species, in the same manner as various soils may react differently to the same tillage practice. Results presented by Nicou and Charreau (1985) (Table 25.2) show the effect of tillage on yields of various crops in the West African semi-arid tropics. Cotton showed the smallest yield increase with tillage within the range of crops tested.

Herbicides are used in all these methods to kill weeds. In no-till systems, the herbicide is applied directly on last season's crop residue. In the other methods, some soil preparation takes place before the herbicide is applied. A common myth is that more herbicide is used with conservation tillage methods but in fact farmers rely on herbicides for weed control under all tillage systems, and the amount used is more or less independent of tillage method.

Table 25.2: Effect of Tillage on Crop Yields in the West African Semi-Arid Tropics

Crop	*Number of Annual Results*	*Yield (kg ha⁻¹)*		*Yield Increase (per cent)*
		Control	*With Tillage*	
Millet	38	1558	1894	22
Sorghum	86	1691	2118	25
Maize	31	1893	2791	50
Rice	20	1164	2367	103
Cotton	28	1322	1550	17
Groundnut	46	1259	1556	24

Equipment Used for Soil Preparation

- *Tractor* - a traction machine that provides mechanical, hydraulic, and/or electrical power to implements to perform a wide range of crop production and handling operations. Tractors are most often used to perform drawbar work (pulling equipment through the field) and PTO (power take-off) (power to rotate equipment components) work.
- *Plow* - an implement used to perform primary tillage. A number of types of plows are in common use including the moldboard plow, chisel plow and disk plow.

  The *moldboard plow* has a large frame that is equipped with a series of "bottoms" each of which consists of a steel coulter to slice through residue followed closely by a steel share that cuts the soil and an attached moldboard that is used to raise and turn over the cut "slice" of soil.
- *Disk plow* works in a similar manner to laterally displace and invert soil through the use of concave steel disk blades.
- *Chisel plow* uses curved shanks to penetrate and "stir" the soil without inverting a soil layer. Chisel plows cause less residue disturbance than moldboard plows and are often used in conservation tillage systems.
- *Disk Harrow (or Disk)* - an implement that uses steel blades to slice through crop residues and soil. Disk blades are mounted in groups or gangs that rotate as they move forward through the soil. Front gangs move soil toward the outside of the disk while rear gangs move soil back toward the center of the disk. A disk can be used for primary or secondary tillage.
- *Field Cultivator* -an implement used to perform secondary tillage operations such as seedbed preparation and weed eradication. Field cultivators are equipped with steel shanks that are typically spring mounted to permit the shank to move within the soil and shatter clods.

Efficient machinery helps in increasing productivity by about 30 per cent (Table 25.3) besides, enabling the farmers to raise a second crop making the agriculture attractive. Raising more crops with high productivity is a path for meeting the future food requirement of population. Development and introduction of high capacity, precision, reliable and energy efficient equipment is the need for judicious use inputs.

Table 25.3: Economic Advantage of Mechanization in Per cent

Effect of Mechanization		*Per cent Increase*
Increase in productivity		12-34
Seed-cum-fertilizer drill facilitates	Saving in seeds	20
	Saving in fertilizer	15-20
Enhancement in cropping intensity		5 - 22
Increase in gross income of the farmers		29-49

Source: Report of the Sub-Group on Agricultural Implements and Machinery for Formulation of 9th Five Year Plan, Govt. of India.

The contribution of different power sources to the total power changed with time (Table 25.4). The share of agricultural workers continuously declined since 1981 and expected to be only 5.09 per cent by 2011-12 and that of draught animal power from 27.23 per cent to 6.37 per cent in same period. The increase in power has been mainly through introduction of tractors, whose contribution has increased from 7.5 per cent in 1971 to 51.08 per cent in 2011-12.

Table 25.4: Contribution of Different Power Sources in India

Year	*Agril Worker (per cent)*	*Draught Animal (per cent)*	*Tractor (per cent)*	*Stationary Engine (per cent)*	*Total Power KW/ha*
1981-82	10.92	27.23	19.95	41.57	0.471
2005-06	5.77	8.02	46.70	41.12	1.502
2011-12*	5.09	6.37	51.08	37.46	1.910

* Estimated

Source: http://www.unapcaem.org/Activities per cent 20Files/A09105thTC/PPT/in-doc.pdf

Leveling of farm land is also a step in preparation of soil in some crops. Traditional methods of leveling land are cumbersome, time consuming and expensive, so more and more farmers are turning to modern methods to level the land. One such method is *Laser Land Leveling*, a process of smoothing the land surface (± 2 cm) from its average elevation using laser-equipped drag buckets. This practice uses large horsepower tractors and soil movers that are equipped with global positioning systems (GPS) and / or laser-guided instrumentation so that the soil can be moved either by cutting or filling to create the desired slope / level. This technique is well known for achieving higher levels of accuracy in land leveling and offers great potential for water savings and higher grain yields.

The major environmental concern related to soil preparation is erosion. Soil erosion is a natural process that occurs when the actions of water and/or wind cause top soil to be removed and carried elsewhere. Best management practices to reduce erosion are:

- ✰ *Conservation tillage*: It leaves at least 30 per cent residue cover on the ground. This simple, low-cost practice can have a huge impact on the amount of soil eroded. Because of energy savings and obvious improvements in soil quality that can result from conservation tillage, it has been widely adopted.
- ✰ *Contour farming and strip cropping*: It is the practice of planting along the slope instead of up-and-down slopes and planting strips of grass between row crops.
- ✰ *Cover crops*: These are crops such as rye that grows in late fall and provide soil cover during winter. By providing a cover to the soil, winter soil erosion from both air and water can be greatly reduced.
- ✰ *Grassed waterways*: It protects soil against the erosive forces of concentrated runoff from sloping lands. By collecting and concentrating overland flow, waterways absorb the destructive energy that would otherwise cause channel erosion and gully formation.
- ✰ *Terraces*: These are structural practices that can reduce erosion by holding back the water and routing it along a channel at a lower velocity to where it can be safely discharged, usually into a grassed waterway.
- ✰ *Windbreaks*: These are the best way to protect soil from wind erosion. They can be in the form of rows of shrubs or trees.
- ✰ *Grass barriers*: They can prevent wind erosion by slowing the wind.

Seed Treatment

In the modern seed industry a lot of attention is paid to protecting plants from different kinds of diseases and pests. This is sometimes accomplished by breeding varieties that are resistant to diseases thereby protecting the crop through its growth cycle. In India, 45 per cent of the pesticides are applied (David, 2008) in cotton alone. Pesticide load in crop ecosystem has culminated in many undesirable effects such as resistance, resurgence, residues *etc.*, disturbing the agro-ecosystem. Sprays and soil application of pesticides are costly and cumbersome to adopt. So it is imperative to find out an eco-friendly and need based use of chemical pesticides as a component of integrated pest management (IPM). Seed treatment is an easy, economic and feasible method for pest control (Mote and Shah, 1993; Murugesan and Annakkodi, 2007). It protects against insect pests and is eco-friendlier to bio control agents like coccinellids and chrysopids under field condition (Murugesan and Annakkodi, 2007).

The new seed varieties that are resistant to diseases give a very good crop. These resistant varieties are developed by biotechnology and known as genetically modified (GM) crops or transgenic/biotech crops. By 2007, biotech crops were grown in 18 countries and were estimated to increase farm income by more than $10 billion. During the decade 1996-07, these increases were valued at more than $44 billion

(Table 25.5). According to ISAAA's (International Service.for the Acquisition of Agri-Biotech Applications) annual report 2009, GM crops occupy 134 million hectares. This is out of a total area of global agricultural land of over 4.9 billion hectares (FAO, 2009).

Table 25.5: Global Farm Income Benefits from Biotech Crops, 1966-07

Global Farm Income Benefits from Growing Biotech Crops, 1996-07, million US $				
Technology	*Increase in Farm Income*		*2007 Income Benefit*	
	2007	*1996-07*	*Share Country Value Per cent**	*Share Global Value Per cent*
Herbicide tolerant soybeans	3935	21814	7.2	6.4
Herbicide tolerant maize	4421	508	0.7	0.4
Herbicide tolerant cotton	25	848	0.1	0.1
Herbicide tolerant canola	3461	439	7.65	1.4
Insect resistent maize	2075	5674	3.2	1.9
Insect resistent cotton	3204	12576	16.5	10.2
Others	54	209	NA	NA
Total	10081	44068	6.9	4.4

* Share of value of production in adopting countries.

Source: Brookes, Graham and Barfoot, Peter, GM Crops: Global Socio-economic and Environmental Impacts, 1996-2007

However, when resistant varieties are not available or the level of resistance is not sufficient under high disease and pest pressure, the grower may decide to use crop protection methods *i.e.*, applying plant protection compounds (chemicals or biological material). These compounds can be used in two ways, either as an application to the soil or sprayed directly on the plant. Such crop protection methods may require the use of significant quantities of active substances.

An excellent alternative is to apply crop protection compounds directly to the seed before sowing. Such a treatment to the seed requires the use of very small quantities of active substances per unit area of land and is a very effective and targeted method of controlling pests and diseases. These seeds are treated by dipping them in certain chemicals like cerosan or agrosan. These chemicals do not allow the microorganisms to attack the seeds and damage them.

Other seed disease prevention measures include:

- ✰ Crop rotation- creating disease-suppressive soil/compost, habitats for beneficial pest predators.
- ✰ Appropriate planting dates, soil temperature and moisture.
- ✰ Selecting disease-resistant varieties.
- ✰ Cleaning and processing methods that control disease.

- ✰ Compost and vermi-compost teas.
- ✰ Biodynamic treatments.
- ✰ Herbal treatments.
- ✰ Hot water bath.
- ✰ Disinfectants

Preparing the Seedbed and Care of the Seedlings

A *seedbed* or *seedling bed* is the local soil environment in which seeds are planted. Often it comprises not only the soil but also a specially prepared cold frame, hot bed or raised bed used to grow the seedlings in a controlled environment into larger young plants before transplanting them into a garden or field. A seedling bed is used to increase the number of seeds that germinate.

In certain crop plants like paddy and some of the vegetables, seeds are not sown directly in the main field. First these seeds are sown in a nursery bed. Once they grow to a certain age they are transferred and planted in the main field. These small plants are called *seedlings*. All weeds or unwanted plants in the field must be removed. This is done by spraying chemicals like Parathion, Sevin, Dimecron and Rogor on the seedlings.

The soil of a seedbed needs to be loose and smooth and without large lumps. These traits are needed so that seeds can be planted easily and at a specific depth for best germination. Large lumps and uneven surface would tend to make the planting depth random. Many types of seedlings also need loose soil with minimal rocky content for best conditions to grow their roots (For example, carrots grown in rocky soil will tend not to grow straight).

Seedbed preparation in farm fields often involves secondary tillage via harrows and cultivators. This may follow primary tillage (if any) by moldboard plows or chisel plows. No-till farming methods avoid tillage for seedbed preparation as well as later weed control. Seedbed preparation in gardens often involves secondary tillage via hand tools such as rakes and hoes. This may follow primary tillage (if any) by shovels, picks or mattocks. Rotary tillers provide a powered alternative that takes care of both primary and secondary tillage.

The preparation of a seedbed may include:

1. *The removal of debris*: Insect eggs and disease spores are often found in plant debris and so this is removed from the plot. Stones and larger debris will also physically prevent the seedlings from growing.
2. *Leveling*: The site will have been leveled for even drainage.
3. *Breaking up the soil*: Compacted soil will be broken up by digging. This allows air and water to enter and helps the seedling penetrate the soil. Smaller seeds require a finer soil structure. The surface the soil can be broken down into a fine granular structure using a tool such as a rake.
4. *Soil improvement*: The soil structure may be improved by the introduction of organic matter such as compost or peat.

5. *Fertilizing*: The nitrate and phosphate levels of the soil can be adjusted with fertilizer. If the soil is deficient in any micro nutrients, these too can be added.

The seedlings may be left to grow to adult plants in the seedbed, perhaps after thinning to remove the weaker ones or they may be moved to a border as young plants.

Cropping Pattern

Cropping systems vary among farms depending on the available resources and constraints, geography and climate of the farm, government policy, economic and social pressures and the philosophy and culture of the farmer (Acquaah, 2002).

Transplanting

The process of removing the seedlings from the nursery bed and planting them in the main field is called *transplanting*. These are sowed at proper distance from each other. The main field must be ploughed and manured before transplanting. Generally rice and vegetables like tomato and brinjal are sown by transplanting. Transplanting of seedlings is a very important practice. This enables us to select good and healthy seedlings and get a better crop. Besides, when we transplant seedlings, their roots are able to go deep into the soil and get more nutrients. When seedlings get good food, they grow into healthy plants and give a better yield.

Monoculture

When one crop is grown alone in a field, it is called a monoculture. Monoculture makes it easier to cultivate, sow seed, control weeds and harvest, as well as expand the size of the farm operation and improve aspects of profitability and cost.

Multiple Cropping

Growing two to four crops one after the other in a year in the same field is called multiple cropping. It is possible when we make use of crop varieties that grow for a short period of time. However, to get best results there must be a properly managed field. In fact, multiple cropping is the best solution for a country with food problem because same piece of land is used to grow different kinds of crops.

Intercropping

When several crops are grown at the same time in a field is commonly known as intercropping. Multiple cropping and intercropping cropping systems are also known as polycultures.

Niche Farming

Niche farming, in which specialized crops are raised for a specialized market for example heirloom tomatoes or exotic herbs sold to food shops and restaurants.

Shifting Cultivation

Shifting cultivation (or slash and burn) is a system in which forests are burnt releasing nutrients to support cultivation of annual and then perennial crops for a period of several years (Chrispeels and Sadava, 1994). Then the plot is left fallow to re-grow forest and the farmer moves to a new plot, returning after many more years (10-20). This fallow period is shortened if population density grows requiring the input of nutrients (fertilizer or manure) and some manual pest control.

In tropical environments, all of these cropping systems are practiced. In subtropical and arid environments, the timing and extent of agriculture may be limited by rainfall either not allowing multiple annual crops in a year or requiring irrigation. In all of these environments, perennial crops are grown (coffee, chocolate) and systems are practiced such as agroforestry. In temperate environments where ecosystems were predominantly grassland or prairie, highly productive annual cropping is the dominant farming system (Chrispeels and Sadava, 1994).

All the above mentioned practices are meant to ensure that plants have a healthy growth, maximize productivity and yield a good crop. Along with these developments in our country we have also brought more and more land under cultivation. The increased cultivation of agricultural crops is in order to meet requirements of a growing population.

Adding Fertilizers

Crops need nutrients like phosphorus, calcium, nitrogen etc. for their growth and pick up these nutrients from the soil. It is very important to add fertilizers to the soil. They provide nutrients to the soil and help to obtain a better crop yield. Modern techniques of soil testing indicate what nutrients are in the soil and what nutrients are needed. After the initial soil test, one should analyze his soil after every three to five years to make sure that his fertilizer programme is on target. The way one uses a fertilizer also depends upon what type of fertilizer is being added to the soil. A fertilizer which contains nitrogen (*nitrogenous fertilizer*) is generally given in two or three doses. Other fertilizers are *phosphatic* and *complex fertilizers*. The effectiveness of fertilizer applications depends on soil properties such as texture, organic matter content, drainage and pH.

Soil particles are grouped by size and designated as sand, silt or clay. Sand is the largest particle size and clay is the smallest. Soils contain different proportions of sand, silt and clay that make up the soils' specific texture. In general, soil textures with a greater portion of silt and clay retain more water and nutrients than those soils composed mostly of sand particles.

Soil organic matter also influences soil productivity. In general, organic matter increases both the water and nutrient holding capacity of a soil. It's addition to soil may also provide nutrients as they decompose or increase the tilth of the soil by amending or modifying the soil structure to promote water infiltration and root penetration.

Soil drainage is critical to plant health. The soil's ability to hold water must be balanced with its ability to retain enough oxygen for plant growth. If soil becomes saturated for a prolonged period of time, the oxygen trapped in soil pore space can be rapidly depleted by the plant and soil organisms. When roots cannot get oxygen, the plant's ability to get nutrients and grow is impaired. Poor drainage causes more problems for landscape plants than any other factor.

Soil pH, a measure of acidity, has a significant impact on the plant's ability to use nutrients. If the soil is too acidic (low pH), the pH can be raised by adding lime and if the soil is too alkaline (high pH), the pH can be lowered by adding sulfur.

Necessary Nutrients

Sixteen elements are essential for plant growth. These are divided into two groups based on the amount of each needed by plants. The nine macronutrients (nitrogen, phosphorus, potassium, calcium, magnesium, sulfur, carbon, hydrogen and oxygen) are required in relatively large amounts. Of these, nitrogen, phosphorus and potassium are the elements most commonly added as fertilizer.

The seven micronutrients (iron, manganese, zinc, boron, molybdenum, copper and chlorine) are required in smaller amounts. Iron and manganese are the micronutrients most often deficient in landscape plants. An adjustment in soil pH usually corrects deficiencies of the micronutrients.

Types of Fertilizers

Fertilizers may be divided into two broad categories, *Natural* and *Synthetic*. Natural fertilizers generally originate from unprocessed organic sources such as plants or animals. Synthetic fertilizers are man-made or processed. Synthetic fertilizers can be organic (for example, urea) or inorganic (for example, superphosphate).

Natural fertilizers commonly misnamed "organic" can also contain inorganic ores such as rock phosphate. Most nutrients from living or once-living organisms are not readily available for plant growth because they are bound in organic molecules such as proteins and amino acids and in structures such as cell walls. These nutrients are released only by microorganisms decomposing the organic matter. Some of the commonly used manures are:

Farmyard manure is a mixture of decomposed cattle dung (excreta) and urine, left over fodder (cattle feed) and litter (bedding provided to cattle in the farm).

Compost is manure made from vegetable and animal refuse collected from domestic waste, straw and weed etc., dumped in a deep pit to decompose.

Vermicompost is compost broken down by earthworms. Like fertilizers, manures also add nutrients to soil.

Slow-release fertilizers may be synthetic or natural. Because nutrients are released over an extended period of time, slow-release fertilizers do not have to be applied as frequently as other fertilizer types. Also, higher amounts of slow-release fertilizer can be added at each application without risking injury to plant roots. Slowly released nitrogen is used more efficiently because a higher percentage is absorbed by plants. The higher efficiency of slow-release fertilizers means less nitrogen is available to contribute to pollution of streams and sub-surface water. The primary disadvantage of slow-release fertilizers is higher cost. When analyzed the cost of nitrogen that is actually absorbed by the plant, however, the unit cost is actually less for slow-release materials. Water-soluble or liquid fertilizers are applied either to the soil or foliage. Many water-soluble formulations are available for almost any specific need, from plant starter, high-nitrogen fertilizers to minor element formulations. Chelated iron is used extensively for prevention and control of iron deficiency in azalea, rhododendron and other popular ornamentals.

Signs of Nutrient Deficiency in Plants

- ☆ *Nitrogen (N)*: Stunted growth, small and fewer leaflets, early leaf drop, dark green to blue-green and slightly smaller leaves. Veins, petioles, or lower surface may become reddish-purple, especially when young.
- ☆ *Potassium (K)*: Partial chlorosis of most recently matured leaves in inter-veinal area beginning at tips, followed by necrosis. Older leaves may become brown and curl downward.
- ☆ *Calcium (Ca)*: Death of terminal buds, tip die-back, chlorosis of young leaves, leaves may become hard and stiff. Root injury is the first apparent sign.
- ☆ *Magnesium (Mg)*: Marginal chlorosis on older leaves followed by inter-veinal chlorosis and the margins may become brittle and curl upward.
- ☆ *Sulfur (S)*: Uniform chlorosis of new leaves, older leaves are usually not affected.
- ☆ *Iron (Fe)*: Interveinal chlorosis of young leaves (sharp distinction between green veins and yellow tissue between veins). Older basal leaves greener; exposed leaves blanched.
- ☆ *Manganese (Mn)*: Interveinal chlorosis of young leaves that begin at margins and progresses toward midribs and further followed by necrotic spots.
- ☆ *Zinc (Zn)*: Young leaves may be yellow, small, deformed or mottled with necrotic spots. There may be a tuft of leaves at shoot tips.
- ☆ *Boron (B)*: Terminal growth dies; later growth that develops has sparse foliage. Young leaves may be red, bronzed or scorched. Leaves may be small, thick, distorted or brittle.
- ☆ *Copper (Cu)*: Rosetting of terminal growth may die. Leaf symptoms not usually pronounced but veins may be lighter than blades.
- ☆ *Molybdenum (Mo)*: Cupping of the older leaves, marginal chlorosis followed by interveinal chlorosis.

Some plants once established may not need additional fertilizer to perform well. Silver maple, willow, ligustrum and forsythia are good examples. Because of fibrous root systems, some ornamentals like azalea, dogwood, hemlock and rhododendron are easily damaged by fertilizers. Split applications of water-soluble nutrients or slow-release formulations are recommended for these plants. So, judicious use of fertilizers is necessary.

Timing Fertilizer Applications

Pre-plant Incorporation

The best time to improve soil fertility is before planting because nutrients can be thoroughly mixed into the soil where the plant roots will be. Phosphorus is the most important element in this category because it moves slowly. Lime or sulfur if needed to correct pH levels, is most effective when incorporated the season before planting. This allows time for the chemical reactions that change pH to take place.

Nitrogen can also be added before planting. Inorganic forms, however, are not recommended at this time, because they leach easily, and at high levels may injure the fine roots of newly-planted ornamentals. The best sources of nitrogen in organic forms are composted materials or fish emulsion.

Roots absorb nutrients most efficiently when they are actively growing in late winter/early spring or fall. The best time to apply fertilizer is about two weeks before these periods so that the nitrogen can move into the root zone in time to be available for plant use.

The following are some modern scientific guidelines for nitrogen application as it relates to the stage of plant growth:

- ✰ At planting time nitrogen needs are relatively low but sufficient nitrogen should be supplied to support the new root generation and top growth of the plant. High rates should be avoided because they will trigger excessive top growth that cannot be supported by a limited root system.
- ✰ Maximum growth of landscape plants is usually desired during the first few growing seasons after establishment. To achieve maximum growth constant rates of nitrogen are needed.
- ✰ As the plant begins to mature and the rapid growth rate is no longer needed or desired lower levels of nitrogen are sufficient to maintain plant vigor.

Plants growing in a restricted root zone should not receive large amounts of nitrogen because restricted roots cannot support excessive top growth.

Application Methods

Five methods–liquid injection, drill hole or punch bar, surface application, fertilizer stakes or spikes, foliar spraying and tree-trunk injection or implants –are most recommended. Each serves a specific role depending on the site and plant health. Regardless of the method selected the soil should be moist at the time of fertilizing to prevent fertilizer injury to the plant.

- ✰ *Liquid Injection*: Through liquid injection into the soil, fertilizer solutions are placed in the root zone. This is an excellent method for correcting nutrient deficiencies. Injection sites should be 2 to 3 feet apart depending on pressure and 6 to 9 inches deep. Fertilizing deeper than 9 inches may place the fertilizer below the absorbing roots thus preventing plant use. When using this method in summer or during periods of drought, water should be added to dry soil.

- *Drill Hole or Punch Bar*: A major advantage of the drill-hole system is the opening of heavy, compacted soils which allows air/moisture and fertilizer to move into the soil. The drill holes should be placed in concentric circles or in a grid system around the main stem beginning 3 to 4 feet from the main stem and extending beyond the drip line. Space the holes 2 feet apart and drill them 6 to 9 inches deep. The recommended rate of fertilizer for the area should be uniformly distributed among the holes and is based on the root-zone space under the tree (and not the trunk diameter). The holes can be filled either with organic materials such as peat moss, compost or inorganic materials such as gravel, sand or calcined clay.
- *Surface Application*: Applying fertilizer to the ground surface is as effective as most other methods. It is okay to place fertilizer on top of mulch in landscape beds.
- *Fertilizer Stakes or Spikes*: Fertilizer stakes or spikes that are driven into the soil contain satisfactory fertilizer materials. Unfortunately, the spacing of spikes is such that very little fertilizer comes in contact with the root system. One or two stakes per inch of trunk diameter do not represent adequate fertilizer distribution because lateral fertilizer movement is limited in soil.
- *Foliar Spraying*: Spraying liquid or water-soluble fertilizer on the foliage is best for correcting deficiencies of minor elements, especially of iron and manganese. This method should not be used as a means of providing all the macronutrients required by plants. To correct chlorosis, yellowing of plants caused by nutrient deficiency, several applications during a growing season may be necessary.
- *Tree-Trunk Injection or Implants*: The infusion of liquid or implants of fertilizer is often the best method for correcting iron and manganese problems in landscape trees. This method is especially useful in areas of adverse soil pH, high moisture or where other means of application are not practical. The wound caused by holes made in the trunk root flare will close within a growing season.

The most important thing to remember when making fertilizer decisions is apply only what your plants will use. The care you take to learn about plant needs will pay off in healthy plants and a protected water supply.

Use of Plant Growth Regulators

Plant growth regulators are certain chemicals which regulate the growth of plants. All plants have growth regulators which determine how tall the plant would be, how big its fruit will be etc. We can now add some plant growth regulators like auxins, gibberellins, cytokinins, abscisic acid etc. to get a better yield of crops (Table 25.6).

Irrigation

Irrigation may be defined as the science of artificial application of water to the land or soil. Irrigation is necessary for proper development of plants. Roots fail to develop and penetrate in the dry soil. The crop is irrigated according to its requirement and soil characteristics. Irrigation is essential during the seedling, flowering and grain filling stages of the crop. Rice crop needs standing water.

Table 25.6: Plant Growth Regulators Associated Function(s) and Practical Uses

Class	*Function(s)*	*Practical Uses*
Auxins	Shoot elongation	Increase rooting and flower formation
Gibberellins	Stimulate cell division and elongation	Increase stalk length, flower and fruit size
Cytokinins	Stimulate cell division	Prolong storage life of flowers and vegetables, stimulate bud initiation and root growth
Ethylene generators	Ripening	Induce uniform ripening in fruit and vegetables
Growth inhibitors	Stops growth	Promote flower production by shortening internodes
Growth retardants	Slows growth	Retard tobacco sucker growth

Several methods of irrigation are used today depending on water availability, the type of crop and the financial investment the grower is willing to make. Various types of irrigation techniques differ in how the water obtained from the source is distributed within the field. In general, the goal is to supply the entire field uniformly with water, so that each plant has the amount of water it needs, neither too much nor too little. The modern methods are efficient enough to achieve this goal.

Surface Irrigation

In surface irrigation systems water moves over and across the land by simple gravity flow in order to wet it and to infiltrate into the soil. Surface irrigation can be sub-divided into furrow, borderstrip or basin irrigation. It is often called flood irrigation when the irrigation results in flooding or near flooding of the cultivated land. Historically, this has been the most common method of irrigating agricultural land.

Localized Irrigation

It is a system where water is distributed under low pressure through a piped network in a pre-determined pattern and applied as a small discharge to each plant or adjacent to it. Drip irrigation, spray or micro-sprinkler irrigation and bubbler irrigation belong to this category of irrigation methods.

Drip Irrigation

Drip irrigation, also known as *trickle irrigation*, functions as its name suggests. In this system, water fall drop by drop just at the position of roots. Water is delivered at or near the root zone of plants, drop by drop. This method can be the most water-efficient method of irrigation (Provenzano, 2007) if managed properly, since evaporation and runoff are minimized (Figures 25.1 and 25.2).

In modern agriculture, drip irrigation is often combined with plastic mulch, further reducing evaporation and is also the means of delivery of fertilizer. The process is known as Fertigation.

Deep Percolation

In deep percolation, water moves below the root zone. It occurs if a drip system is operated for too long or if the delivery rate is too high. Drip irrigation methods range from very high-tech and computerized to low-tech and labor-intensive. Lower water pressures are usually needed than for most other types of systems, with the exception of low energy center pivot systems and surface irrigation systems and the system can be designed for uniformity throughout a field or for precise water delivery to individual plants in a landscape containing a number of plant species. Although it is difficult to regulate pressure on steep slopes, pressure compensating emitters are available, so the field does not have to be level. High-tech solutions involve precisely calibrated emitters located along lines of tubing that extend from a computerized set of valves.

Figure 25.1: Drip Irrigation–A Dripper in Action

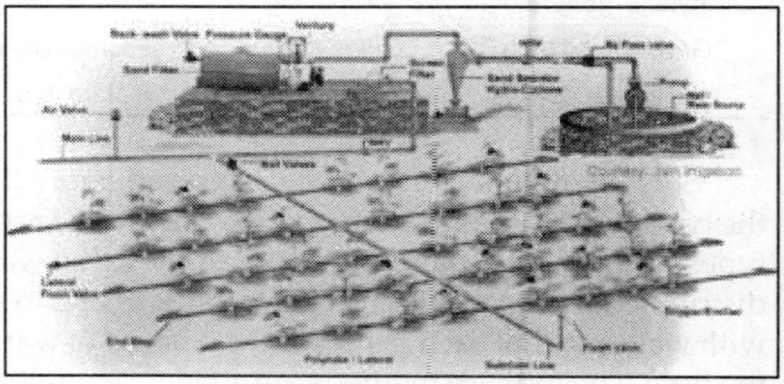

Sprinkler Irrigation

In sprinkler or overhead irrigation water is piped to one or more central locations within the field and distributed by overhead high-pressure sprinklers or guns. A system utilizing sprinklers, sprays or guns mounted overhead on permanently installed risers is often referred to as a solid-set irrigation system (Figure 25.3). Higher pressure sprinklers that rotate are called rotors and are driven by a ball drive, gear drive, or impact mechanism. Rotors can be designed to rotate in a full or partial circle. Guns are similar to rotors, except that they generally operate at very high pressure (275 to 900 kPa) and flows of 3 to 76 L/s usually with nozzle diameters in the range of 0.5 to 1.9 inches (10 to 50 mm). Guns are used not only for irrigation but also for industrial applications such as dust suppression and logging.

Figure 25.3: Sprinkler Irrigation of Blueberries

Sprinklers can also be mounted on moving platforms connected to the water source by a hose. Automatically moving wheeled systems known as traveling sprinklers may irrigate areas such as small farms, sports fields, parks, pastures and cemeteries unattended. Most of these utilize a length of polyethylene tubing wound on a steel drum. As the tubing is wound on the drum powered by the irrigation water or a small gas engine, the sprinkler is pulled across the field. When the sprinkler arrives back at the reel the system shuts off. This type of system is known to most people as a "waterreel" traveling irrigation sprinkler and they are used extensively for dust suppression, irrigation and land application of waste water. Other travelers use a flat rubber hose that is dragged along behind while the sprinkler platform is pulled by a cable. These cable-type travelers are definitely old technology and their use is limited in today's modern irrigation projects.

Center Pivot Irrigation

Center pivot irrigation is a form of sprinkler irrigation consisting of several segments of pipe (usually galvanized steel or aluminum) joined together and supported by trusses mounted on wheeled towers with sprinklers positioned along its length. The system moves in a circular pattern and is fed with water from the pivot point at the center of the arc (Figure 25.4).These systems are found and used in all parts of the world and allow irrigation of all types of terrain.

Figure 25.4: A Small Center Pivot System from Beginning to End

Figure 25.5: Center Pivot with Drop Sprinklers

Most center pivot systems now have drops hanging from a u-shaped pipe attached at the top of the pipe with sprinkler heads that are positioned a few feet (at most) above the crop thus limiting evaporative losses (Figure 25.5). Drops can also be used with drag hoses or bubblers that deposit the water directly on the ground between crops. Crops are often planted in a circle to conform to the center pivot. This type of system is known as LEPA (Low Energy Precision Application). Originally most center pivots were water powered. These are replaced by hydraulic systems and electric motor driven systems. Many modern pivots feature GPS devices.

Wheel Line Irrigation System

A series of pipes, each with a wheel of about 1.5m diameter permanently affixed to its midpoint and sprinklers along its length, are coupled together at one edge of a field (Figure 25.6). Water is supplied at one end using a large hose. After sufficient water has been applied the hose is removed and the remaining assembly rotated either by hand or with a purpose-built mechanism so that the sprinklers move 10m across the field. The hose is reconnected. The process is repeated until the opposite edge of the field is reached. This system is less expensive to install than a center pivot but much more labor intensive to operate and it is limited in the amount of water it can carry. Most systems utilize 4 or 5-inch (130mm) diameter aluminum pipe. One feature of a lateral move system is that it consists of sections that can be easily disconnected. They are most often used for small or oddly shaped fields such as found in hilly or mountainous regions or in regions where labor is inexpensive.

Figure 25.6: Wheel Line Irrigation System

Sub-irrigation

Sub-irrigation also sometimes called *seepage irrigation* has been used for many years in field crops in areas with high water tables. It is a method of artificially raising the water table to allow the soil to be moistened from below the plants' root zone. Often those systems are located on permanent grasslands in lowlands or river valleys and combined with drainage infrastructure. A system of pumping stations, canals, weirs and gates allows it to increase or decrease the water level in a network of ditches and thereby controlling the water table.

Sub-irrigation is also used in commercial greenhouse production, usually for potted plants. Water is delivered from below, absorbed upwards and the excess collected for recycling. Typically, a solution of water and nutrients floods a container or flows through a trough for a short period of time (10–20 minutes) and is then pumped back into a holding tank for reuse. Sub-irrigation in greenhouses requires fairly sophisticated, expensive equipment and management. Advantages are water and nutrient conservation, labor-saving through lowered system maintenance and automation. It is similar in principle and action to subsurface drip irrigation.

Manual Using Buckets or Watering Cans

This conventional system has low requirements for infrastructure and technical equipment but need high labor inputs.

Automatic, Non-Electric Using Buckets and Ropes

Besides the common manual watering by bucket, an automated and natural version of this also exists. Using plain polyester ropes combined with a prepared ground mixture can be used to water plants from a vessel filled with water. The ground mixture would need to be made depending on the plant itself, yet would mostly consist of black potting soil, vermiculite and perlite. This system would (with certain crops) allow to save expenses as it does not consume any electricity and requires only little water (unlike sprinklers, water timers etc.). However, it may only be used with certain crops (probably mostly larger crops that do not need a humid environment. *e.g.* paprikas).

Using Water Condensed from Humid Air

In countries where at night, humid air sweeps the countryside, water can be obtained from the humid air by condensation onto cold surfaces. This is for example practiced in the vineyards at Lanzarote using stones to condense water or with various fog collectors based on canvas or foil sheets.

Harvesting

Harvesting machines have now replaced the back breaking job of hand harvesting with the sickle and scythe. Harvesting machines cut or dig out the plant or its parts as required. The machines gather the plant parts, separate desired parts and eliminate parts not needed. Certain harvesting machines may even load the crop for transport. However, the above mentioned functions of harvesting machines depend on kind of crop, plant parts to be harvested, crop use, stage of maturity etc.

Harvesting methods differ according to the part of the plant to be used. As regards forage crops, the whole plant is cut but for underground crops (eg, groundnuts, roots and tubers), the crop is lifted while the soil sticking to it is removed. With cereals, the crop is first cut either as a whole or partially (ears), threshed and cleaned to separate the grain from the ears and straw. In the latter case two main alternatives exist: separate harvesting and threshing or combined harvesting and threshing.

Attempts have been made since the beginning of the 20th century to devise machines which would both harvest and thresh grain so as to reduce the labour requirements involved. Combine harvesters (combines) which can cut, convey, separate and thresh the grain were the product of this development work. They are in widespread use and have been used already on large grain production schemes in a number of developing countries.

Traditional Threshing

The traditional threshing of rice is generally made by hand: bunches of panicles are beaten against a hard element (*e.g.* a wooden bar, bamboo table or stone) or with a flail. The outputs are 10kg to 30kg of grain per man-hour according to the variety of rice and the method applied. Grain losses amount to 1-2 per cent or up to 4 per cent when threshing is performed excessively late. Some unthreshed grains can also be lost around the threshing area.

Mechanized Threshing

They consist of a threshing device with pegs, teeth or loops and (in more complex models) a cleaning-winnowing mechanism based upon shakers, sieves and centrifugal fan. The rates of work can vary from 2 hours per hectare with a 3-row self-propelled harvester to 5 hours per hectare with a tractor-drawn or -mounted single row unit. Generally speaking, harvest losses range from 3 per cent to 5 per cent but they may be up to 10 per cent -15 per cent under adverse conditions. Depending on the situation, a single-row harvester can be employed effectively on up to 20 hectares or more but the use of a multi-row machine demands several tens of hectares to be economically effective.

Timeliness of harvest is of prime importance. During harvesting season, often rains and storms occurs causing considerable damage to standing crops. Rapid harvest facilitates extra days for land preparation and earlier planting of the next crop. The use of machines can help to harvest at proper stage of crop maturity and reduce drudgery and operation time. Considering these improved harvesting tools, equipment, combines are being accepted by the farmers.

Combined Harvesting and Threshing Methods

Combine–Harvester–Thresher is designed for harvesting threshing, separating, cleaning and collecting grains while moving through the standing crops. Bagging arrangement may be provided with a pick up attachment, it may be self propelled or tractor operated.

Strippers

Because of their size, conventional harvesters and combine harvesters prove unsuitable for many rice growing areas with small family farm holdings. In response to this problem research services have developed small-sized machines for harvesting the panicles without cutting the straw. Such machines are known as strippers. However, the harvested grain has to be threshed and cleaned in a separate thresher. Since harvesting unthreshed produce results in frequent stoppages for emptying the machine, this constitutes the main drawback to the progress of the prototype.

Grain Cleaning

Threshing operations leave all kinds of trash mixed with the grain. They comprise both vegetable (*e.g.* foreign seeds or kernels, chaff, stalk, empty grains etc.) and mineral materials (*e.g.* earth, stones, sand, metal particles etc.) and can adversely affect subsequent storage and processing conditions. The cleaning operation aims at removing as much trash as possible from the threshed grain.

Traditional Cleaning

The simplest traditional cleaning method is winnowing, which uses the wind to remove light elements from the grain.

Mechanized Cleaning

The most rustic equipment is the winnower a fan-originated current of air passes through several superposed reciprocating sieves or screens. It can be either manually powered or motorized; capacities range from a few hundred kilogrammes to several tonnes per hour.

Storage of Agricultural Products

Once harvested, food grains have to be safely stored. They have to be saved from being attacked and eaten up by rodents, birds or insects. Also they have to be protected from spoilage due to improper temperature and moisture in the storage place or due to growth of fungi. Some of the methods to prevent loss and spoilage of agricultural products are as follows:

- ☆ *Drying*: The grains can be dried in the sun or by blowing hot air on them.
- ☆ *Maintaining storage containers*: Godowns or gunny bags or tanks or earthen pots used for storage should be free of the cracks and holes and should be clean.
- ☆ *Chemical treatment*: Spraying or fumigation (insecticide solution converted into fumes) of godowns and containers with insecticides and fungicides should be done before storage. Care should be taken to ascertain that the grains for consumption by human beings are not treated with chemicals poisonous to human beings. Grains are often treated with neem kernel powder or pepper or mineral oil which prevents laying of eggs by insect pests.
- ☆ *Use of improved storage structures*: Modern structures which are airtight, rat proof, moisture proof and can maintain a steady temperature are now used for storage. Few of them are named Pusa bin, Pusa cubicle and Pusa kothar.

Conditions for Storage

When dried to moisture contents below the safe moisture level, cereals and pulses can be stored for periods of a year or more under a wide range of temperatures, provided that during storage the moisture level does not rise and precautions against insects are taken. Insects may still develop at a relative humidity of approximately 35 per cent and temperatures of around 15°C.

Storage Methods

Depending on the weather during the harvesting period and the duration of storage there are four possible combinations:

1. Dry harvesting period and dry storage period.
2. Dry harvesting and wet storage period.
3. Wet harvesting and dry storage period.
4. Wet harvesting and wet storage period.

Conclusion

Sustainable agriculture integrates three main goals - environmental health, economic profitability and social and economic equity. Making the transition to sustainable agriculture is a process. For farmers, the transition to sustainable agriculture normally requires a series of small, realistic steps. Family economics and personal goals influence how fast or how far participants can go in the transition. It is important to realize that each small decision can make a difference and contribute to advancing the entire system further on the "sustainable agriculture development." The key to move forward is the will to take the next step. It is important to point out that reaching towards the goal of sustainable agriculture is the responsibility of all participants in the system including farmers, laborers, policymakers, researchers, retailers and consumers. Each group has its own part to play, its own unique contribution to make to strengthen the sustainable agriculture community. With years of practice in agriculture, there have been new inventions and modern techniques adopted by farmers in agriculture. To spread and encourage knowledge of agriculture among the youth generation, government has even launched new programmes and courses with specialization in agriculture. Modern Agricultural comprises of improved farming techniques and the using of irrigation and high-yield grains resulting into increased production. Modern agricultural systems have been developed with two related goals in mind: to obtain the highest yields possible and to get the highest economic profit possible. Each practice is used for its individual contribution to productivity but when they are all combined in a farming system each depends on the others and reinforces the need for using the others. The work of agronomists, specialists in agricultural production, has been keys to the development of these practices.

References

Acquaah, G. (2002). Agricultural Production Systems. pp. 283-317 in "Principles of Crop Production, Theories, Techniques and Technology". Prentice Hall, Upper Saddle River, NJ.

Allmaras, R.R., Schomberg, H.H., Douglas C.L. and Dao, T.H. (2000). Soil organic carbon sequestration potential of adopting conservation tillage in U.S. croplands. *Journal of Soil and Water Conservation*. 55: 365–373.

Altieri, M.A. (1992). Agroecological foundations of alternative agriculture in California. Agriculture. *Ecosystems and Environment*. 39: 23-53.

Barrionuevo, A. and Bradsher, K. (2005). "Sometimes a Bumper Crop Is Too Much of a Good Thing". *The New York Times*, December 8.

Chrispeels, M.J. and Sadava, D.E. (1994). "Farming Systems: Development, Productivity, and Sustainability". pp. 25-57 in *Plants, Genes, and Agriculture*. Jones and Bartlett, Boston, MA.

David, B.V. (2008). Biotechnological approaches in IPM and their impact on environment. *Journal of Biopesticides*. 1(1): 1-5.

FAO (2009). *FAO Statistical Yearbook 2009*, FAO Rome.

Lal, R. (1980). Crop residue management in relation to tillage techniques for soil and water conservation. In: Organic recycling in Africa 74-79. *Soils Bulletin* 43. FAO, Rome.

Lal, R. (1991). Tillage and agricultural sustainability. *Soil and Tillage Research*. 20: 133-146.

Matson, P.A., Parton, W.J., Power, A.G. and Swift, M.J. (1997). Agricultural Intensification and Ecosystem Properties. *Science*. 277: 504-509.

Mote, U.N. and Shah, J.M. (1993). Efficacy of insecticides against stem fly, *Ophiomyia phaseoli* Tryon infesting french bean. *Plant Protection Bulletin*. 45(1): 24-27.

Murugesan, N. and Annakodi, P. (2007). Seed Treatment with Insecticides, botanicals and antagonistic organisms against the leafhopper, *Amrasca devastans* (Distant).PP 183-188 In: *Proc. National Seminar on Applied Zoology*, (Ed.) S. Baskaran. Ayya Nadar Janaki Ammal College, Sivakasi.

Nicou, R. and Charreau, C. (1985). Soil tillage and water conservation in semi-arid West Africa. In: *Appropriate Technologies for Farmers in Semi-arid West Africa*, (Eds.) H. Ohm and J.G. Nagy. pp. 9-32. Purdue University Press, West Lafayette.

Provenzano, G. (2007). Using HYDRUS-2D Simulation Model to Evaluate Wetted Soil Volume in Subsurface Drip Irrigation Systems. *Journal of Irrigation and Drainage Engineering* 133(4): 342–350.

Thomas Malthus, *An Essay on the Principle of Population*, 1798, Chapter 1.

Unger, P.W. (1984). Tillage and residue effects on wheat, sorghum, and sunflower grown in rotation. *Soil Science Society of America Journal*, 48: 885-891.

Chapter 26

Organizations Perspective on Human Resource Management for Sustainable Development

A.K. Makwana

Assistant Professor, SMC College of Dairy Science, Anand Agricultural University, Anand – 388 110, Gujarat

ABSTRACT

Organization carryout many roles in the society. This role tends to change over a period of time. Organizations right from inception of human race have been regarded as constructive partners in the communities in which they operate. Though organizations have been instrumental in creating employment, wealth, products, and services yet the pressure on them to play a role in social issues involving employees (HR), society, environment, government, other stakeholders etc is growing. They are called upon to take sustainable development activities. Organizations are comprised of people and building a sustainable organization should consider the human as well as the physical dimensions of organization's actions. Today, organizations are struggling with a new role, *i.e.* to meet the present generations without compromising the ability to meet the needs of next generations. Organizations are required to take responsibilities for their actions and policies which impact societies and natural environment. They are also being required to apply sustainability principles to the ways in which they conduct their business. In this context role of human resources and HRM department becomes crucial. Human resource professionals have a key role to play to help organizations achieve its key sustainable development activities. Human resource managers have the tools and the opportunity to leverage employee commitment to and engagement in the organizations effort towards sustainable development. This paper discusses the issue of organizations perspective on human resource management for sustainable development. It defines sustainable development, human resource management, some specific HR issues, and steps to be undertaken by HRM departments of the organization to achieve their sustainable development targets. The paper also highlights the challenges faced by present day organizations with respect to sustainable development, major environmental force affecting HR function and gives brief about key questions to be addressed by HR function of the organization for environmental forces.

Keywords: Organizations, HR, HRM, Sustainable development.

Introduction

The term sustainable development connotes different meanings to different people. But the basic concept of sustainable development is to meet the needs of people of present generations today without compromising on the ability of future generations to meet their own needs. Sustainable development thus involves (*i*) A broad view of social, environmental and economic outcomes. (*ii*) A long term perspective concerned with the interests and rights of future generations as well as of people today. (*iii*) An inclusive approach to action which recognizes the need for all people to be involved in the decisions that affect their lives.

The concept of sustainable development was coined first time in 1972 at the United Nations conference on human development. It gained popularity in 1987 with the release of seminar report 'Our Common future by the United Nations World Commission on Environment and Development (WCED).

A working definition for sustainable development based on social and ecological considerations is given by Rees (1989) as : "sustainable development is positive socioeconomic change that does not undermine the ecological and social systems upon which communities and society are dependent. Its successful implementation requires integrated policy, planning, and social learning process; its political viability depends on the full support of the people it affects through their governments, their social institutions and their private activities."

The term human resources can be defined as the total knowledge, skills, creative abilities, talents and aptitudes of an organizations labourforce (employees) as well as the values, attitudes and beliefs of the individuals involved. Organizations require different resources in the form of man, machines, materials and money to produce output in the form of goods and services. Thus manpower (Human Resource: HR) is one of the important resource for the organization. It can also be equated with any natural resource. Therefore it can be concluded that the organizations can utilize the skill, knowledge, ability etc of their employees by developing a positive attitude among them.

The term human resource management (HRM) is the organizational function that deals with recruiting, managing, developing and motivating people, including providing functional and specialized support and systems for employee engagement and managing systems to foster regulatory compliance with employment and standards relevant for human rights. The important techniques of HRM are: recruitment and selection, training and development, performance appraisal, motivation etc.

HR and Sustainable Development

Organizations carryout many roles in the society which tend to change over a period of time. If the organizations do not perform well on economic aspects, it will not have resources to fulfil its responsibilities to the society. Of late, organizations have realized that they are born out of society and must therefore serve it for their own enlightened sustenance.

From the organizations viewpoint, sustainable development is concerned with how business organizations can contribute to some of the most significant challenges faced by the world today ranging from climate change and bio diversity to working conditions and health amongst the poorest in the world. HR managers often hear the terms stakeholder engagement, corporate social responsibility (CSR), sustainable development etc. But HR mangers find it difficult to introduce sustainable development in their organizations. For example, how should they integrate sustainable development issues into programmes for recruitment and staff development and how can they use their knowledge of the business to bring critical issues to the attention of top management.

Literature Review

Nancy (2004) has discussed role of HR in forming and then grounding culture of sustainable development in organizations. She has emphasized that human capital acts as a success factor in today's competitive world wherein role of HR leadership has become more critical in leading and educating stakeholders of organizations on the issue of sustainable development and how best to strategically implement the sustainable development practices.

Rupp *et al.* (2006) emphasis that sustainable development practices of organizations plays a role about fostering positive social relationships between organizations and society. They highlighted that employees will adopt sustainable development practices which will meet employees need for belongingness with the organization and the society.

A survey by sirota survey intelligence (2007) found out that employees who are satisfied with their organizations, commitment to sustainable development practices are likely to be more positive, more engaged and more productive than those working for less responsible employers.

Murray (2008) has conducted a survey and based upon that reported that more than one third of respondents pointed that working for a caring and responsible employer was more important than the salary they earned and nearly half would turn away from an employer that does not adopt sustainable development practices.

Agrawal K (2007) while discussing the aspects of corporate social responsibility writes that with adoption of right HR policies, organizations can reap the full business benefits and become successful to the great satisfaction of all stakeholders.

Krishan and Balachandran (2004) highlight the role of HRM in incorporating responsible practices within the organization. According to them, lack of involvement of employees and failure to embed the socially responsible values into the organizational culture leads to failure of many corporate social responsibility initiatives.

The above mentioned research findings highlight the fact that HR function can be a powerful agent in effecting organization wide progress in its sustainable development performance.

Emerging Issues and Challenges for Sustainable Development

The issues of sustainable development are growing more and more. The capacity building needs of developing countries like India are growing rapidly. The demands placed on their natural resources are increasing due to their own development as living standards, aspirations, and resource consumption increase and also due to external factors such as climate change.

As the world progresses, a number of existing and emerging environmental issues present themselves as key issue for concern. These include globalization, coping with climate change and variability, poverty, health and environment, emergence of megacities, human vulnerability to environment change, freshwater depletion and degradation, marine and coastal degradation, population growth, rising consumption in developing countries and those with economies in transition, biodiversity depletion, bio safety and many more such issues.

Some Specific HR Issues and Challenges

Recruiting and retaining top talent: Organizations can only succeed in long run if they recruit and motivate people who are able to respond to and shape challenges of future. Although there is little evidence that people apply for jobs on the basis of ranking of organizations but the aspects of job security, respect and reward, learning opportunities and the organizations contribution to society are important criteria for increasing number of people. Skilled people are the organizations most valuable resource and a sustainable development perspective is also becoming crucial in today's globalized world. But where can such flexibility and creativity be found? One rarely considered source is those people with a strong awareness of social and environmental issues. Some HR managers believe that such people have a better knowledge of the challenges faced by today's business organizations and frequently command the skills to engage with a wide diversity of institutions and people including other businesses, governments, universities and non government organizations as well as cross functional teams in their organizations.

Motivation: Employees are motivated by complex mix of rewards. These range from financial security and bonuses to promotion and peer recognition and from taking on new challenges to making a difference to the things they care about. If organizations and employee values are aligned such that the organization is operating ethically providing a suitable worklife balance and clearly demonstrating a concern for health and safety, will increase the staff morale and commitment will also increase reducing staff turnover and boosting productivity.

Employee targets and incentives: They need to be aligned with the goals of sustainable development. Organizations should adopt an approach which recognizes the role of social and environmental issues in enhancing commercial performance. To make such approach successful, employees need to visualize that sustainable development objectives are clearly reflected in their personal targets and rewards. Performance appraisal criteria must consider contribution of individuals and teams to long term social and environmental goals as well as short term financial objectives. For example if all the bonuses for a procurement team are based upon obtaining lowest cost contracts, then employees will definitely display less interest in the working conditions of supplier base.

Challenges

1. HR has to understand how sustainability development is linked with business and HR practices. HR has to obtain yes signal for sustainable development strategies from inside and outside the organization and communicate consistently.
2. HR has to make sure that their suggested sustainable development strategy can stand up to the inevitable scrutiny by stakeholders.
3. HR has to implement sustainable development strategy as a strategic opportunity which should be market led.
4. HR has to develop appropriate code based laws and regulations and also ensure that reporting systems are accountable and transparent.
5. HR has to find out ways to leverage social responsibility initiatives internally, communicate the organizations contribution in the society and involve the employees in all its related efforts.
6. HRM department should effectively measure and evaluate sustainable development activities. The value added by such activities in the form of direct results such as economic savings and indirect results like increase in employee satisfaction, less employee turnover measured by staff attitude survey should be related with improved organization performance. It is also necessary to conduct periodic review of sustainable development activities.

Sustainable HR Policy

Human resource managers are responsible for adoption of sustainable development practices by their organization which reduces its negative and enhances its positive impacts on society and the environment. The following steps should be adopted by HR professionals on behalf of their organizations as sustainable development strategy.

1. Vision, mission, values and sustainable development: Successful implementation of sustainable development practices requires a clearly articulated vision, mission and values. The HR practitioner should initiate or support the development, or upgrade, of a vision, mission and values foundation of the organization covering sustainable development actions.

The mission of a socially responsible organization will specify that it will engage in ethical and responsible business practices and seek to make decisions which balance the needs of key stakeholders, employees and society at large. As an ideal, key stakeholders including society representatives should be involved in formulating organizations vision, mission and value statements.

2. Code of conduct of employees: Employees code of conducts is drafted by HR department. Compliance with code of conduct articulate ethical values of the organization. It is important to avoid rhetoric and undefined terms but to enunciate the conduct standards expected of employees.
3. Manpower planning and recruitment: Manpower planning is concerned with forecasting future manpower needs and ensuring availability of manpower in required number and of required type. For an organization which has policy of assuring sustainable development, the manpower planning function consists of evaluating the need for skill sets and competencies central to the emergent sustainability economy and economy of resource and energy scarcity, human and environmental security constraints, changing social norms and government expectations. Organization should highlight their sustainable development activities in their recruitment branding and marketing programmes. Campus recruitment policy is ideal for sustainable development oriented organizations emphasizing on questions on ethics and sustainable development in recruitment interviews.
4. Orientation and training : During orientation, all the new employees regardless of level should be given through overview of organization's vision, mission and core sustainable development policy adopted by the organization. If the organization is facing any key sustainable issue, the employee should be made aware with it in right and simple manner indicating performance requirement of the employee. Regular training programmes should be arranged in the concerned area for all the employees, whether one has direct responsibility (*e.g.* energy manager) or indirect responsibility (*e.g.* store clerk). The direct responsible persons should be given specialized technical training and indirect responsible persons should be given training in top priority sustainable development issues of a more general nature.
5. Compensation and performance management: Compensation and performance management are important HR tools. HR managers should integrate sustainable development elements into job descriptions, individual performance plans and team goals. HRM Practitioners know very well that 'you get what you pay for'. Traditionally organizations reward the employees based upon financial performance, which foster profit maximizing behaviour, overlooking sustainability factors. In the changed scenario, organizations compensation policy should be aligned with the much required sustainable development values and strategy. Although the strategic direction of the organization is formulated by the board of

management and chief of the organization, however HRM department can help the organization establish sustainable development targets and develop performance evaluation systems by providing right tools and counsel.

6. Change management and corporate culture: Change management experts realize that people come in different states of readiness for advances for sustainability, or any change for that matter. People can be grouped by state of readiness and then the organization should adopt suitable approach appropriately to each group. The employee can be categorized as (a) those who have the value and the behaviour (greens), (b) Those who have value but not the behaviour (sprouts), (c) Those who do not have the value or the behaviour (browns). The change strategy for greens is to encourage them to continue it. Promote, incent and reward sprouts for behaviour changes. Specific tools should be designed for this group so that the benefits are meaningful and the barriers to change are removed from this group. Leave browns alone. Many browns will change their behaviour once the sprouts have changed their behaviour so that they do not standout as minority.
7. Employee involvement and participation: Employee are one of the key stakeholders of any organization. It will be beneficial if the employees are consulted and engaged in the development of new programmes and approaches. This has more likelihood to be followed through with their implementation.
8. Sustainable development policy and programme development: It is also possible for HRM department to drive policy development and programme implementation in HR areas that directly support sustainable development activities *e.g.* wellness, diversity, work life balance, flexi time polices etc. organizations which strive to reduce their carbon footprint, HR practitioners can develop programmes enabling employees to use alternative transportation to get to work. (*e.g.* bus passes, common company vehicle etc.) Wellness programmes can become a platform for engaging employees in discussions about 'personal sustainability' and provide support for the employees in the areas of stress management, spirituality at work, health and fitness, healthy lifestyle etc. HR can also initiate employee volunteering programmes which can help build out the employee value proposition and employer brand while simultaneously delivering on the organizations goal for community engagement and investment.
9. Employee communications: Every sustainable development strategy necessitates development and implementation of an employee communication programme to communicate the organizations direction, objectives, innovation and performance on its sustainable development efforts. Intranets, websites, blogs, wikis, social networking sites, videos, forms, outdoor meetings, regular team briefings, webcasts, voicemails, print and electronic newsletters and other forms of social media need to be used by HR to bring the sustainability development message to the employees, with ultimate aim to engage employees in the sustainable development efforts of the firm.

10. Measurement, reporting and celebrating successes along the journey:It is said that 'what gets measured gets managed'. Thus it is important that both sustainable development performance of the organization and employee engagement be actively measured and reported to public and all other stakeholders. This can be achieved by annual sustainability report. It is also necessary to celebrate the successes, large and small along the journey. From pats on the back to specific articles on organizations intranet, to celebratory events, make sure that concerned persons are congratulated and achievements celebrated on formal and informal basis.

HR Implementation Strategy

The strategy and tactics of sustainable development adopted by the organization must be achievable by the HR team within its budgetary constraints and the skills, experience and interests of its people. But more than this, the HR strategy must respond to the wider environment in which the organization is functioning. The HR environment of the company is shown in Figure 26.1 and the key questions to be addressed by HR specialist are shown in Table 26.1.

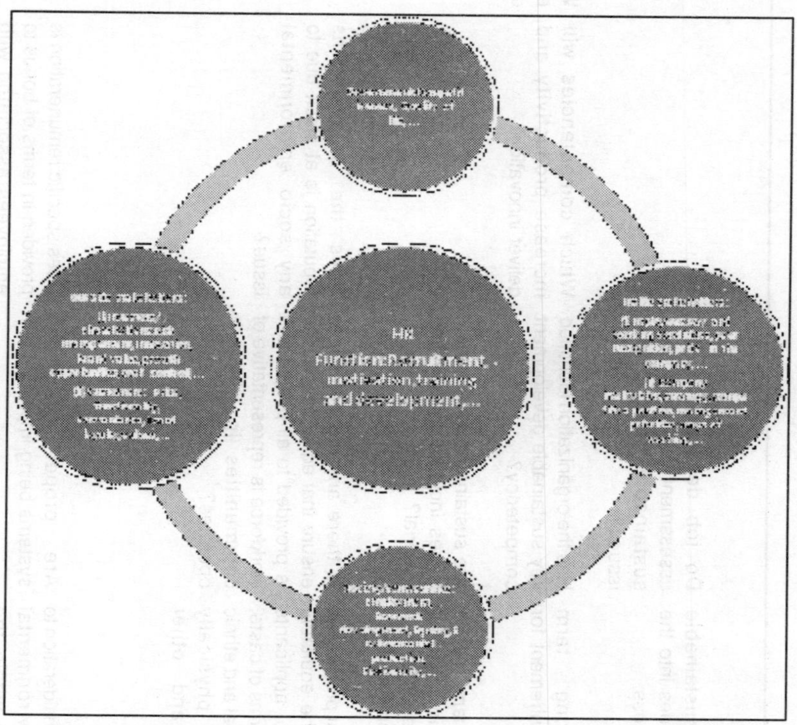

Figure 26.1: Major Environmental Forces Affecting HR Function

***Source*: Driving Success, Human resources and sustainable development, www.cpsl.ca.ac.uk**

Table 26.1: Key Questions to be Addressed by HR in Relation to Environmental Forces

	Employees and Company	*Investors and Shareholders*	*Customers*	*Society/Comunities and Government*
Recruitment				
Integrating sustainable development issues into the recruitment process	Do job descriptions and assessment materials contain sustainable development issues?			
Monitoring long term competency requirement for the organization	Has the organization identified key sustainable development competency?	Which competencies will increase productivity and deliver innovation?	Which competencies will enhance the value of customer proposition?	Which competencies will deliver value to society?
Providing new employees with information about sustainable development policies and commitments	Are sustainable development issues included in induction material?			
Using recruitment procedures which support the equitable representation of applicants and recruits in terms of caste, gender, age, racial and ethnic groups, physically handicapped and other relevant groups.	Is there any mechanism to ensure that equal opportunity is provided to all people and workforce is representative of communities in which it operates?	Does the organizations reputation is at stack due to any socio environmental issue?		What are the key diversity issues?
Incentives				
Providing due consideration to social and environmental considerations within the organizations performance appraisal and management systems	Are proper monitoring systems being implemented?	Does specific remuneration is provided in terms of bonus to employees associated with social and environmental performance?		

Contd...

Table 26.1–Contd...

	Employees and Company	*Investors and Shareholders*	*Customers*	*Society/Comunities and Government*
Identifying and supporting an appropriate work/life balance	Does any procedure exist to promote an appropriate work iife balance within the organization?	How do employee package contribute to financial performance?		What are the key employee benefit issues?
Encouraging employees to participate in corporate community involvement programmes	Whether employees are involved in organizations choice of community programmes?			What are the important social and environmental benefits, the organization can contribute?
Competancy development				
Including social and environmental issues within employee training and development programmes at all levels, from technical-health and safety considerations on the shop floor to strategic sustainable issues at top management and board level	Are sustainable development related training programmes provided?		Which training programme will enhance the value of customer proposition?	
Identifying specific training programmes and ensuring that they consider the business risks and opportunities from sustainable development	Whether training programmes include full range of social, environmental and economic risks and opportunities?	What are the financial opportunities from enhanced training in sustainable development area?		
Identifying the most effective means of building sustainable development competencies for all business functions and for all levels of employees	Which learning approaches are most effective in diverse parts of the organization?			

Source: Driving Success, Human resources and sustainable development, www.cpsl.ca.ac.uk

Conclusion

Successful implementation of sustainable development practices by organizations depends much on enlightened human resource management practices. In this context, HRM department activities of the organization in getting the employment relation right which is necessary condition for establishing effective relationships with external stakeholders and thus orient the human resources of the organization and thus the whole organization towards a socially responsible unit. It is important to link HR with external environment to chart out a better future for the society. Sustainable development activities definitely enhance the reputation, respect and the brand image of the company in the place they operate. This would lead to higher profits and establishment of their strong foothold in their sphere of operation. In this way they will ensure a better world for generations to come. As the issue of sustainable development enters the business mainstream, the responsibility for managing social and environmental issues is slowly transferring from the corporate fringe to important business function. To carry out this function efficiently to the satisfaction of different stakeholders of any business organization, the concerned manager which may be a human resource manager has to incorporate key questions related to sustainable development and key HRM functions, as a key HRM policy of the organization.

References

Agrawal, Kalpana (2007). Corporate Excellence as an outcome of corporate governance: Rethinking the role and responsibility of HRM. The Icfai Journal of corporate governance. 6(1) pp 6-16.

Ahmed, Mohammad Khalil. The role of human resources in corporate social responsibility. http://www.mmbgims.com assessed on 9-12-2011.

Anon. Driving Success, Human resources and sustainable development, www.cpsl.ca.ac.uk assessed on 1-12-2011

Krishnan, K., Sandeep and Balachandran, Rakesh (2004). Corporate social responsibility as a determinant of market success: An exploratory analysis with special reference to MNCs in emerging markets, paper presented at IIMK-NASMEI international conference.

Lockwood, Nancy R. (2004). Corporate social responsibility: HR's leadership role. http://www.shrm.org assessed on 9-12-2011.

Murray, Norm E. (2008). Corporate social responsibility is the number one criteria for job holders today. http://normmurray.org assessed on 9-12-2011.

Subbarao, P. (2004). Personnel and Human Resource Management, Himalaya Publishing House, Mumbai.

Rupp, Deborah E, Ganapathi Jyoti, Aguilera Ruth V, Williams Cynthia A. (2006). Employee reactions to corporate social responsibility: An Organizational justice framework, Journal of Organizational Behaviour. 27. pp 537-543.

Sharma, S., Sharma, J. and Devi, A. (2009). - Corporate Social Responsibility: The Key Role of Human Resource Management Business Intelligence Journal - January, 2(1): 205-213.

Sirota Survey Intelligence (2007). Social responsibility boosts employee engagement. http: / / www.management-issue.com assessed on 10-12-2011.

www.corostrandberg.com accessed on 9-12-2011.

Chapter 27

Emerging Trends of NGOs in Providing Livelihood Security and Empowering the Rural Women in Manipur

Ch. Victoria Devi[1] and Sagar Mondal[2]*

[1]Ph.D. Scholar, [2]Associate Professor, Department of Agricultural Extension, BCKV, Mohanpur, District Nadia, West Bengal

ABSTRACT

Women Empowerment by formation of "Joint Liability Group" is progressing with the leadership of different NGOs in a number of districts in Manipur. Organization for Rural Improvement (ORI) is one such NGO, which is actively engaged in improving the living standards of the rural poor, especially women, in the state.ORI's use of JLG models is concentrated in rural areas of Manipur, a backward and hilly state where the fruits of development have not percolated in the same measures as is in the rest of the country. The primary objective of ORI for creating such models is to promote and undertake different activities to ensure an integrated rural development through optimum utilization and full development of human material resources available locally.The study was conducted on the Joint Liability Groups (JLGs) of Rural Women in Bishnupur District of Manipur with some specific objectives like the extent of success of income generation and empowerment of the members. Purposive as well as simple random sampling method was adopted for the investigation. Data were collected by interview schedule method and both null hypothesis and alternative hypothesis were formulated. Moreover, Paired t-test was employed to

* *Corresponding Author:* E-mail: ch_venus2005@yahoo.com

ascertain any significant change in the mean income of JLG members before and after joining the groups.It was found that most of the women have earned higher income with the help of micro-finance given by ORI. Some of them have even started new enterprises. Today, ORI is playing a key role of providing secured livelihood of rural women in particular and women empowerment in general, in the rural areas of Bishnupur, Manipur.Some recommendations have been made with reference to (*i*) proper training for JLG members, (*ii*) enhancement of loans (micro-finance), (*iii*) arrangement of marketing facilities outside the state for indigenous products iv) organising fair/exhibition to encourage the women members of JLG.

Keywords: Women empowerment, Livelihood security, NGOs, Manipur.

Introduction

Empowerment is a multi-dimensional social process that helps people gain control over their own lives, communities and in their society. Empowerment occurs within sociological, psychological, economic spheres and at various levels, such as individual, group and community and challenges our assumptions about the status quo, asymmetrical power relationships and social dynamics.

Acharya and Ghrime (2005) identified three dimensions of women's empowerment-economic, social and political. According to them each component of these dimensions reinforces the other. The economic aspects include increasing women's access to and command over tangible and intangible resources, such as wealth, property, employment, knowledge and information. Social aspects include changing discriminatory ideology and culture, which determine the environment for women's existence.

Women Empowerment

Empowerment of women includes many things - economic opportunity, property rights, political representation, social equality, personal rights and so on. Empowerment is a holistic concept. It is multidimensional in its approach and covers different aspects of life. In recent years, empowerment of women has been recognized as a central issue in determining the status of women. Empowerment covers aspects such as women's control over material and intellectual resources.

Objectives of the Study

The present study was conducted on the JLGs of rural women in Bishnupur district of Manipur with some objectives, which include:

1. To study the extent of success of JLG members in relation to income generation. And the extent of success of the NGO.
2. To study some JLG members who have successfully started their own enterprises and the reasons for their success.

Research Methodology and Data Sources

Women empowerment by formation of JLGs is progressing with the leadership of NGOs in a number of districts in Manipur. For the present investigation the

Bishnupur Community Development block under Bishnupur district Manipur was purposively selected. In this block, ORI plays a pioneer role in rural development work with formation of large numbers of JLGs.

A total of 61 JLG members were selected randomly from 220 women members in 13 villages (centres) and interviewed. The data were collected by the researcher herself with interview schedule and a few case studies in the month of December 2009. For analysis of data some statistical tools like frequency, percentage, paired't'- test were adopted and income of the respondents were measured in a ratio scale. PRA tools like "seasonality diagram" were also used.

Figure 27.1: The Researcher and the Respondents during Data Collection

Hypothesis Testing

In the present investigation both null and alternative hypothesis were formulated. These are: (*i*) H: There is no difference between the mean income of JLG members before and after joining into the group. (*ii*) H: The mean income of JLG members differ significantly after joining into the group. Level of significance taken was 5 per cent.

Results and Discussion

The activities and monthly income of 61 Joint Liability Groups members before and after joining into JLG along with their enhanced income are presented in Table 27.1.

Table 27.1: Extent of Income Generation through Different Activities of the Joint Liability Group (JLG) Members

(N=61)

Respondents (Members of JLG)	*Activities before Joining JLG*	*Monthly Income before Joining JLG Rs)*	*Activities after Joining JLG*	*Monthly Income after Joining JLG (Rs.)*	*Increase in Percentage in Income*
1.	Maroi Nakuppi	5000	Maroi Nakuppi+Poultry	10000	50
2.	Cattle Rearing	2000	Cattle+Poultry	10000	80
3.	Weaving	2000	Weaving+Silk Yarning	4000	50
4.	Weaving	2000	Weaving	3000	33
5.	Weaving	3000	Weaving+Silk Yarning	5000	40
6.	Weaving	1000	Weaving	2000	50
7.	Tea Stall	2000	Tea Stall+Pan Shop	4000	50
8.	Smoked fish	5000	Smoked Fish	10000	50
9.	Handloom Products	5000	Handloom Products	10000	50
10.	Seasonal Vegetables	10000	Seasonal Vegetables	15000	33
11.	Vegetable Vendor	1000	Vegetable Vendor	2000	50
12.	Bamboo Basket (not commercially)	0	Bamboo Basket (done commercially)	1000	100
13.	Tea Stall	3000	Tea Stall, Food items	9000	66
14.	Pan Shop	1000	Pan Shop	2500	60
15.	Tailoring	1000	Tailoring	2000	50
16.	Weaving	1000	Weaving	2000	50
17.	Blouse Stitching and Embroidery	6000	Blouse Stitching and Embroidery	10000	40
18.	Potato cultivation	1000	Potato cultivation	2000	50
19.	Mat Making	1000	Mat Making	5000	80
20.	Fishery	1000	Fishery	3500	50
21.	Nil	0	Cattle (for milk purpose.	1500	100
22.	Nil	0	Weaving	2000	100
23.	Cane Works	2000	Cane Works	5000	60
24.	Vegetable Vendor	1000	Vegetable Vendor	5000	80
25.	Nil	0	Rice processing unit (purchase a dehusking and cleaning machine)	2000	100
26.	Vegetable Farm	1000	Vegetable Farm	1700	41
27.	Mushroom Cultivation	2000	Mushroom Cultivation	5000	60
28.	Nil	0	Cattle (purchased milk cow)	1500	100
29.	Pottery	2000	Pottery	5000	60
30.	Bakery	5000	Bakery	10000	50

Contd...

Table 27.1–Contd...

Respondents (Members of JLG)	Activities before Joining JLG	Monthly Income before Joining JLG Rs)	Activities after Joining JLG	Monthly Income after Joining JLG (Rs.)	Increase in Percentage in Income
31.	Juice Making (Pine-apple, orange)	2000	Juice Making (Pine-apple, Orange, Passion fruit)	4000	50
32.	Nil	0	Weaving (Dhoti, bought a traditional weaving machine)	2000	100
33.	Embroidery	1000	Embroidery	2000	50
34.	Bori Making	2000	Bori Making	5000	60
35.	Papar Making	1000	Papar Making	2000	50
36.	Duckery	2000	Duckery	5000	60
37.	Nursery (vegetables)	1000	Nursery (vegetables)	2000	50
38.	Smoked fish	6000	Smoked fish	15000	60
39.	Pickle Making	5000	Pickle Making	10000	50
40.	Nil	0	Bamboo Basket	1000	100
41.	Tea Stall	3000	Tea Stall, Food Items	9000	66
42.	Mat Making	1000	Mat Making	5000	80
43.	Cattle Rearing	1000	Cattle Rearing	2400	58
44.	Weaving	1000	Weaving	2000	50
45.	Nil	0	Rice processing (purchased a dehusking and cleaning machine)	2000	100
46.	Smoked Fish	1000	Smoked Fish	2000	50
47.	Mushroom Cultivation	2000	Mushroom Cultivation	9000	77
48.	Muri Making	1000	Muri Making	2000	50
49.	Bamboo Basket	900	Bamboo Basket	1500	40
50.	Agarbati Making	500	Agarbati Making	1500	66
51.	Muri Ladu	1000	Muri Ladu	2000	50
52.	Silk Yarning	1000	Silk Yarning	3000	66
53.	Kouna Craft	1000	Kouna Craft	3000	66
54.	Fermented Fish	4000	Fermented Fish	6000	33
55.	Poultry	5000	Poultry	15000	66
56.	Weaving	500	Weaving	1500	66
57.	Weaving (silk sarees.	5000	Weaving (silk sarees.	10000	50
58.	Agarbati Making	500	Agarbati Making	2000	75
59.	Maroi Nakuppi	10000	Maroi Nakuppi	30000	66
60.	Mixed Farming (poultry, vegetable)	10000	Mixed Farming (poultry vegetable, piggery, fishery)	30000	66
61.	Cherra/Chempak Making	10000	Cherra/Chempak Making	30000	66

Discussion

Table 27.1 reveals that most of the women members have generated income for earning their livelihood after joining into the group. For testing the null hypothesis, a 5 per cent level of significance was selected. Paired t-test was done to know whether the monthly income of the JLG members has increased significantly after joining as a member of JLG

The main findings of paired t test are presented below:

Mean (+ SD) income before being JLG member = 2383.61(+ 2614.01)

Mean (+SD) income after being JLG member = 6026.23(+ 6642.24)

$|t| = 6.586$

DF =60

Table value at 1 per cent level of significance level with 60 D.F = 2.660(two-tailed test)

The level of significance was selected at 5 per cent. But after analysis of data it was found that the calculated value of paired t test ($|t| = 6.586$) was higher than the statistical table value (2.660) at 1 per cent level of significance with n-1(61-1) =60 degrees of freedom (DF), the null hypothesis was rejected. So the mean income of Joint Liability Group members increased significantly after joining as JLG. ORI is playing a pioneer role in the rural areas of Bishnupur District, Manipur for economic improvement of the rural women.

Case Study on some Joint Liability Group Members who have Established their Own Enterprises and Reasons for Success

Yin (1984) defines a case study as an empirical inquiry that investigates a contemporary phenomenon within its real-life context; when the boundaries between phenomenon and context are not clearly evident; and in which multiple sources of evidence are used. Detailed data were collected and long interactions done during the case studies along with the JLG members (beneficiaries of ORI) in this study. Three such case studies are presented below:

Case Study–1: Product → Silk Threads

M. Manileima is one of the beneficiaries of ORI since 2004. Her family is known for yarning silk in the locality and this is her main source of family's income. *In the year 2006 she was given the privilege to attain a free 15-days training programme organised by Sericulture department, Manipur*. Her husband is a member of a mulberry and oak tassar plantation farm, Mantripukhri. So she has the privilege to get the tassar/ mulberry leaves as feed for the worms. They also planted around 1 hectare of their homeland with mulberry.

Ningol Chakouba is a popular festival in Manipur and during this festival there is a great demand for silk items. With the practice, she earns Rs. 3000/- extra per month, but during the festival time she manages to earn around Rs 6000/-. With the help of loans and her savings she planned to buy a yarning machine by August, 2010. Following is a seasonality diagram of her sell of the product

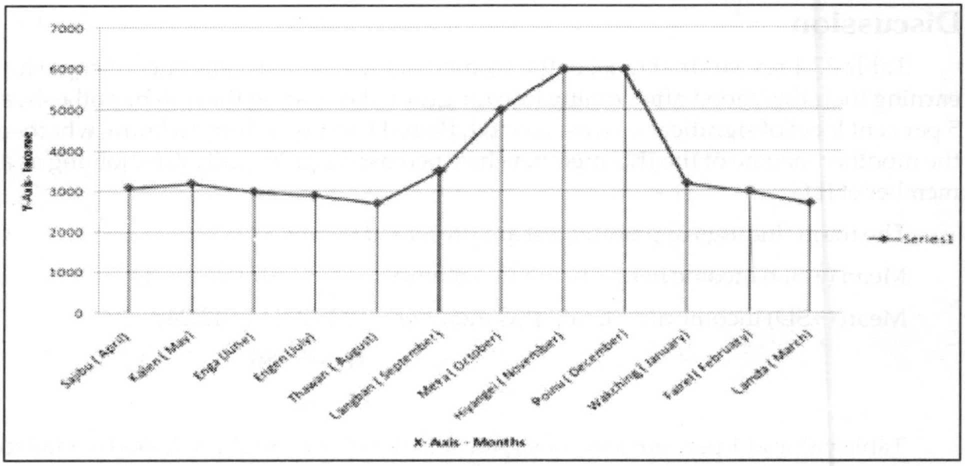

Figure 27.2: Y Axis Shows the Amount of Money She Earns in a Month and X Axis Shows the Months of a Year

Figure 27.3: The Researcher and the Respondent (Practicing Silk Yarning) during the Case Study

Case Study–2: Mixed Farming

Thoibi Devi is a beneficiary of ORI since 2004. She efficiently practices one of the best enterprises of agriculture, *i.e.* "Mixed farming". The enterprises she adopted are:

1. *Piggery:* The particular breed she rears is Hampshire breed, which she purchased at the rate of Rs. 2250 per two months old piglet and *for feeds of the pigs she buys the factory bi-products from Food Factories.* Generally, a full grown pig weighs around 80kg and fetches her Rs. 8000/-. For *health coverage and prevention from diseases she consults the veterinary surgeon from the nearby KVK.*
2. *Fishery*: This is one of the enterprises which give her maximum income because fish is a popular dish in Manipur. The best concept she applies in her enterprises is *the waste recycling process.* The faecal matters of the pigs are washed into the pond, to be served as good feeds for the fishes.
3. *Poultry*: Wastes from the vegetable garden are utilised as feeds of the poultry. She sells the chickens (which she bought at the rate of Rs 15 per chick) after 45 days at the rate of Rs 100 per kg. when they attain one kg.of weight. The poultry faecal matters are used as manures for vegetables.
4. *Vegetable Garden*: She grows different varieties of vegetable like potato, cabbage, pea, cauliflower, brinjal, onions and other indigenous vegetables, all meant for sale.

One of her remarkable achievements is that she managed to purchase a car (Maruti Alto) with her savings. She said that every month she manages to earn not less than Rs. 30000 (Figure 27.4).

Case Study–3: Weaving-Silk Manipuri Saree

T. Ibecha Devi is a beneficiary of ORI. She told that her family has been weaving silk manipuri sarees since time immemorial and she has inherited the skill. Before she joined the JLG she used to weave simple cotton clothes. But after her joining into the group and started taking loan, she started weaving silk sarees, which are in good demand all over India.

Before she could earn around Rs. 4000 – 5000/- with which it was very hard to manage her joint family. Now, as she could get loan and also repay, she manages to weave 10 silk sarees (silk chaddar) in a month. To weave 1 saree she takes 3 days and the cost of silk thread for weaving 1 saree is Rs. 1500/-. She sells at the rate of Rs. 2500 per saree. So in a month she manages to earn Rs. 10000/- easily and it seldom goes below Rs. 8000/- per month (Figure 27.5).

Conclusion and Recommendations

The following conclusion and recommendations may be drawn from the study on Joint Liability Group in Bishnupur district who are working under the guidance and supervision of an NGO, "Organisation for Rural Improvement" (ORI).

Figure 27.4: During the Case Study of a Respondent Practicing Mixed Farming

Along with the respondent at her piggery enterprise

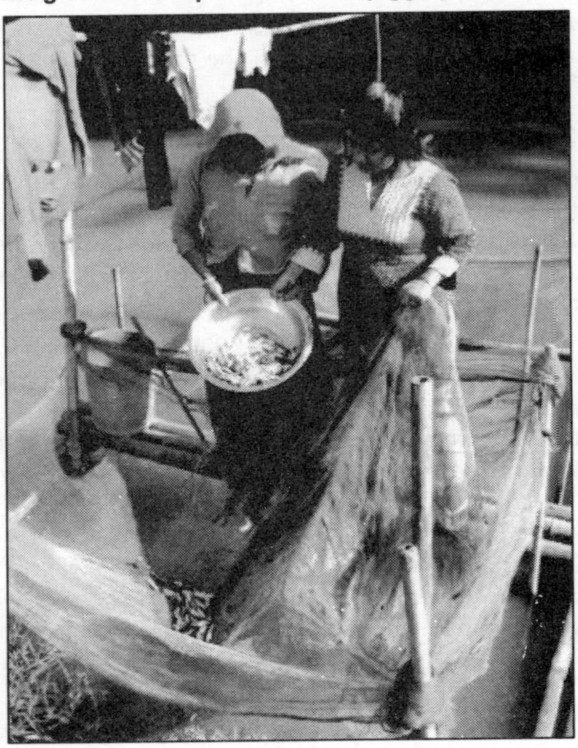

At her fishery enterprise

Figure 27.4–Contd...

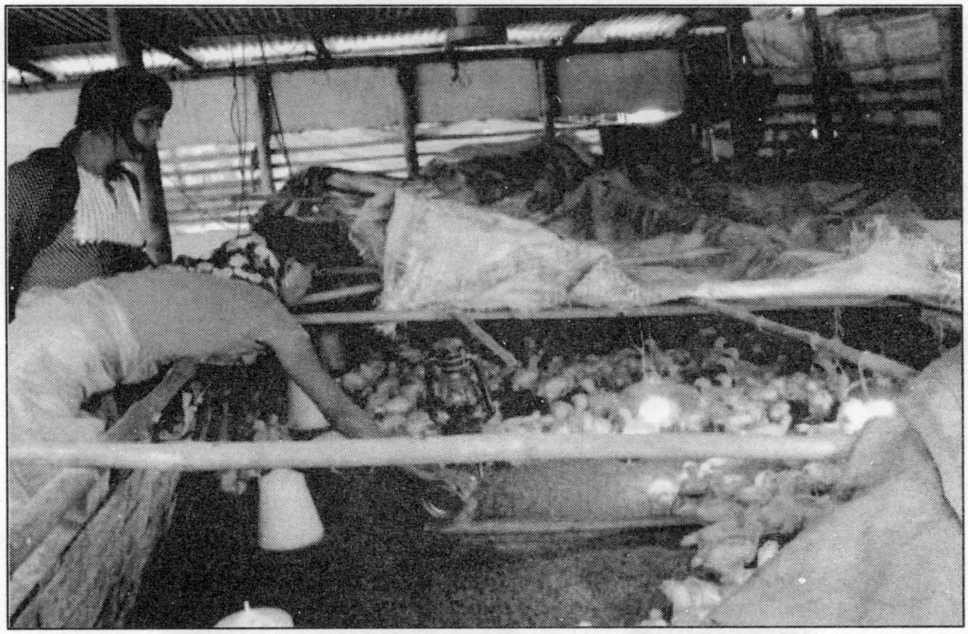

At her poultry enterprise

At her vegetable garden

Figure 27.5: The Woman Respondent and her Products (Manipuri Silk Sarees) seen during the Case Study

Conclusion

Employment Generation by the NGO, "Organisation for Rural Improvement" (ORI)

Table 27.1 reveals that before joining as JLG members, some women had no income generated activities. They were unemployed (Respondents Sl. No. 12, 21, 25, 28, 32,40,45).

The ORI has played a great role of employment generation by providing income generation activities.

Income Generation by Additional Activities

Before joining as JLG members the respondents used to engage in only one activities but after being JLG members they have included extra income generating activities under the guidance and supervision of ORI.(respondent no.1,2,3,5,7,13,31,41,60 in Table 27.1).

Manipuri Rituals and Festivals

It reveals from the case studies that Manipuri rituals and festivals are closely associated with the income generation of JLG members. Activities like silk yarning, weaving and their income generation are closely related with the rituals and festivals of Manipur. This leads to significant increase in the income of some JLG members. With the coming/onset of some festivals the workload as well as the income of the women members increases.

Adoption of New Technology and the Inherited Skills of the Women Members

Some of the women members are very much skilled with their activities. Women practicing weaving, cottage industries are very much skilled and make very good quality products. Some of them were well aware of the recent technology. Mention may be made of the respondent no. 60 of Table 27.1 who practiced "Mixed Farming" and was found to adopt many good concept on "Integrated Farming", "the idea of waste recycling" etc.

Economic Empowerment

The study reveals that not only the NGO, "Organisation for Rural Improvement" had succeeded in empowering the rural women economically through income generating activities but the Manipuri rural women have been empowered in all spheres of rural life *viz.* in decision making, family size decision making and even freedom of movement.

Success of "Joint Liability Group"

The mean income of 61 women members before and after JLG and subsequent application of statistical paired t- test scientifically proves that the concept "Joint Liability Group" as evolved by NABARD and tested in a pilot basis is a successful endeavour particularly in the Bishnupur district of Manipur.

Recommendations

1. Since the women practicing handicraft and handloom works are extraordinarily skilled, ORI may help in the promotion of such works, specially kouna craft and silk Manipuri sarees.
2. The NGO, ORI may take necessary steps for loan enhancement for the women members to increase their production.
3. Arrangement of marketing facilities outside the State for kouna products and silk Manipuri sarees may be made as the products are of high demands outside the State.

4. Fair/Exhibition may be organized once a year to let the women showcase their products. This will be a good step to encourage the women in increasing their products further. Awards can also be given to those hardworking and successful women. ORI may take a leading role for this purpose.

References

Acharya and Ghrime (2005). 'Gender indicators of equality, inclusion and poverty reduction: Measuring programmes and project effectiveness', *Economic and Political weekly* 40(44-45), March 20th 2009 issue.

Alexander, S. Kritikos and Denitsa Vigenina (2005). The empirical Analysis "Key Factors of Joint-Liability Loan Contract".KYKLOS, 58(2), 213–238. (*www.gdrc.org/icm/country/georgia-grouploankyklos.pdf*).

Batiwala, Srilatha (1994). 'The meanings of women's empowerment: New Concepts from action. In: *Population Policies Reconsidered: Health, Employment and Rights,* (Eds.) G. Sen, A. Germin and L.C Chen. Harvard University Press.

Cheston, Susy and Kuhn, Lisa (2000). 'Empowering women through microfinance', A draft Report, UNIFEM.

Chinnadurai, R. (2005). 'Women Entrepreneurship and Service Sector'. *Kurukshetra,* 54(1):19-20.

Yin, R.K. (1984). Case Study Research: design and methods. *Applied Social Research.* Prentice-Hall of India Pvt. Ltd., New Delhi.

Chapter 28

Issues of Sustainable Development of Jhum Centered Land Use System in Mizoram

Vanlalruali[1], Ch. Victoria Devi[2] and Sagar Mondal[3]***

[1]M.Sc. Agril. Extension; [2]Ph.D. Scholar; [3]Associate Professor
Department of Agriculture Extension, BCKV,
Mohanpur, District Nadia – 741 252, West Bengal

ABSTRACT

Jhuming is a way of social life in Mizoram. A study was conducted on jhumias and factors associated with jhuming in Mizoram with some specific objectives viz to delineate the major activities undertaken during jhum cultivation and festivals associated with jhuming in Mizoram; to study the perception of the jhumias relating to environmental impact of jhuming and to delve into some agro- economic, socio- psychological factors contributing to jhuming in Mizoram. The field investigation was carried out in village Chhingchhip under Serchhip district and Airfield area near Aizawl under Aizawl district. Forty respondents have been selected by non- probability sampling method and interviewed. Data were collected in the month of May, 2007 and December, 2007- January, 2008 by the researchers with the help of Interview Schedule. The jhumias are well aware of the adverse effect of jhuming on environment and ecology but some important factors like community integration with some festivals, food security by way of mixed cropping, food habit, topography of the State, low cost practice, inheritance of jhuming from their forefather etc. playing a vital role for continuance of jhuming over the centuries. Since jhum cannot be completely replaced by terrace cultivation in Mizoram, some recommendations have been made for modernization of jhuming and for upliftment of the economic condition of jhumia in the study area.

E-mail: *ch_venus2005@yahoo.com; **sagarmondal.bckv@gmail.com

Introduction

The origin of shifting cultivation can be traced back to 9000 years ago on the basis of archeological data. Shifting cultivation in the history of cultivation is regarded as the first transitional step towards agriculture from food gathering and hunting to domestication and cultivation of useful cultivars for human survival.

The system has been variously termed as primitive agriculture, nomadic agriculture, slash-and –burn agriculture, migratory agriculture, primitive horticulture, band tillage, hoe and burn, bush-fallow cultivation, swidden agriculture and shifting cultivation or jhum cultivation etc.

About 2 million people cultivate nearly 11 million hectares of land following these practices in India. It is known in different areas by a variety of local names. In India, it is known as "jhuming" in north-east; "khallu" or "kurwa" in Bihar; "dahiya" in Madhya Pradesh.; "podu" in Andhra Pradesh and Orissa and as "kumri" or "kumari" in Western Ghats. There are differences in their mode of operations, cropping pattern and duration of cultivation on the same land from place to place.

Shifting Cultivation in the North-East Region

Shifting cultivation, locally named as "Jhum" as a system of land use and agriculture is widespread in the mountainous areas of the north eastern region of India. The ideal conditions for shifting cultivation prevail in the region because of extensive forest resources and sparse population. The region which constitutes less than 8 per cent of the country's geographical area comprises little more than 25 per cent of the country's forest cover of 63.34 million ha.

Table 28.1: Shifting Cultivation in North-East Region

States	*Annual Area Under Shifting Cultivation (Sq. Kms.)*	*Fallow Period (in years)*	*No. of Families Practising Shifting Cultivation*
Manipur	900	4–7	70,000
Arunachal Pradesh	700	3–10	54,000
Assam	696	2–10	58,000
Meghalaya	530	5–7	52,290
Mizoram	630	3–4	50,000
Nagaland	190	5–8	116046
Tripura	223	5–9	43,000
Total	**3869**	–	**4,43,336**

Source: Basic Statistics of NER, 2002, Govt. of India, North Eastern Secretariat, Shillong, p. 42.

Thus, mountain topography, hot humid climate, deep and fertile soils, scanty indigenous population and communal ownership of land constitute the ideal conditions for the rise and continuation of shifting cultivation in north-east India. These estimates, however, make one thing clear that shifting cultivation is widespread in all the states of north-east India where it is a way of life.

Shifting Cultivation in Mizoram

Mizoram is a small state located between 21° 58′ N to 24° 35′N latitude parallel and 92°15′ E to 93°29′ E meridians longitude. It encompasses about 21087 sq.km. and population of 888573 persons (2001 census, Statistical Handbook, Mizoram, 2006).

As per the statistics of North-Eastern Hill Region the average annual area under shifting cultivation in Mizoram was 530 Sq.km. involving as many as 50,000 families operate minimum 63,000 ha. under this method of cultivation. However, the area under this system of agriculture as well as the number of families have come down drastically every year as it is no longer a paying proposition..

Jhum Cycle

Jhum is a method of cultivation in the hilly areas of Mizoram. Because of the rapid loss of fertility, the jhum land is abandoned after one or two years of cultivation and a new plot is selected for fresh cultivation. In the process a plot once used is left fallow for a number of years to regain fertility. This is known as *jhum cycle*. The farmers in the Mizoram who are cultivating using the methods of jhum are called *jhumias*.

The jhumias cultivate patches of land on the hill slopes for one year only and shift to a new patch to return to the same land after some year. The jhum area lies around the village, rotating the cultivation which makes the jhum cycle. The cycle of cultivation varies from village to village depending on the availability of jhum land within the village jurisdiction and number of families dependent on it in the village. Logically villages with high population density should have short jhum cycles. The important parameters of jhum cycles are: i) rural population density (*ii*) number of villages per 100 sq. km and (*iii*) number of cultivators per sq. km.

System of Jhum

A brief resume of the practice of jhuming existing now-a-days will not be relevant. Earlier the village chief and the village council headed by him had much power and control over the villages, which is lost to a large extent now. Instead, lately formed Mizo Hills District Council has made a number of Acts, regulations and rules pertaining to various economic practices including jhuming, collection of forest produces and other agricultural uses of land, which are executable through the Village Council of the region. According to these rules, each village council has to present a written report by September regarding allotments of jhum lands, number of families and amount of land to be allotted, etc, to the Executive Committee of the District Council.

On receiving such reports instructions are issued for distribution of land which rests to the power of village council, and usually accomplished by early January. Mode of distribution of jhum land now is either (*i*) by draw of lots to distribute previously demarcated jhum lands, or (*ii*) by draw of lots to divide the order of preference by which selection of jhum shall be done

Though shifting cultivation has long been viewed as a problem, in the absence of any workable alternative, people continued "jhuming" on the slopes much as they have been doing for generations. At the same time, farmers have started feeling pressure of growing population and shortening shifting cultivation cycle. As a result, shifting cultivation is widely practiced in the whole state.

Adverse Effect of Jhuming

Although the system of jhuming was good for a time when it emerged, yet, it is considered to be wasteful for the present time. The various adverse effects of jhuming are summarized below:

1. Some experts are of the opinion that this system of economy is very harmful as it destroys the forest cover and lowers the fertility of the soil.
   (*a*) In areas of extensive exploitation by shifting cultivation, it has been noted that the soil is rapidly washed away and trees can no longer grow there.
   (*b*) There has been large scale deforestation resulting denudation of hill tops and slopes.
   (*c*) Further, since the hill tops are the source of water, deforestation of hill tops leads to elimination of the sources of water.
2. There is large scale soil erosion due to deforestation and cultivation on hill slopes without effective soil conservation practices. This in turn leads to several effects.
   (*a*) Erosion of soil in catchment areas resulting in shifting of reservoir and streams leading to floods in plains (Goswami, 1968);
   (*b*) Removal of top soil leads to loss of fertility which cannot be easily restored or replenished. This leads to low productivity and subsequent pressure on land.
   (*c*) Productivity from the point of view of land use is low under shifting cultivation. (Barkakoti 1990)
3. Continuance of primitive form of agriculture offers very little scope for introduction of modern technology. This along with loss of fertility leading to such a low level of productivity that the farmers live in near famine (bear subsistence level) (Saha, 1973 and 1976).
4. The system is labour-intensive with low technology resulting in low productivity (Roy Burman and Sharma 1970). Further, there is no scope for development of sources of subsidiary income for the farmers. Besides these, which are directly related to agriculture, there are other aspects which adversely affect social welfare. The public health, education, communication and such other basic facilities are difficult to be developed when there is no permanent settlement.

Advantages of Jhuming

Inspite of all these problems or adverse effects of jhuming, there are a number of advantages of shifting cultivation as envisaged by the jhumias. These are presented below:

1. Low labour cost, as no labour outside family is employed. The amount of the land allotted to each family essentially depends upon the size of it, *i.e.*, number of able bodied members.
2. Inputs required for cultivation are much less which include nothing but seeds and a few simple agricultural implements.
3. The problems of attacks from insects are almost nil if the firing is done effectively.
4. The jhum double cropping or multiple cropping is also possible with the application of some local improved variety seeds. *e.g.*,"Phulbun" which, if sown 6 feet apart, can produce much more yield. These seeds can also be sown in terraced field of dry condition with good result.
5. Damage of crops by birds is more when small area is under cultivation. This argument of the farmers justifies for larger area under jhum.
6. Combination of different crops not only helps in optimizing the use of natural processes but it also contributes in a significant way to soil and water conservation and enrichment and also in disease and pest control.

Objectives of the Study

1. To delineate the major activities undertaken during jhum cultivation and festivals associated with jhuming in Mizoram
2. To study the perception of the jhumias relating to environmental impact of jhuming.
3. To delve into some agro- economic, socio- psychological factors contributing to jhuming in Mizoram.

Methodology

Pilot Study

A pilot study is a preliminary study conducted on a limited scale before the original studies are carried out in order to gain some primary information, on the basis of which the main project would be planned and formulated.

Pilot study enables the researcher to gain some systematic knowledge of universe and its population. It plays a vital role in identifying he variables involved in research, the nature of the problem, the nature of the respondents, kind of responses that are likely to be available etc. On the basis of the information gathered or experience gained during such a study, the schedule/questionnaire is then framed.

A pilot study was undertaken in Chhingchhip village under Serchhip district in order to access the feasibility of the research study. For this purpose, the researchers stayed with her relatives in Chhingchhip. A Pilot study was conducted in month of January 2007.

Informal discussions were made with few villagers and village council personnel during the period. Exhaustive list of information was collected from different sources. The information collected was used for preparation of the schedule for data collection.

Selection of District

As per Statistical Handbook Mizoram 2006, this state Mizoram has 24 Rural Development blocks under 8 Districts. The study was conducted in the village Chhingchhip under Serchhip district of Mizoram which is about 80 km away from Aizawl, the capital of Mizoram. The Tlangnuam Rural Development block in the district of Aizawl was also selected purposively for conducting the study.

Selection of Villages

The village Chhingchhip under Serchhip district and the village Airfield area under Aizawl District were selected purposively. The village Chhingchhip was consisted with three paras (wards) which is called *'veng* 'in local language. The name of this veng was *Venghlun, Mualpui,* and *Venglai,*. The study was conducted in venghlun veng (para). According to the book', *chhingchhip khaw chanchin'* (About the Village Chhingchhip) by Upa. Haukhama published in 2008, the total number of houses at the village Chhingchhip were 709, total number of families were 708, total no. of population were 4148; out of which total no. of male persons (*mipa*) were 2018 and that of female persons (*hmeichhia*) were 2130. The study was conducted in *Venghlun Veng* of the village Chhingchhip.

Selection of Respondents (Jhumias)

For selection of respondents Non- probability sampling method was followed. Since, the total number of jhumias families in Venghlun para of village Chhingchhip could not be ascertained, and also due to limitation of time and other constraints, instead of adopting random sampling method, non- probability sampling method was followed. In this type of sampling, the items (respondents) for the sample are selected deliberately instead of using the techniques of random sampling.

For the present investigation total 40 (forty) respondents (jhumia) have been selected. Thirty Jhum cultivators (Jhumias) were identified in the village Chhingchhip and were selected and interviewed. Only 10 Jhumias have been selected deliberately by the researchers from Airfield area under Aizawl District.

Empirical Measures of Variables

A variable is something which varies. More specifically variables are those attribute of objects, events, things and beings which vary and can be measured (Ray and Mondal, 2004)

For selection of appropriate variable, the experts in the Department of Agriculture, Government of Mizoram were consulted by the researchers and available literature on jhum cultivation in the North East in general and particularly in Mizoram were collected and gone through. By following specific objectives of the study, the theoretical basis for selection of variables and their empirical measures are presented.

Major Activities Undertaken during Jhum Cultivation

Jhuming as practiced by Mizo's involves some specific activities which are to be performed within one full cycle. The activities are as follows:

- Selection of land
- Cutting of forest
- Natural drying
- Burning of dried forest
- Cleaning
- Sowing of seeds
- Weed control
- Construction of watching house (Thlam)
- Harvesting
- Threshing/winnowing/drying
- Storing of crop, and
- Marketing

In the present investigation it was intended to find out the calendar year (month of operation), time required (in no. of days), whether it is performed by male or female and some festivals associated with these activities.

Perception of Jhumias Relating to Environmental Impact of Jhuming

The system of jhum cultivation involves cultivation of crop on hill slopes after clearing the land by cutting down the jungle and burning the dry debris on which seeds of different crops are sown by dibbling or broadcasting before the onset of rain,

Though shifting cultivation has long been viewed as a problem, in the absence of any workable alternative, people continued "jhuming" on the slopes much as they have been doing for generation. At the same time, farmers have started feeling pressure of growing population and shortening shifting cultivation cycle. As a result, shifting cultivation is widely practiced in the whole state.

Although the system of jhuming was good for a time when it emerged, yet it is considered to be wasteful for the present time. Some experts are of the opinion that this system of economy is very harmful as it destroys the forest cover and lowers the fertility of the soil. There has been large scale deforestation resulting in denudation of hill tops and slopes. Removal of top soil leads to loss in fertility which cannot be easily restored or replenished. This leads to low productivity and subsequent pressure on land. Further, since the hill top are the sources of water, deforestation of hill tops lead to elimination of the sources of water.

The present investigation was intended to study the perception of the jhumias relating to environmental impact of jhuming. According to Cambridge Advance Learners Dictionary (2005), perception is an awareness of things through the physical senses especially sight. It is the belief or opinion often held by many people and based on appearance.

To study the perception of the jhumias relating to environmental impact of jhuming, the following steps were gone through.

Collection and Initial Selection of Items

The items (statement) relating to the environmental impact of jhuming were collected from different sources such as books, literature, Mizo Farmers Union, Cooperative Society, scientist of Agricultural University, expert in the Department of Agriculture, Govt. of Mizoram and the researchers own experience.

Initially 18 items relevant to the study were collected. These were edited as per criteria set up by Edwards (1969). The general guidelines for editing the statements following Edwards are presented below:

1. The statements should be short, simple, clear, direct and easily understandable.
2. Each statement should contain only one idea. Their should be no insignificant idea or repetition of idea
3. Statements that are irrelevant or may be interpreted in more than one way should be avoided.
4. The use of double negatives should be avoided

Final Selection and Scoring of Items

The items were pre tested with the jhumias to judge their relevance in field situation. After discussion with expert in the Department of Agriculture, Govt. of Mizoram, Village Council of Mizoram, 11(eleven) items were finally selected for the administration of the respondents (jhumias).

The responses were obtained through nominal level of measurement (Yes/No) which means numbers are assigned to objects or events which can be placed into mutually exclusive and exhaustive category.

Agro-Economic, Socio-Psychological Factors Contributing to Jhuming in Mizoram

The present investigation was aimed to identify some important factors which are responsible for jhum cultivation in Mizoram. The different factors of jhum cultivation are as follows:

1. Socio- psychological factors of juming
2. Agronomic factors of juming
3. Economic factors of juming

Apart from agronomic factors for jhum cultivation there are also some sociological and psychological factors that are contributing for jhuming since time immemorial. Some socio- psychological aspect like inheritance of jhuming from their fore fathers, maintenance of harmonious relationship among the community, involvement of women in jhuming and some important festivals in Mizo community namely *Chapchar-kut, Mimkut* etc. have been associated with jhum cultivation and these factors cannot be ignored or obliterated

Collection and Initial Selection of Items

The items relating to the agro- economic, socio-psychological factors contributing to jhuming were collected from different sources, such as books, literature, jhumias, village council members, experts in the Department of Agriculture, Govt. of Mizoram and the researchers own experience. Initially 17(seventeen) items relating to socio-psychological aspect, 11(eleven) items relating to agronomic aspect and 10(ten) items relating to economic aspect were collected. These were edited as per criteria set up by Edwards. The general guidelines for editing the statement have been mentioned earlier.

Final Selection and Scoring of Items

The items were pre tested with the jhumias to judge their relevance in field situation. After discussion with the expert of the Department of Agriculture, Govt. of Mizoram the items (activities) were divided into three categories *viz.* socio-psychological aspect where 10(ten) items out of 17(seventeen) were finally selected for the administration to the respondents. Under Agronomic aspect 6(six) items were selected out of 11(eleven) items and for economic aspect 6(six) items out of 10(ten) items were selected finally for administration to the respondents (jhumias)

Each item was provided with a two point scale, agree or disagree and the jhumias were asked to indicate their agreement or disagreement in respect of each statement under agronomic, socio-psychological and economic aspect.

Statistical Methods Used

In the present study all the data are nominal data and it has been measured by nominal level of measurement which means numbers are assign to objects or events which can be placed into mutually exclusive and exhaustive categories. The statistical tools frequency and percentage has been used for the present study.

Field Data Collection

Relevant data from the jhumias were collected on the basis of personal Interview method with the help of Interview Schedule. Field data were collected in 2 phases *viz.*(1) In summer season i.e, May 2007 and in winter season, December 2007. The researchers went to the village Chhingchhip, 80 km away from Aizawl city and stayed there for a number of days.

Responses from 30 jhumias were obtained at village Chhingchhip under Serchhip district. During data collection the researchers visited the houses of the jhumias either in the evening or in the morning as because during day time they used to go to their jhum field.

Responses from ten jhumias were obtained from Airfield area adjacent to Aizawl. The data were collected in the end of December 2007 to 1st week of January 2008.

Results and Discussion

The results of the investigation are presented and discuss in this chapter after statistical analysis. These are presented according to specific objectives of the study:-

Figure 28.1: Researchers Interviewing the Jhumias During Data Collection

Major Activities Undertaken During Jhum Cultivation and Festival Associated with Jhuming

Shifting cultivation that is known as jhuming in the state of Mizoram and the rest of the North-East India is an agricultural practice evolved as a reflex to certain inherent difficult physiological characteristics of land that renders hoeing or ploughing undesirable on hilly and slopy land. Jhuming is more than an occupation in Mizoram. It is a life style of the mizo people in rural area. It is also an integral part of the socio-cultural life of Mizo's. In the present investigation it was intended to find out the major activities of the jhum cultivation along with the important festivals associated with jhuming in Mizoram. This is presented in Table 28.2.

The Village Council allots the forest land for cultivation. The individual has no ownership over the land. Cutting of forest was done in the month of November and December. Small trees and bamboos are cut using indigenous implements. And the fallen vegetations are left in the field to dry for one to two months.

The Village Council declares a suitable day for burning on each of the selected slopes where a consolidated block of plots is located. All jhum lands of a village are burned on a single day. Cleaning is done after burning; they collect the unburned material for re-burning. Jhum shelter (*Thlam*) is constructed on the field where members of the family stay during long periods of work. Sowing is done after burning where different crop seeds like vegetables, paddy, maize, cotton, chillies etc. are sown together. All the activities are performed by both male and female.

Table 28.2: Major Activities Undertaken During Jhum Cultivation and Festival Associated with Jhuming

N= 40

Activities	*Calendar Year*	*Time Required (In Days)*	*Performed by (Male/Female)*	*Festivals*
Selection of land	October–November	10 days	Male	–
Cutting of forest	November–December	24 days	Male	Pawlkut
Natural drying	January–February	50- 60 days	–	–
Burning of dried forest	20 February–15 March	1 day	Male	Chapchar kut
Cleaning	March	7 days	Male and Female	–
Sowing	April–June	20- 25 days	Male and Female	–
Weeding	i) May–Aug	30- 50 days	Male and Female	–
	ii) Sept–Oct	121 days		
Construction of watching house (Thlam)	March	7 days	Male and Female	–
Harvesting	October–November	60 days	Male and Female	Mimkut
Threshing/Winnowing	November	30 days	Male and Female	–
Storing	November–December	30 days	Male and Female	–
Marketing	August–October	100 days	Male and Female	–

The calendar of operations of shifting cultivation controls their social life, festivities and religious observances. Almost all operations are connected with socio-religious functions. There are many dances, songs, folk tales associated with this type of cultivation in the study area. Some of the festivals are *Chapchar kut*, *Mimkut* and *Pawlkut* which are associated with jhuming but after the advent of Christianity most of these festivals disappeared.

Perception of Jhumias Relating to Environmental Impact of Jhumming

In this chapter it was intended to study the perception of jhumias relating to environmental impact of jhuming. For this purpose 11 items or statements relating to environmental impact of jhuming were presented to the respondents in order to ascertain whether the jhumias are well aware of the adverse effects of jhuming. The perception includes both ecology and economy perception and the results have been presented in Table 28.3.

Ecology Perception

The study reveals that the jhumias are well aware of the harmful ecological impact of jhuming. Most of the jhumias agreed with all these items relating to harmful ecological aspects of jhuming. When forest areas were large and the population was small, the practice of shifting cultivation created no adverse effects on forest and the

Figure 28.2: Chapchar Kut Festival in Mizoram

Figure 28.3: Jhum Implements

environment and worked in perfect harmony with nature. But with the increase of population in the last centuries and reduced forest area shortened the cultivation cycle, which has adversely affected forests and the environment. Shifting cultivation causes large- scale damage to the forests and has resulted in deforestation and denudation of hill slopes.

Table 3 : Perception of Jhumias Relating to Environmental Impact of jhuming

N = 40

Sl.No.	Items	Response Category	Frequency	Percentage
	Ecology Perception Do you know that			
1.	Jhum cultivation (due to deforestation) helps to endangered wild animals and birds	Yes No	40 0	100 0
2.	Jhum cultivation destroy the flora and fauna	Yes No	39 1	97.5 2.5
3.	Jhum cultivation causes in loss of soil fertility	Yes	38	95
	No		2	5
4.	Jhum cultivation reduces the water holding capacity of the soil	Yes No	38 2	95 5
5.	Jhum cultivation indirectly causes to increase the temperature in your locality	Yes No	38 2	95 5
6.	Jhum cultivation causes in soil erosion	Yes No	37 3	92.5 7.5
7.	Jhum cultivation causes destruction of forest	Yes No	37 3	92.5 7.5
8.	Jhum cultivation increases the occurrences of drought/flood	Yes No	37 3	92.5 7.5
				N= 40
	Economy Perception Do you know that:			
1.	Jhum cultivation reduces the yield of crops	Yes No	20 20	50 50
2.	Indiscriminate cutting and burning of timber, forest trees, bamboo etc., means loss of money of the Govt	Yes No	23 17	57.5 42.5
3.	Jhum cultivation gradually reduces the quality of seed	Yes No Could not reply	18 16 6	45 40 15

Shifting cultivation reduces the diversity of species in the environment.. Several species of trees have already disappeared due to continuous cutting and burning. Birds and animals are more difficult to sight and sing today than was the case several decades ago. The jhum is also liable to soil erosion caused by the direct impact of raindrops and are cited as causes of flood in the plain areas. Burning and high temperatures reduces the organic matter and leads to rapid mineralization and volatilization of inorganic nutrients.

Economy Perception

The jhumia in the study area also perceived that jhuming also causes loss of property and money for the Government of Mizoram. The production of agricultural crops under shifting cultivation areas is usually low. Their resistance to modern inputs and their reluctance to produce for the market labels them as primitive farmers. The system of shifting cultivation provides no opportunity for infrastructural development which could be helpful in increasing agricultural production and providing basic amenities, such as public health, roads, education, communication, etc. since there is no permanency of settlements and agriculture. Valuable trees are gradually replaced by useless shrubs and more resistant tree species. Already trees of larger size are almost absent. Cutting and burning of such valuable timber species is a national waste.

Agro-Economic, Socio-Psychological Factors Contributing to Jhuming in Mizoram

In the previous discussions it was found that jhumias in study area are well aware of the harmful consequences of jhuming on ecology and environment. Still they are practicing jhum generation after generation. The present investigation was aimed to identify the important factors which are contributing the jhum cultivation in the state, Mizoram. The important factors of jhuming have been identified as follows:

1. Socio- psychological factors of jhuming
2. Agronomic factors of jhuming
3. Economic factors of jhuming

The findings have been presented in Table 28.4.

Table 28.4: Important Factors of Jhuming in Mizoram

N= 40

Sl.No.	*Items/Statements*	*Response Category*	*Frequency*	*Percentage*
A.	**Socio-Psychological Factors of Jhuming**			
	Community Integration			
1.	I practice jhum because jhum can help to maintain a harmonious relationship among the community	Agree Disagree	34 6	85 15
2.	I practice jhum because our community is habituated with jhuming	Agree Disagree	31 9	77.5 22.5
3.	I practice jhum because I want to save my identity from the non-jhumias	Agree Disagree	26 14	65 35
4	I practice jhum because some important festivals in our community *viz.* (chapchar kut, mimkut etc.) are performed with jhuming	Agree Disagree	23 17	57.5 42.5
5.	If I do not practice jhum I may be isolated or detached from my community	Agree Disagree	73 3	17.5 82.5

Contd...

Table 28.4–Contd...

Sl.No.	Items/Statements	Response Category	Frequency	Percentage
	Inheritance of Jhum			
1.	I practice jhum because I have learnt it since childhood	Agree Disagree	34 6	85 15
2.	I practice jhum because my father/forefather use to practice it	Agree Disagree	33 7	82.5 17.5
3.	I practice jhum because our women are also efficient in jhum agricultural operation	Agree Disagree	31 9	77.5 22.5
	Tlawmngaihna			
1.	I practice jhum because I get mental satisfaction through jhuming	Agree Disagree	35 5	87.5 12.5
2.	I practice jhum for promoting the spirit of selfness service to needy person (Tlawmngaihna)	Agree Disagree	32 8	80 20
B.	**Agronomic Factors of Jhuming**			
	Mixed Cropping and Food Habits			
1.	I practice jhum because from jhum I can get variety of crops which cannot be raised by settled cultivation/terrain cultivation	Agree Disagree	40 0	100 0
2.	I practice jhum due to non- availability of irrigation water	Agree Disagree	39 1	97.5 2.5
3.	I practice jhum because the taste and flavour of jhum rice is better than other crop	Agree Disagree	28 12	70 30
4.	I cultivate jhum because burning of forest helps in control of pest and diseases	Agree Disagree	14 26	35 65
	Topography			
1.	I practice jhum because limited flat land is available	Agree Disagree	30 10	75 25
2.	I practice jhum because enough jhum land is available in Mizoram	Agree Disagree	22 18	55 45
C.	**Economic Factors of jhuming**			
	Food Security			
1.	I practice jhum because by jhumming I can store foodgrains which are under my control	Agree Disagree	40 0	100 0
2.	I practice jhum because other wage income is uncertain and only jhum can give me food security	Agree Disagree	39 1	97.5 2.5
3.	I practice jhum because mixed cropping in jhum helps to increase food production	Agree Disagree	38 2	95 5

Contd...

Table 28.4–Contd...

Sl.No.	Items/Statements	Response Category	Frequency	Percentage
	Low Cost Practice			
1.	I practice jhum because jhum is performed by my family members	Agree Disagree	391	97.5 2.5
2.	I practice jhum because agricultural inputs *viz.*, chemical fertilizers, pesticides are not required which are costly inputs	Agree Disagree	38 2	95 5
3.	I practice jhum because jhum cultivation requires only indigenous agricultural implements	Agree Disagree	35 5	87.5 12.5

Socio-Psychological Factors

Community Integration

Jhuming has been equated to the way of life of the jhumias because it was considered to be so important in their lives that all other activities of the group as well as their values, attitude and beliefs were supposed to revolve around it. It was claimed that their society, polity and culture were all based on this form of agricultural practices.

From table 4, it is found that 85 per cent jhumias agreed with the statement that jhuming can help to maintain a harmonious relationship among the community. The community life of the mizo's (jhumias) is deeply embedded in agricultural activity which invites the whole community to share both pleasure and pain. The community life is comprehensive and free from any rank or status consciousness. They participate in all social or religions functions without any consideration of social hierarchy.

The community festivals associated with jhuming have promoted community feeling and helped in influencing the attitude of the jhumias. Every festival is accompanied with dance and music which is the symbol of their love and fellow feeling. Among the community festivals called *Kut* in mizo. *Pawlkut, chapchar kut* and *mimkut* are most important. These are celebrated on specific occasions of the agricultural cycle. *So community integration is an important aspect of jhum cultivation.*

Inheritance of Jhum

Jhuming is a socially learned behavior which affects social organizations, and on the higher level upon the organization of the entire community and subsequently the populations.

It was found that about 85 per cent of the jhumias agreed that they practice jhum because they have learnt it since childhood. About 82 per cent respondents agreed with the statement that they practice jhum because their father/forefather used to practice it.

Women labour force can be fruitfully utilized, if shifting cultivation were to be practiced side by side with settled agriculture. Jhum operations can be performed by women also. Sowing, weeding and harvesting are done by both men and women, assisted by children. About 77 per cent of the jhumias agreed with this statement that women are also efficient in jhum agricultural operation.

Tlawmngaihna

In Mizoram this term *tlawmngaihna* means helping a needy person who is unable to perform his job due to sickness, or any other reason. According to the principles of tlawmngaihna, "It is a well- established convention for people, to assist families lacking male members in the task of jhuming. Opportunities for such service were frequently found in the event of sickness, death or other misfortune.

Eighty seven per cent of the jhumias responded that they get mental satisfaction from shifting cultivation and it is more of a convenience and compulsion rather than a choice to fulfill the basic needs of food. The stock of grain by jhuming which is under the control of the farmer can provides one type of psychological security.

Their code of morals and good form *'Tlawmngaihna'* (prescribing courtesy, consideration, selflessness, courage, and industry) was manifest in a high degree of mutual support and cooperation. *According to this principle, the group of people jhuming on a particular slope carried out a number of activities jointly such as construction of jhum road, clearing vegetation to a height of about five feet. A group of ceremony was performed halfway through operation and a feast was held on completion.*

Agronomic Factors

Apart from socio- psychological factors of jhum cultivation there are also other factors for which the Mizo people are practicing jhum cultivation generation after generation. The agronomic factors such as advantages of mixed cropping in the jhum field and topography also play an important role for continuance of jhum cultivation in the North- East States.

Mixed Cropping and Food Habits

The main interest of the cultivators of Mizoram is to produce almost everything that he needs. Therefore he grows not only a variety of cereals but also vegetables, oilseeds, fruits, spices, cotton etc. The requirements of timber and fuel are easily met and therefore, no importance is usually attached to trees.

It was found that most of the shifting cultivation plots were being used for growing vegetables, fruits and other cash crops. For jhum cultivation the forest in the hill slope is cleared and burnt. The land is then roughly prepared with crude implements such as wooden digging stick and hoe and the crops are raised in mixture. The usual mixtures are paddy, maize, and vegetables of various kinds. Only a few crops are grown as single crop by the tribal people.

Climatic factors such as the timing and quantum of rainfall, and the incidence of storms, site factors like aspect, attitude, and slope, decisions on crops, varieties, mixtures and placement, and the application of human effort and skills all add up to influence the productivity of a jhum.

In settled/terrain cultivation, irrigation is not available, so only a single crop can be grown. Ninety seven per cent of the jhumias responded that they practice jhum due to non-availability of irrigation water. They also responded that settled cultivation incurred high costs which require not only high investment but also high recurring expenses.

Thus the mixed cropping which is characteristic future of shifting cultivation regime once considered primitive is now being propped up as a way of increased food production in Mizoram.

Seventy per cent of the jhumias agreed with the statement that the taste of the flavour of jhum rice is better than other crop; it is purchased by non-jhuming families who can afford to do so. There are numerous indigenous varieties of each crop jhumias favoured. In maize, there is *'sticky mimban'* and *'non sticky mimpui'* varieties which are locally grown. In chillies, varieties like large *'hmarchapui'* the bird eye *hmarchate*; and in paddy 23 varieties are grown *e.g., Buhbial* (early medium duration, roundish grain), *Buhpui* etc.

Thirty five per cent of the jhumias agreed with the statement that burning of forest helps in control of pest and diseases. Burning must take place at a stage when the felled bamboos or trees are completely dried so that they burn well. The role of fire and mixed cropping together make jhuming one of the few surviving methods of organic farming. *This has enabled jhumias to reach the brink of twenty first century without feeling the need for chemical fertilizer, weedicides, pesticides or insecticides.*

Topography

Shifting cultivation occurs generally in areas where extensive hilly habitat with steep slopes and no flat areas or valleys, plenty of forest around the hills, low population density, humid climate and with little development of communication facilities are available.

Though shifting cultivation has long been viewed as a problem in the absence of any workable alternative people continued 'jhuming' on the slopes much as they have been doing for generations.

Seventy five per cent of jhumias agreed with the statement that they practice jhum because limited flat land is available. So they prepare proper seed beds in steep slopes where any disturbance of the surface by hoeing and ploughing will result in washing away the fertile top soil.

Each year the Village Council select consolidated blocks of land to be jhumed, which may be located on one or more slopes. The jhuming regulations laid down the process by which the council was to distribute plots among the village families. It is not always that the same household would cultivate the same plot of land. Very frequently individual household do not have absolute ownership right over the lands cultivated by them. They can hold the land so long as they make effective use of the same. As soon as they stop their operations, their right ceases. Fifty five per cent respondents agreed with the statement that due to long fallow period, the jhum lands are far from the village. The jhum lands are being increased from 1 tin area to 5 tin areas in the present situation(1 tin Area= 3 Bighas).

By and large the permanent fields (homestead, irrigated fields, orchards etc.) are owned individually while the bulk of the land under shifting cultivation is communally owned. There are two distinct sections of population in each village. One who claim ownership on large chunks of land and therefore, own some permanent fields besides claiming rights over the jhum land. The second category made of subsequent settlers have access to land for only shifting cultivation whatever is allotted by the village council. In such cases also only right of use is enjoyed by members undisturbed and continuously but they can not sell the land at their will though it is heritable.

An exceptional feature of shifting cultivation in Mizoram, which has survived generations of cultivators, is the freedom of each family to jhum as much land as it wants. This can only be feasible in a situation where land is perceived as plentiful.

Economy Factors

The present finding also reveals that not only socio- psychological and agronomic factors are responsible for jhuming, the economic condition of the Mizo people in particular and the people of North- East in general are also responsible for jhum cultivation in the region. The two parameters food security and low cost practice have been identified which are discussed below:

Food Security

According to FAO (2002), food security implies, "a situation that exists when all people, at all times, have physical, social and economic access to dietary needs and food preferences for an active and healthy life".

According to United Nations, a household is food secure when it has access to food need for a healthy life for all its members (adequate in terms of quality, safety and culturally acceptable) and when it is not at undue risk of losing such access. Food security should have three specific aims: namely, ensuring production of adequate food supplies; maximizing stability in the flow of supplies; and securing access to available supplies on the part of those who need them.

A perusal of table 4 reveals that most of the jhumias agreed with the statement that they used to practice jhum because by jhuming they can store food grains which are under their control. The stock of grain a jhumias maintain provides him a psychological security. Ninety eight per cent respondents agreed with the statement that other wage income is uncertain and only jhum can give them food security. Experiences show that money income cannot be fully trusted for physical subsistence in the far flung mountain areas. *Therefore the jhumias is driven to shifting cultivation so that he may replenish his stock of food grains which is visible, within his reach and under his control. He is therefore; psychologically not prepared to opt for an uncertain wage income however attractive it may be in lieu of shifting cultivation.*

Combination of different crops not only helps in optimizing the use of natural processes but it also contributes in a significant way to soil and water conservation and enrichment and also helps in pest and disease control. The multiple choice principle also helps in the management of agriculture because harvesting of crops is not one time operation which is generally followed in settled cultivation. The choice of crops is so considered that the gestation period of the crops is any where between two to eight months. *Thus, harvesting of products under shifting cultivation continued for a longer period of the year and it generally starts from July to January which is not available in settled cultivation system.*

The wide stretch of sowing and harvesting season and the variety of crops grown are an insurance against the uncertainties of rain both in terms of quantity and timing. Thus, the mixed cropping which is characteristic feature of shifting cultivation regime once considered primitive is now being propped up as a way of increased food production.

Low Cost Practice

The labour requirement for sowing, rearing and harvesting for the variety of crops is evenly spread over a long period of time and not peaked seasonally. It helps in intensive use of manpower in the sparsely populated mountain areas. The labour is manual and tools are of indigenous agricultural implements.

Land being abundant for jhum cultivation in Mizoram, the jhumias having more working hands in command can produce more crops. Under the jhum economy the size of a family farm is directly related to the number of working members in a family.

Jhum operations like clearing of the jungles, burning, and construction of jhum hut (*thlam*) are performed by man. Men, women as well as children performed in various operation like sowing, weeding, harvesting and marketing.

Figure 28.4: Thlam (Jhum Shelter) Constructed by the Jhumia

The act of burning has many benefits other than building up soil fertility. The destruction of weeds and improvement of tilth necessary for a proper seed bed are achieved with the help of fire which jhuming alone made to improve the fertility of the jhum land. Ninety five per cent of the jhumias do not used agricultural inputs *viz.*, chemical fertilizers, pesticides which are costly inputs.

The desire to invest in agriculture is greatly hampered due to lack of opportunities. The jhumias generally do not prefer high yielding varieties of seeds for different reasons. They fear of losing soil fertility and production in the second cropping if they use high yielding varieties (HYV's), Use of indigenous seeds and implements is, therefore, a measure of self reliance and food security in a non-intensive system of land use in the interior hill areas. The application of chemical fertilizers in hilly slopes is considered useless as these will be washed away with heavy rain.

Conclusion and Recommendations

The whole issue of "sustainable development" of jhum- centered land- use system has to be viewed from a variety of different perspectives *viz.* population pressure on land, social organization and function as part of a given cultural landscape, conservation and management of biodiversity linked natural resources, and land degradation issues.

Since jhum is a way of social life in Mizoram where many factors like food security by way of mixed cropping practices, taste and flavour of the jhum food, culture and festivals associated with jhuming, community integration and inheritance of jhuming from forefathers and even the topography of the North- East in general and Mizoram in particular-- all have been intermingled with jhuming over the last few centuries which is very difficult to dissociate these from the Mizo community.

Whereas the jhum cultivation of the village would surely persist in jhuming, cultivation of low- volume high- value crops in the jhum lands rather than cereals (paddy) may be encouraged as an immediate measure, to start, with keeping the topography and landscape of the village in view, plantation of horticulture crops may be taken up as long- term measure which has caught the imagination of a great majority of jhumias. As the system is multidisciplinary in nature, separate education and training for development, management and extension of agroforestry system at different level are required.

Whilst access to markets is difficult for a large number of communities in view of the lack of infrastructure, alternative solutions to the remoteness of communities should be considered.

Processing and small manufacturing facilities such as Cottage industries, notably, cane- works may also be a viable proposition. There is an abundant reserve of this natural resource in the nearby forests for the cane- works industry which remains to be tapped and exploited fully. To start with, provisions may be made for financial assistance and incentives for undertaking such ventures. Promoting value addition through processing of farm and homestead products particularly, fruits, vegetables, ginger, tea etc. through the indigenous entrepreneurs, in order to increase returns and to retain wealth within the region.

Inputs such as fertilizers, seeds and plant protection chemicals and equipment etc. may be made available within the easy reach of the farmers to encourage cultivation of vegetables on commercial scale.

Jhumia's Co- operative Society can play a pivotal role in accelerating the tempo of rehabilitation of the jhum land. The Co- operative Society will create awareness among the farmers and motivate them to implement rehabilitation programmes of jhum land. They should arrange timely disbursement of short, medium and long term credit. The jhumias will get technical guidance from the experts through the society. Processing and marketing of the produce of the members by the society will also facilitate the timely recovery of loans and hence the problem of overdues will be automatically solved. Thus, it is apparent that Co- operative Society can solve the problems of jhumia's and help in improving the overall quality of life of jhum cultivation, leading to the formation of egalitarian social order.

Recommendations

In order to improve the socio- economic condition of the jhumias in the study area, the following recommendation are made which may be examined for adoption by the concerned leaders and planners of Village Councils and by the Department of Agriculture and Horticulture, Govt. of Mizoram.

Jhuming with Plantation Crops and Orchards

Since jhuming is a way of life and ensures *food security* by way of mixed cropping for the Mizo people, total replacement of jhuming in Mizoram is undesirable and impossible also.

In order to enhance their economic condition, a portion of the jhum field allotted by the Village Council to the jhumia may be utilized for orchards or plantation crops which is a profitable enterprise and can harvest good return. Appropriate rules and regulation may be framed up by the Village Councils who has the responsibility to distribute the jhum land and provision of high yielding varieties (HYV) of planting materials of orchards should be ensured for the jhumias by the Department of Agriculture and Horticulture, Govt. of Mizoram.

Organic Jhum Farming

It is evident from the study that continuous jhuming with reducing the jhum cycle accelerates the environment degradation and extinction of flora and fauna to a great extent.

In each and every household of Mizoram, particularly among the jhumia family, rearing of poultry and piggery is a common practice. Although the excreta of the animals are used in the kitchen garden in a limited scale, there is an immense scope of application of these animal excreta in the jhum field as an organic manure which not only enhance the production and productivity of crops but also improve the soil health.

The Department of Horticulture and the representatives of Village Council should take necessary steps for changing the knowledge and arousing the interest regarding the application of organic manure through use of mass media.

Value Addition and Marketing Facilities of Horticultural Crops

Horticultural crops like ginger, passion fruits, banana, lemon, pineapple etc. are abundantly growimg in Mizoram. A food processing unit (*MIFCO, food processing and training center*) near Chhingchhip has been established and different vegetables and fruits are processed in this plant. There is an immense scope of marketing facilities of this bottled juice specially the passion fruits, bamboo shoots pickles etc, in other states in India.

Development of Cottage Industries

The forest areas of Mizoram are blessed with plenty of Bamboo's and trees, herbs and shrubs. Different cottage industries, notably, cane- works may also be a proposition, provided that marketing facilities are given top priority by state Govt. there is abundant reserves of this natural resources in the nearby forest which remains to be tapped and exploited fully.

References

Barkakoti, S. (1990). Alternatives Plan for Jhum Area Development. In: *Shifting Cultivation in NE India*, (Ed.) D.N. Majumdar. Om Sons Publications, New Delhi.

Basic Statistics of NER (2002). Govt. of India, North Eastern Secretariat, Shillong p. 42

Edwards A.L.(1969). Techniques of attitude scale construction, Vakils, Feffer and Simonds Pvt. Ltd. Mumbai

Goswami (1968). Shifting Cultivation in the Hills of North- East India. *Indian farming* XXI (4) pp. 10-13

Nunthara, C. (1981). Social Dimension of Rural Development in Mizoram. *North-Eastern Hill Regions of India- Problems and Prospects of Development*, Agricole Publishing Academy, New Delhi pp. 105-111.

Ray, G.L. and Mondal Sagar (2004). Research Methods in Social Sciences and Extension Education: second edition, Kalyani Publishers Ludhiana.

Roy Burman, B.K. and Sharma, B.D. (1970). Quoted from Tripathy, R.N. Agricultural Problems in Tribal Areas. Perspective on Tribal Development and Administration, National Institute of Community Development, Hyderabad

Saha, N. (1973). *Economics of Shifting Cultivation in North-East India*. Agro- Economic Research Centre, Jorhat.

Saha, N . (1976). Levels of Production and Income under Shifting Cultivation in North- East India. *Proceedings of the Seminar on Shifting Cultivation in North- East India*. North- East India Council for Social Science Research, Shillong.

Statistical Handbook of Mizoram (2006). Mizoram at a glance: (2006), Directorate of Economics and Statistics Government of Mizoram.

Upa Haukhama (2008). *Chhingchhip Khaw Chanchin*. Mizoram Publication Board, Aizawl.

Chapter 29

Underexploited Fruits of North West Himalaya: A Way to Sustainability

Shailesh Tripathi[1]*, *V.K. Sah*[2]**, *V.K. Yadav*[3]*** and *Rahul Dev*[4]****

[1]*Assistant Professor (Horticulture);* [2]*Professor (Ecology)*
College of Forestry and Hill Agriculture,
G.B. Pant University of Agriculture and Technology
Hill Campus, Ranichauri (Tehri Garhwal) Uttarakhand – 249 199
[3]*Principal Scientist (Plant Breeding), IGFRI, Jhansi, U.P*
[4]*Research Scholar, IARI, New Delhi*

ABSTRACT

India, known for its excellent climate and fertile land, is a reservoir of many fruit species, most of which are not commercially cultivated but provide a significant source of livelihood support for many rural communities. There is a wide variety of wild tropical fruits in India which are underexploited and underutilized. As the definition goes, underutilized fruits are those generally with high nutritional and medicinal value, having less economic importance than the more popular ones, and also those usually grown in limited or very limited areas familiar only to very few people. Underutilized fruits are also the source of sustainable income to the native community living in the forests, degraded lands, wastelands and scarcity areas and proper exploitation of these minor fruits had great potential to provide sustainable income to the people with the least risk of the ecological disorder. These

E-mail: *shaileshgbpuat@gmail.com; **sahvk_111@yahoo.co.in; ***vijayyadav@gmail.com; ****rahul2iari@gmail.com

underutilized fruits occupy a prominent place in India; enter a limited way in the interstate trade. A large number of lesser known fruit species that have a rich potential are yet to be utilized fully. Besides, these fruits are an important source of minerals and vitamins and have high therapeutic values they also have the ability to grow successfully even under adverse agro-climatic conditions, marginal and wastelands, making it a business worth looking. Thus, these underexploited fruits play an important role in the security of food and nutrition. These fruits are known for their typical flavor and taste. The variety of processed products can also be prepared from the underutilized fruits. The north-West Himalayan region which includes foothills, sub-temperate, temperate and dry temperate or cold arid region has rich diversity in different fruits and their wild relatives. The species of importance varied as per altitude and there are some species *viz. Ampelocissus latifolia, Berberis asiatica, Zizyphus mauritiana, Syzygium cumini, Pyrus pashia, Rubus ellipticus, Punica granatum, Prunus armeniaca, etc.* which are important in all altitude from tropical to dry temperate. Native information about fruit plants, their local names, family, brief botany, season of flowering and fruiting, distribution and method of utilization of some important fruit plants are discussed in the chapter.

Keywords: ***Underutilized fruits, Underexploited, Himalaya, Nutritional security, Sustainability.***

Introduction

Nature bestowed man with about 2.50 *lakh* species and perhaps the same number of unknown species. Out of this plant wealth, 80,000 plant species are edible but man exploited only 3000 of these plant species for food, shelter and clothing. Commercial scale cultivation is limited to 150 species and consumption for food is limited to 20 different kinds of crops, the world over. Four crops; rice, wheat, maize and potato alone supply about 60 per cent of total calories intake. The food basket base is shrinking consequent to monocrop culture and reduced area under cropping. There is reduction in the consumption of organic compounds making ourselves vulnerable to stresses and blockage of immune regulators in the body system. Awareness to enlarge food basket incorporating edible underutilized and underexploited crops has come up as a positive development in food and nutrition.

Underutilized fruits (UUF) are those species with underexploited potential in contributing to food security, health, income generation and environmental services (Haq and Hughes, 2002). The UUF species have a potential to contribute to food security and poverty alleviation including the value addition aspects to diversify use and shelf life. This has not been fully exploited. Therefore, the demands and their promotion should be done in a sustainable manner (Ravindran *et al.*, 2002). The range of these fruit species covered is wide, including plants that provide edible fruits, grains, leaves, nuts, oils, roots and tubers, fibres, medicines, spices, stimulants and their derived products. Medicinal values of these fruits become more vital at the moment due to high poverty levels and HIV/AIDS pandemic in the developing countries (Sah and Tripathi, 2009; Tripathi and Sah, 2010).

Species are underutilized or less utilized because they are not competitive with other crop species in the same agricultural environment. UUF species differ from one country to another. A species which may be considered exotic in one country may be quite common in another. It is perhaps better to define underutilized fruits as those generally having less economic importance than the more popular ones, and also those usually grown in limited or very limited areas and is familiar only to very few people. Further, many crops considered neglected at a global level are staples at a national or regional level, contribute considerably to food supply in certain periods or are important for a nutritionally well balanced diet (Padulosi and Hodgkin, 1999).

There is increasing interest throughout the world for neglected and underutilized crop species (NUS) to identify and develop new crops for export and domestic market. From an international perspective/emerging global and national attention on neglected and underutilized species originated basically from there major areas of interest and concern: (*i*) new market and uses (*ii*) the environment (*iii*) food security and nutrition (Datta, 2007).

Concept/Interest developed slowly on neglected and underutilized crop species due to variety of factors, concerns and needs. True novelty of underutilized fruit species depends on the following factors:

- Species that are wild and for which there is an existing market
- Species that are suitable for introduction into cultivation
- Species that are already cultivated or semi-domesticated on a small scale (underutilized species) and for which there is potential for more extensive or more productive cultivation
- Their contribution towards agricultural diversification
- Better land use (marginal land and changing environment)
- Food security and a more balanced diet
- Better safeguard of our agro biodiversity
- Their economic potential
- Opportunities they provide for market diversification
- Additional source of income to farmers and employment opportunities

Why Underutilized Fruits are Unpopular?

The reasons for poor popularity of underutilized fruits crops are:

- Lack of awareness about the economic benefits
- Non-availability of good quality planting materials
- Lack of technology to reduce the gestation period and enhance the fruit production
- Poor marketing network
- Lack of technology for value addition through processing

Scope of Underutilized Fruit Crops

- The average productivity of horticulture crops is just half of the national productivity. As grain farming is un-remunerative in the undulating topography of hilly tracts, which is deprived of irrigation facilities, despite government of India's has been putting for the endeavors to uplift the region, vast potential remains unexploited. It becomes possible to exploit the untapped potential of the region through location specific horticulture and subsequently expanding the area under horticultural crops. Production of UUF can also be increased through adoption of scientific technologies (Rai *et al.*, 2005).

☆ There are large areas of marginal and waste lands which are not suitable for cultivation of the above species either due to poor quality soil or lack of water resources, can easily be put to use for low inputs crops (UUF) in order to diversify the present day agriculture. In this way, UUF can play a significant role in maintaining the ecological balance and environmental conservation (Patel *et al.*, 2010).

Table 29.1: Suitable Underutilized Fruit Crops for Wastelands

Wastelands	*Suitable Underutilized Fruit Crops*
Degraded forest lands	Jackfruit, Custard apple, Bread fruit, Hogplum, Governor's plum, Kaphal, west Indian cherry
Waterlogged marshy lands	Alligator's apple, Bael, Local Mahua, *Salacia edulis, Salacia chinensis*
Sandy waste lands	Bread fruit, Ber, Star apple, Bael, Bullock's heart, Custard apple, Cherimoya, Sour soap
Salt affected lands	Rambutan, Ber, Local Mahua, Alligator's apple, Jamun
Industrial waste lands	Cashewnut, Governor's plum, Jamun, Tamarind, Ber, Custard apple
Undulating uplands	Custard apple, Jamun, Aonla, Kendu, Cherimoya
Degraded pastures and grazing land	Aonla, Tamarind, Jamun, star apple, Katta phal

☆ Underutilized fruit crops are particularly important for their medicinal properties and famous for the retentive value in *Ayurvedic* medicine. Many of wild fruit species origin possesses laxative and other curative properties and mostly people are familiar with medicinal properties of locally grown fruit crops but these are yet to be utilized fully for their medicinal and therapeutic value (Ponnuswamy *et al.*, 2009).

Table 29.2: Underutilized Fruits having Medicinal Properties

Fruits	*Medicinal uses*
Bael	Appetizer, stomachic, cooling
Mahua	Cough, cold, bronchitis
Kaith	Fruit astringent, stomachic, stimulant
Bahera	Fruits purgative, astringent, useful in piles, dropsy, dying and tanning
Bread fruit	Resin and gum
Chalta	Stomachic, cooling
Wild Jamun	Stomachic, improves hemoglobin in blood
Governor's plum	Young leaves astringent and stomachic, bark used as gargle

☆ Apart from medicinal properties, UUF are also known for their nutritive values. In comparison to majority of other common cultivated fruits, underutilized fruits are very rich source of vitamins, proteins, carbohydrates, minerals, fat etc. which are needed to develop resistance against diseases, normal vision, development of bones, proper growth and maintenance of human body etc.

Table 29.3: Sources of Vitamin A

Fruits	*Vitamin A (IU)*
Diospyrus kaki (Persimmon)	2000-2710
Artocarpus heterophyllus (Jackfruit)	175-540
Cyphomandra batacea (Tree tomato)	150-540
Moringa oleifera (Drumstick tree)	184

Table 29.4: Sources of Vitamin C

Fruits	*Ascorbic Acid (mg/100 g pulp)*
Seabuckthorn (*Hippophae* spp.)	2000
Barbados cherry (*Malpighia glabra*)	1540
Aonla (*Emblica officinalis*)	600
Kiwi fruit (*Actinidia chinensis*)	300
Manila tamarind (*Pithecellobium dulce*)	138
Indian jujube (*Zizyphus mauritiana*)	39-166
Sugar apple (*Annona squamosa*)	42

Table 29.5: Sources of Protein

Fruits	*Protein (mg/100 g pulp)*
Chironji (*Buchanania langan*)	21.60
Pistacia vera (*Pistacia vera*)	19.10-25.40
Indian almond (*Terminalia cattappa*)	19.10-24.40
Indian lotus (*Nelumbo nucifera*)	16.60-24.20
Water chestnut (*Trapa nutans*)	12.20

☆ Several underexploited fruits provide raw material for processing industry. Commercial production of food products from underexploited fruit species has to reduce wastage and promote widespread marketing for these items.

☆ Underutilized fruits are the source of sustainable income to the native community living in forests, degraded lands, wastelands and scarcity areas. The Indian National Commission on Agriculture has suggested that collection, processing and marketing of the minor forest produce can generate gainful employment for over 10 million persons every year. Proper exploitation of these minor fruits had great potential to provide sustainable income to the people with the least risk of the ecological disaster.

Table 29.6: Underutilized Fruits Used for Processing Purposes

Products	*Fruits*
Jam	Jamun, Karonda, Aonla, Jackfruit
Jelly	Tamarind, Jamun, Karonda
Preserved	Ber, Aonla, Bael, Ker, Karonda
Candy	Aonla, Karonda
Glazed fruits	Tamarinds, Annanas
Juice/Syrup/Sherbat/ Beverage	Tamarind, Bael, Jamun, Karonda, Phalsa, Jujube
Wine	Mahua, Jujube
Chutney	Karonda, Wood apple
Sauce	Karonda, Tamarind
Canning	Jujube, Jackfruit
Pickle	Aonla, Amra, Karonda, Ker
Dehydration	Bael, Jackfruit, Karonda, Ker, Phalsa

☆ Wild fruit flora is a genetic reservoir to draw genes for the resistant to biotic stresses and abiotic stresses like, cold, drought, salinity etc. Besides, the genes for higher productivity and distinctive quality traits could also be located from this gene pool *e.g.*, in case of *Annona* species, *Annona atemoya* has low acid content and smooth red surface. *Annona reticulata* and *Annona cherimoya* fruits have better self life. These traits can be utilized for improving existing *Annona* varieties.

Underutilized Fruits Crops of Himalayan Region

The Himalayans region is one of the hot spot for genetic variability for different characters in underutilized crops. India is a primary centre of diversity of various fruit crops such as jack fruit, banana, mango, jamun, large cardamom, black pepper etc.

The horticulture based food basket is limited to less than 100 crops, ornamentals, medicinal and aromatic plants. In India alone, there are around 36,000 species of plants including microorganisms, of which 17,500 species are higher plants. India possesses 11.9 per cent of the world flora with 5725 endemic species belonging to 141 endemic genera and 47 families of higher plants. Among the domesticated species, about 166 crop species originated and diversified in this part of the world. The species of horticultural importance include 54 in vegetables, 104 in fruits and 27 in species. As per one estimate there are more than 3000 edible fruit and nut species. Many have unique taste and are good source of minerals and vitamins. Several other worlds' fruits hold great promise and yet to be recognized and exploited for their potential

(Jalikop, 2006). Very little work has been done to exploit genetic resources of these unexploited fruits of Himalayan region and they are facing severe genetic erosion and disappearance as result of activities related to urbanization, construction of hydro-electric power projects and intensive agriculture.

The North-West Himalayan region which includes foothills, sub-temperate, temperate and dry temperate or cold arid region have rich diversity in different fruits and their wild relatives. The species of importance varied as per altitude and there are some species *viz. Ampelocissus latifolia, Berberis asiatica, Zizyphus mauritiana, Syzgium cumini, Rubus biflorus, Rubus ellipticus, Pyrus pashia, Punica granatum, Prunus armeniaca, Morus serrata, Opuntia elatior, Ficus auriculata, Ficus palmata, Juglans régia, Debregeasia longifolia, Pyracantha crenulata, Coriara nepalensis, Cotoneaster microphyllus etc.* (Negi, 1986) which are important in all altitude from tropical to dry temperate. Native information about fruit plants, their local names, family, brief botany, season of flowering and fruiting, distribution and method of utilization of some important plants are presented in Table 29.9.

Important Underutilized Fruit Crops

Rubus ellipticus

It is also called Hisalu or Yellow Raspberry in Uttarakhand. It is a tall sub-erect shrub. The stem and branches are shaggy with long tawny bristles. There are numerous prickles on the stem. Flowers are white in colour; fruit is yellow and 0.5 inch in diameter. It is found in all types of forests but is typical of open chirpine and ban oak forests. It is commonly eaten raw or jelly is prepared from ripe fruits. The nutritional composition per 100 g of ripe fruit is:

Moisture (g)	94.2	Energy (Kcal)	25
Total ash (g)	0.19	Ascorbic acid (mg)	53.77
Crude fat (g)	0.69	β-Carotene (mg)	1867
Crude fibre (g)	0.16	Iron (mg)	24.33
Crude protein (g)	0.52	Calcium (mg)	473
Carbohydrate (g)	4.23		

Pyrus pashia

It is a medium sized deciduous tree belongs to the family Rosaceae, locally called *Mehal* or wild pear in Uttarakhand. Flowers are white; fruits 1.5-4 cm across, globose, depressed at top, some what rough with raised white specks, dark brown when ripe, endocarp having grit cells. Fruits do not have much edible value. Seedlings are used as rootstock for pear. The nutritional composition per 100g of ripe fruit is:

Moisture (g)	17.21	Energy (Kcal)	85
Total ash (g)	0.68	Ascorbic acid (mg)	57.3
Crude fat (g)	0.04	β-Carotene (mg))	104.0
Crude fibre (g)	3.23	Iron	0.71

Crude protein (g)	1.04	Calcium	73.4
Carbohydrate (g)	16.4		

Ficus roxburghii

Ficus roxburghii Wall. is again one of popular fruits found growing wild in the hills. In local dialects it is called as *trimmal, timal, timla, Bedu* in different Himalayan region. It is a deciduous woody tree. It is found in the Himalayan region up to an elevation of 1,700 m. Fruit is syconoid, globose, having a 4.5-cm-long stalk, 4.5 cm in diameter. The fruit is, in fact, a fleshy receptacle, enclosing a number of true fruits and average yield of a tree recorded to be 32.4 kg.

The fruit contains 87.1 per cent moisture and 7.5 per cent total soluble solids. The total soluble solids of the sweet jelly-like substance, however, are 9.9 per cent. The fruit is almost devoid of acidity; the total sugars are 6.15 per cent; the reducing sugars are 6.12 per cent; the non-reducing sugars are 0.03 per cent and pectin is 0.48 per cent. The vitamin C content of the fruit is only 3.35 mg per 100 g and this content is rather very low. The fully ripe fruit contains 1.068 per cent total minerals, as represented by its ash. The percentage content of some of the mineral elements in the fruit, *viz.*, phosphorus, potassium, calcium, magnesium, and iron is, 0.039, 0.331, 0.039, 0.045 and 0.003, respectively.

The fruits are very good for eating and the jelly-like substance contained in them makes them very tasty. The fruits should be very suitable for bottling, and work should be initiated on this aspect. The possibility of jam-making from this fruit should also be explored.

Myrica esculenta

Myrica esculenta Buch. – Ham. Ex D. Don., belongs to family Myricaceae is called "Bay Berry" and "Box Myrtle" in English and popularly called "Kaphal" in Hindi. It is native to China and Japan and now found in subtropical Himalayas, from River *Ravi* eastwards to Assam. It is an evergreen tree and grows wild in the temperate regions of India at an altitude of about 900-2100 m msl. The trees are dioecious. The fruit is purplish red in colour. The stone is much bigger when compared to the size of the fruit and the amount of pulp is very less. The potential yield at different sites is 2.0-4.2 tones/ha of which 2.8-7.2 per cent is harvested for income generation. The nutritional composition per 100 g of ripe fruit is:

Moisture (g)	71.21	Energy (Kcal)	115
Total ash (g)	0.67	Ascorbic acid (mg)	70.11
Crude fat (g)	1.46	β-Carotene (mg)	306.66
Crude fibre (g)	1.11	Iron (mg)	3.33
Crude protein (g)	4.58	Calcium (mg)	19.04
Carbohydrate (g)	20.96		

Figure 29.1: Tree and Fruits of *Myrica esculenta* (Kaphal)

Hippophae spp.

Among various underutilized plant genetic resources, seabuckthorn possesses a very important place with respect to the species of the dry temperate regions of high Himalayan region. Genus *Hippophae*, commonly known, as seabuckthorn, is a thorny shrub or a medium size tree, grows widely between 1900 to 3000 m altitudes mostly in dry temperate regions of North-Western Himalaya of India. It belongs to the family Elaeagnaceae and has an ability to fix atmospheric nitrogen, thus helps in soil amelioration. In North-Western Himalayas, *Hippophae* has 3 main species, *viz., H. rhamnoides, H. salicifolia* and *H. tibetana*. *H. rhamnoides* which is a common seabuckthorn or Indian seabuckthorn alone has a share of 80 per cent of the total area under seabuckthorn. It is distributed in drier areas of Jammu and Kashmir, Himachal Pradesh, Uttarakhand and some parts of Arunachal Pradesh and Sikkim (Dwivedi *et al.*, 2006; 2007).

Due to its wider adaptability, drought resistance capabilities, deep root system and capacity to fix atmospheric nitrogen through Frankia symbiosis, the plant has been widely planted in China to control soil erosion and reclamation of degraded lands. The leaves, berries and seeds of seabuckthorn have high nutritional and medicinal value and are excellent source of vitamins C, B_1, B_2, E, F, K, P, pro-vitamin A, sugars and organic acids. High content of vitamin C and E and other bioactive substances makes it a suitable species for pharmaceutical, cosmetic and food industry (Yadav, *et al.*, 2005; 2006; 2011 and Rao, *et al.*, 2006). On the basis of survey of districts

Figure 29.2: Naturally Growing Plants of *Hippophae salicifolia* (Seabuckthorn)

Uttarkashi and Chamoli of the Uttarakhand state 10 plus stands selected for further multiplication.

Table 29.7: Distribution and General Characteristics of *Hippophae* spp.

Species	*Distribution*	*Growing Altitude (m)*	*Plant Height (m)*	*Flowering Time*	*Fruit Ripening Time*
Hippophae rhammoides	India, China, Tibet, Kyrghystan, Kazakistan, Uzbegistan	600-4200	5-6 m	May	September–October
Hippophaesa licifolia	India, Tibet, Bhutan, Nepal	2700-3700	3-10 m	June	October
Hippophae tibetana	India, Tibet, China	3000-5200	0.8-1.2 m	May	August–September

·runus armeniaca

Wild apricot is one of the most important under-utilized fruit of mid hills and ry temperate regions of the country. It is called as *chulu, chulli, sahara* and *zardalu* in ıe local dialect found in different regions of Uttarakhand, Himachal Pradesh and ımmu and Kashmir (1200-3500 m amsl). Botanically the plant is same as cultivated pricot but lower content sugar make it unfit for table purpose. The wild apricot pulp

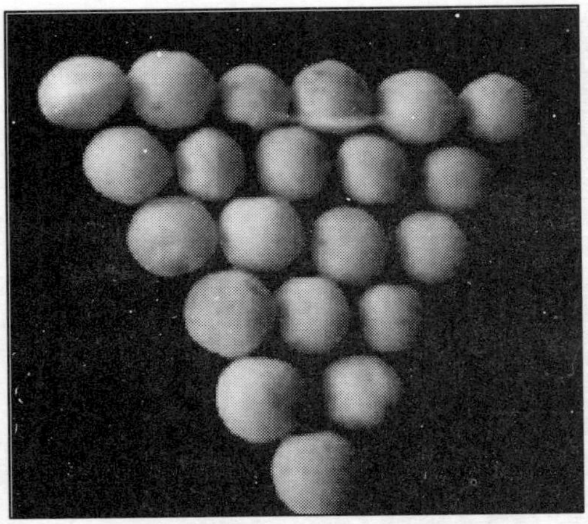

Figure 29.3: Fruits of Wild Apricot *(Prunus armeniaca)*

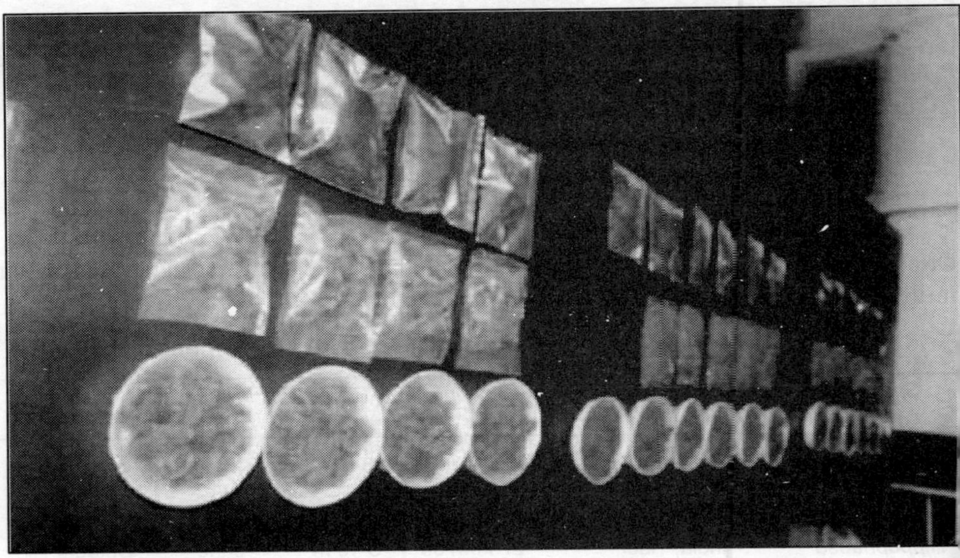

Figure 29.4: Osmotically Dried Fruits of Wild Apricot

contains many important nutrients. The mature 100g fruit reported to have wate 85.3g, protein 0.9 g, sucrose, ascorbic acid 6.36 mg, citric acid 2130 mg, carotene 61 mg, potassium 350 mg, calcium 16.0 mg, iron 0.3 mg, niacin 1.4 mg and thiamin an riboflavin in traces. The kernels contain 40-45 per cent of oil and are used for extractio of oil. The apricot oil consists of unsaturated fatty acids, such as oleic acid (74.3 pe cent) and linoleic acid (21.6 per cent) which contributes up to 90 per cent of total o yield.

Table 29.8: Chemical Characteristics of Selected Plus Stand/Trees

Sl.No.	Plus Stand/Trees	TSS	Per cent Acidity	Ascorbic Acid (mg/100gm)	Sugar Per cent
1.	Gangotri Selection 1	8.0	5.57	409.20	1.42
2.	Harsil Selection 1	7.5	5.38	228.80	1.50
3.	Dharali, Selection 1	8.0	5.70	484.00	1.61
4.	Badrinath Selection 1	6.0	4.42	365.20	0.88
5.	Badrinath (Mana) Selection 2	8.0	6.46	391.60	1.25
6.	Yamnotri Selection 1	10.0	6.72	462.00	1.02
7.	Jhelum Selection 1	9.5	6.72	400.40	0.97
8.	Suraithota Selection 1	6.0	4.67	316.80	0.62
9.	Yamunotri Selection 2	6.5	4.74	330.00	1.19
10.	Badrinath Selection 3	7.5	5.57	224.40	1.58

Rubus niveus

It is called as Mysore raspberry, Ceylon Raspberry, Hill Raspberry in English and Kala Hisalu in Hindi, is a species of *Rubus* native to Southern Asia, from Afghanistan East through India and China to Taiwan and the Philippines. In India,

Figure 29.5: Fruits of *Rubus niveus* (Kala Hisalu)

Table 29.9: Some Important Potential Fruit Crops of North-West Himalaya

Sl.No.	Botanical Name/ Family/Vernacular	Brief Description	Season	Distribution	Uses/Method of Utilization
1.	*Aegle marmelos* (L.) Corr./Rutaceae/Bel, Bengal quince holy fruit	Tall tree, flowers greenish-white, fragrant, fruit globose, 4.5-20.5 cm diam., grey or yellow	Flowering and Fruit.: April-November	Uncommon associated with Oak-Rhododendronforest	Sweet aromatic orange pulp of ripe fruits edible and diluted with water, it makes refreshing drink or sherbet and considered good of patients suffering from bacillary dysentery
2.	*Aesandra butyracea* (Roxb.) Baehni/ Sapotaceae/Cheura, Bhulel, Indian butter tree, Honey flowers	Small delicious tree, flowers white, fragrant, berries 0.8-3.2 cm, ellipsoid, green shining with milky juice	Flowering and Fruit.: October-July	Not common, along river bank, semi domesticated	The juice extracted from the flower is made in to jaggery (gur) after prolonged streaming. It is mixed with milk, tea and with other preparations. Ripe fruits are eaten, sweet in taste. Edible oil is extracted from the seeds
3.	*Aesculus indica* Colebr. ex Cambess/Hippocastanaceae/Bankhor, Paamal, Panger, Himalayan horse chestnut	Large deciduous tree, flowers white or yellow, capsule rough outside, ovoid, 2.5-4.5 cm long, dark brown	Flowering and Fruit.: May-November	Common, associated with Oak-Rhododendron forest, cultivated	Kernels are made into flour, fried with better oil (ghee) and prepared in to chapattis during famine
4.	*Ampelocissus latifolia* (Roxb.) Planch./ Vitacecae/Panibel, Jungle angoor, Jungle grape wine	Large stemmed colour climber, flowers red, fruits black, globose, 6-9 mm long, succulent	Flowering and Fruit.: August-December	Associated with *Shorea robusta, Anogeissus latifolia* forests	Ripe fruits are edible, sweet-sour in taste
5.	*Artocarpus lakoocha* (Ham.)/Moraceae/ Dhau, Monkey jack, lakoocha	Large deciduous tree, flower heads short peduncle, pubescent, fruits loculate, velvety, yellow or orange when ripe, 4.5-10 cm in diameter	Flowering and Fruit.: February-August	Common, associated with *Shorea robusta* and *Anogeissus latifolia* forests	Male inflorescences (flowers/balls) are cooked as a vegetable or pickled. Ripe fruit is eaten raw. Unripe and mature fruits are pickled or cooked as vegetable. Fruits are sold In the market

Contd...

Table 29.9–Contd...

Sl.No.	Botanical Name/ Family/Vernacular	Brief Description	Season	Distribution	Uses/Method of Utilization
6.	*Berberis asiatica* DC./ Berbericaceae/Kashnoi, Kingora, Indian berberry, tanner's berberry	Robust deciduous shrub, flowers yellow, berry blue-black, 0.8 cm diameter, ovoid	Flowering and Fruit.: April-November	Not common, associated with Oak-Rhododendron forest and wastelands	The sour tasting fruits are eaten after mixed with salt and mustard
7.	*Berberis chitria* Lindl./ Berberidaceae/Chotar, Kingor, Phachang, Indian bergerry	Deciduous shrub, berry red, oblong arrowed both ends, 0.6-0.9 cm long	Flowering and Fruit.: May-October	Common, associated with *Quercus* sp. forest	Ripe fruits are eaten raw, also taken after drying in sun
8.	*Berberis lycium* Royle./ Berberidaceae/Chatroi, Daruhaldi, kingor, Indian berberry	Evergreen, rigid shrub, flowers pale-yellow, berry violet, about 5 mm long, ovoid, glaucous	Flowering and Fruit.: March-July	Common	Fruits are edible, sweet-sour in taste
9.	*Berberis pachyacantha* Kohne/Berberidaceae/ Chotar	Deciduous shrub, flowers yellow, berry blue-black, about 1 cm long, ovoid, edible, sweet-sour in taste	Flowering and Fruit.: April-June	Common, associated with Quercus semecarpifolia forest	–
10.	*Careya arborea* Roxb./ Barringtoniaceae/Kumji, Kalikhatai, Patna Oak	Small deciduous tree, flowers pale-greenish, tinged with pink, fragrant, fruits 4-5 cm in diameter globose, green, glabrous	Flowering and Fruit.: March-July	Common, associated with *Shorea robusta* and *Anogeissus latifolia* forests	Ripe fruits are eaten raw and seeds taken after roasting. Bark and roots are also eaten
11.	*Carissa opaca* Stapf./ Apocynaceae/Karonda	Evergreen shrub, flowers white, often tinged with pink, fragrant, berry ellipsoid, purplish-black, 0.4-1.8 cm long	Flowering and Fruit.: March-June	On dry slopes	Ripe fruits are eaten raw or made in to tarts; unripe fruits are pickled
12.	*Castanea sativa* Mill./ Fagaceae/Hinguri; Sweet chestnut	Tree, fruits densely spinous, spilitting into 2-4 valves. Seeds 1-3	Flowering and Fruit.: March-October	Common, cultivated	Boiled and eaten

Contd...

Table 29.9–Contd...

Sl.No.	Botanical Name/ Family/Vernacular	Brief Description	Season	Distribution	Uses/Method of Utilization
13.	*Castonopsis tribuloides* (J.E. Smith) A. DC./ Fagaceae/Ghiseri, Ghogsa, Katonj, Tumari	Small evergreen tree, nut acorn like, 10 mm long, smooth, enclosed in scaly cup with branching spines, 1-3 together	Flowering and Fruit.: June-December	Common, often cultivated	Fruits (kernel) are eaten either raw or after roasting
14.	*Citrus medica* L./ Rutaceae/Limbba, Madkakar, Citron	Evergreen small tree, flowers white, fruits yellow when ripe, ovoid-oblong or globose, 4.5-7.5 cm long, pulp edible	Flowering and Fruit.: December-April	Common, sometimes cultivated	The peel and pulp is made into preserves and marmalades
15.	*Cordia dichotoma* Forester/Ehretoaceae/ Lisora, Clammy cherry, Indian cherry, Sebestin	Deciduous tree, flowers white, fragrant, Drupe globose or ovoid, yellowish or pinkish, 1.0-1.8 cm diam.	Flowering and Fruit.: March-July	Common, in ravines along the streams, occasionally planted	Ripe and unripe fruits are eaten raw, sweet in taste. Fruits are also pickled
16.	*Coriara nepalensis* Wall./Coriariaceae/ Gogsa, Gangroo, Masuri, Markauli, Varul, Mussoorie berry	Large deciduous shrub, flowers greenish-yellow. fruits red or dark blue at first, finally blue-black, 3.5 mm in diameter with 5 carpels	Flowering and Fruit.: February-July	Associated with Oak-Rhododendron forest	Ripe fruits are eaten through not very good in taste. If taken in a large quantity causes intoxication
17.	*Cornus macrophylla* Wall./Cornaceae/ Khaysa, Khagshi	Small deciduous tree, flowers pale yellowish white. fruits globose, black when ripe, 3-7 mm in diameter	Flowering and Fruit.: February-April	Common, riverain and shady moist places	Ripe fruits are occasionally eaten raw
18.	*Corylus jacquemoniti* Decne./Corylaceae/ Bhotiya badam, Kabasi, Datheloo, Turkish hazelnut	Small deciduous tree, nut 1.2-1.8 cm long, globose, very hard, enclosed in glandular hairy lobate scales and bracts	Flowering and Fruit.: April-September	Common, semi-domesticate	Kernels are eaten. Edible oil extracted from the nuts

Contd...

Table 29.9–Contd...

Sl.No.	*Botanical Name/ Family/Vernacular*	*Brief Description*	*Season*	*Distribution*	*Uses/Method of Utilization*
19.	*Cotoneaster microphyllus* Wall. ex Lindl./Rosaceae/ Gheri	Procumbent rigid much-branched evergreen shrub, flowers white, fruits scarlet when ripe, globose, 6-12 mm in diameter	Flowering and Fruit.: April-November	On rocky and alpine slopes	Ripe fruits are astringent, frequently eaten by shepherds
20.	*Cucumis sativus* L. var. *hardwickii* Royle/ Cucurbitaceae/Ailaru, Elaru, Jangli kheera	Climber, flowers yellow, fruits oval-oblong, rounded at both ends, 4.5-8.0 cm long and 2.0-3.5 cm in diameter marked with narrow white stripes, ripening to yellow	Flowering and Fruit.: August-October	Common, shady and moist forests	Although fruits and its flesh are very bitter in taste, seeds are consumed. Fruits are used to treat pneumonia
21.	*Debregeasia longifolia* (Burm. F.) Wedd./ Urticaceae/Sigar, Tussara, Wild rhea	Evergreen small tree, berry clusters round, 3-7 mm in diameter, orange-yellow	Flowering and Fruit.: February-July	Near by ravines and shady forests	Ripe fruits are eaten raw, sweet in taste
22.	*Debregeasia salicifolia* (D.Don) rendle./ Urticaceae/Siaru, Syanru, Tusarra	Evergreen shrub, berry globose, about 6 mm in diam., fleshy orange-yellow	Flowering and Fruit.: February-July	In riverain places	Ripe fruits are eaten raw, sweet in taste
23.	*Dendrobenthamida capitata* (Wall.) Hara/ Cornaceae/Bhamora, Himalayan strawberry	Small green tree, flowers white, fruits depressed globose, dark purple or partly orange when ripe, 0.7-1.5 cm long, glabrous, succulent	Flowering and Fruit.: June-November	Common, associated with Oak-Rhododendron forest	Ripe fruits are eaten raw and made into preserved
24.	*Dillenia indica* L./ Dilleniaceae/Chalta, Elephant apple	Small tree, flowers white, scented, fruits fleshy, 6.5-11.5 cm in diameter, enclosed by flashy calyx	Flowering and Fruit.: May-November	Common in *Shorea robusta* forest. Often cultivated	The flashy calyx is eaten raw or cooked, sour in taste. The juice of the fruit mixed with sugar and water is used as cooling drink in fever

Contd...

Table 29.9–Contd...

Sl.No.	*Botanical Name/ Family/Vernacular*	*Brief Description*	*Season*	*Distribution*	*Uses/Method of Utilization*
25.	*Elaeagnus latifolia* L./ Elaeagnaceae/ Leuchadia, Bastard Oleaster	Small evergreen tree, fruits 1.5-3.5 cm long, ovoid-oblong, succulent, red or yellow when ripe	Flowering and Fruit.: March-October	Common	Ripen fruits sweet-sour in taste and pulpy, also made into pickles
26.	*Elaeagnus umbellata* thunb./Elaeagnaceae/ Gewaii, Giyii, Silverberry	Deciduous shrub, flowers dull yellowish white, fragrant, fruits avoid, 0.5-1.2 cm long, succulent, red or orange	Flowering and Fruit.: April-September	Common, near streams and swampy places	Fruits are eaten, sour in taste and also made into pickles
27.	*Emblica officinalis* L./ Euphorbiaceae/Aonla, Amla; Indian goose-berry, Emblic, Myrobalan	Tree, flowers greenish-yellow, drupe globose, 1.5-2.5 cm in diameter, obscurely smooth, 6-lobed with verticle lines, fleshy containing a large stony seed	Flowering and Fruit.: March-June	Common, associated with *Shorea robusta* and *Anogeissus latifolia* forests	Ripe fruits are eaten raw and also made into pickles; fruits are taken as sauce in headache; used as purgative and also in hot summers, to quench thirst
28.	*Ephedra garardiana* Wall. ex Stapf. var. *gerardiana*/Ephredraceae/ Somlata, Ephedrene	Densely tufted shrub, fruits ovoid, red, succulent	Flowering and Fruit.: July-August	Common, open rocky places	Ripe fruits are eaten raw
29.	*Ficus auriculata* Lour/ Moraceae/Timla, Tirboi	Small tree, fruits strongly ribbed, hairy when young but becoming hairless, russet- brown and tinged with red or purple when ripe, 2.5-5.0 x 5.0-7.5 cm	Flowering and Fruit.: May-July	Planted near villages and margins of agricultural fields	Ripe fruits are eaten raw and also cooked as vegetable
30.	*Ficus palmata* Forsk./ Moraceae/Bedu	Small tree, figs axillary, stalked, solitary or in pairs, globose or pear-shaped, 0.8-69.2.4 cm in diameter, narrowed to the base, usually tomentose, purple when ripe	Flowering and Fruit.: June-October	In dry and waste places	Ripe fruits are eaten raw, sweet in taste, and also taken with curd with salt and mustard. The boiled fruits mixed with curd and condiments and preparation for remedy for dysentery is made

Contd...

Table 29.9–Contd...

Sl.No.	Botanical Name/ Family/Vernacular	Brief Description	Season	Distribution	Uses/Method of Utilization
31.	*Ficus racemosa* L./ Moraceae/Gular, Umnari, Cluster fig	Deciduous tree, fruits globose, pyriform or turbinate, smooth or downy, 2.0-4.5 cm in diameter, reddish or orange when ripe	Flowering and Fruit.: March-July	In open and shady forests, often planted near villages	Ripe fruits are eaten raw. Roasted fruits are taken with milk and sugar considered to increase strength. Unripe fruit is pounded, mixed with flour, made into cakes, taken during famine
32.	*Ficus semicordata* Buch-ham. Ex. Sm./ Moraceae/Jangli anjeer, Khainu	Small tree, fruits small and in clusters on long leafless scaly shoots up to 4.5 cm in diameter, arranged on main stem near the base	Flowering and Fruit.: May-November	Common, open and dry places	Ripe fruits are eaten raw and also used as curry. Unripe fruits are made into vegetable
33.	*Flacourtia indica* (Burn. F.) Merr./Flacourtia-ceae/Kandai, Madagas-car plum Bilangara	Deciduous small shrub, flowers yellowish-green, berry red or dark brown or pink, 6-10 mm in diameter	Flowering and Fruit.: February-June	Common, mixed forests	Ripe fruits are eaten, somewhat sweet in taste and also made into jellies, tarts. Rarely leaves are also eaten raw
34.	*Gardenia turgida* Roxb./ Rubiaceae/Thanera, Thunel	Small deciduous tree, flowers dimorphic, fragrant, berry grey or reddish-brown, mealy, ovoid or ellipsoid, 2.5-7.0 cm in diameter	Flowering and Fruit.: March-December	Uncommon, associated with *Shorea robusta* forest	Unripe fruits are eaten after thoroughly boiling and cooking
35.	*Garuga pinnata* Roxb./ Burseraceae/Titmer, Kharpat	Small deciduous tree, flowers yellow, slightly hairy. drupe greenish-yellow, globose, 1.8-2.8 cm in diam., smooth, fleshy	Flowering and Fruit.: March-December	Common, associated with *Shorea robusta* and *Anogeissus latifolia* forests	Fruits are eaten raw, sour in taste, pickled and cooked as a vegetable
36.	*Gaultheria trichophylla* Royle/Ericaceae/ Bhuinala, Jheri, Bher, Wintergreen, Checker berry	Prostrate under shrub, flowers light pink or white, capsules pulpy or succulent, sky blue, 8-12 mm in diam.	Flowering and Fruit.: June-September	Abundant, open grassy slopes along with mosses	Ripe fruits are eaten raw, sweet and aromatic in taste. Rarely mature leaves are steeped for tea

Contd...

Table 29.9–Contd...

Sl.No.	Botanical Name/ Family/Vernacular	Brief Description	Season	Distribution	Uses/Method of Utilization
37.	*Glycosmis cochinchinensis* Pierre ex Engler/ Rutaceae/Bannimboo, Ghurbheli, Wild lemon	Evergreen shrub, flowers white or yellowish, fragrant, berry yellow or pale orange or white or blue, depressed-globose, 6-12 mm long, juicy	Flowering and Fruit.: February-September	Common, associated with *Shorea robusta* forests	Ripe fruits are eaten, sweet in taste
38.	*Gmelina arborea* Roxb./Verbenaceae/ Gambhari, Gumari, Malaya bush beech, White teak	Deciduous tree, flowers brownish-yellowish, drupe ovoid or pyriform, orange-yellow when ripe, smooth, 1.8-2.5 cm long	Flowering and Fruit.: March-June	Common	Ripe fruits are eaten raw, sweetish aromatic in taste; fruits are thoroughly rubbed by hand to remove the rind, dried in sun, boiled and eaten
39.	*Hippophae rhamnoides* Serv. ssp. *salicifolia* (D. Don) Serv./Elaeagnaceae/Amali, Aameel, Tarwa, Seabuckthorn, Sallow thorn	Deciduous thorny shrub or small tree, flowers monoecious, appearing just before or with the leaves, fruits avoid, 0.8 cm in diam., juicy, ranged along the stems	Flowering and Fruit.: March-December	Fairly common, river bank, wetland sides and damp places	Fruits are edible, acidic in taste and preserved in boiling mustard oil, used as pickles and marmalades
40.	*Hippophae tibetana* Schlecht/Elaeagnaceae/ Ames, Chuk	Densely branched, deciduous shrub, flowers stalkless, yellow, fruit oblong, globose, orange-yellow scarlet, arranged along the stem	Flowering and Fruit.: August-December	Common, dry inner ranges bordering streams, screes	Fruits are much acidic, boiled with sugar or jaggery to form palatable preserves
41.	*Holboellia latifolia* Wall. Var. *angustifolia* (Wall.) Hook. f./Lardizabalaceae/Gumphal	Climber, flowers purplish-green, scented, berries 1-3 funnel like, first green, black after ripening, 4.0-10.0 cm, oblong, fleshy	Flowering and Fruit.: April-November	Uncommon, climbing on *Rhododendron arboreum* and *Pyrus pashia* trees	Mature fruit are eaten, sweet in taste
42.	*Juglans regia* L./ Juglandaceae/Akhorot, Himalayan walnut	Tree, flowers green, drupe globose or ovoid, 2.5-4.0 cm, pubescent, green often tinged with pink pericarp, thick, fleshly, enclosing a woody in a single seed	Flowering and Fruit.: February-November	Associated with Oak-Rhododendron forests and cultivated	Kernels are caton, an article of diet and in some villages pressed for edible oil

Contd...

Table 29.9–Contd...

Sl.No.	Botanical Name/ Family/Vernacular	Brief Description	Season	Distribution	Uses/Method of Utilization
43.	*Mahonia acanthifolia* G. Don/Berberidaceae/ Bakil, haldia, ban-totar, Guram, Khoru	Evergreen shrub, flowers yellow, scented. berry dark purple or blue-black, about 7 mm long, ovoid, glaocous	Flowering and Fruit.: February-June	Not common, riverine and shady places	Fruits are eaten, sour in taste, considered diuretic and demulcent in dysentery
44.	*Malus baccata* (L.) Borkh. var. *himalaica* (Maxim) Scheneidr/ Rosaceae/Banmehal, Leik-mehal, Moli, Siberian crab apple	Small deciduous tree, flowers white, fruits red, ovoid or globose, 0.6-1.0 cm in diameter, fleshy with 2-5 celled stone	Flowering and Fruit.: April-November	Not common, associated with Oak-Rhododendron forest	Ripe fruits are eaten raw, acrid in taste; unripe fruits are made into sauce and jam
45.	*Melothria heterophylla* (lour.) Cogn./Cucurbitaceae/Ban kakri, Gwal-kakri	Perennial scandent climber, flowers pale yellow or white, fruits glabrous, pubescent or tomentose, oblong, ovoid, cylindric, 2.5-6.0 cm long, bright red when ripe, tapering towards the apex	Flowering and Fruit.: March-September	Common, climbing on *Berberis* sp., *Pyrus* sp. and other shrubs in Oak-Rhododendron forest	Fruits are eaten raw or cooked as vegetable. Roots are boiled and eaten
46.	*Miliusa velutina* Hook. f and Thomas/Annonaceae/Guasal, Gidar-rukh, Domsal	Deciduous tree, flowers greenish-yellow, fruits blackish-purple or black, 1.5-1.8 cm in diameter, ellipsoid	Flowering and Fruit.: March-October	Not common, associated with *Anogeissus latifolia, Diospyros melanoxylon* and *Shorea robusta* forests	Ripe fruits are eaten
47.	*Moringa oleifera* Lamk./ Moringaceae/Sahajan, Saingina, Sundan; Drumstick tree, Horse radish tree	Small deciduous tree, flowers white or pink, pods 22.5-45.0 x 1.2-1.8 cm, pendulous, 9-ribbed and slightly constricted between the seeds	Flowering and Fruit.: February-June	Common, mixed forests	Young leaves, flowers and green pods are thoroughly washed, boiled and cooed as vegetable. Unripe pods or green pods are also preserved as pickle. Incisions are made in the trunk and stem, from which exudes a reddish gum used as adulterant in many sweets. Gum is sold in market

Contd...

Table 29.9–Contd...

Sl.No.	Botanical Name/ Family/Vernacular	Brief Description	Season	Distribution	Uses/Method of Utilization
48.	*Morus alba* L./ Moraceae/Shahtut, White mulberry	Medium sized deciduous, monoecious tree, fruits white or red	Flowering and Fruit.: March-June	Common, grassy localities	Fruits are eaten, sweet in taste
49.	*Morus serrata* Roxb./ Moraceae/Kimu, Himalayan mulberry	Tree, flowers greenish, fruits white or dark purple, 1.5-2.5 cm in long, mucilaginous, fleshy containing many seeded fruitless	Flowering and Fruit.: April-June	Associated with Oak-Rhododendron forests	Ripe fruits are eaten raw, sweet in taste and also made into sauce
50.	*Murraya koenigii* (L.) Spreng./Rutaceae/ Gandalu, Curry leaf tree	Deciduous, strongly scented pubescent shrub, flowers white, berry black when ripe, sub-globose, 0.5 cm in diameter, ripe fruits are edible, aromatic, taste sweet	Flowering and Fruit.: January-April	In dry places	Leaves are much used for flavouring curries and other food dishes
51.	*Myrica esculenta* Buch.-Ham. ex D. Don/Myricaceae/Kaphal, Bey berry, Box myrtle	tree drupe stalk less, ovoid, red, succulent, granulate, about 2 cm in diameter	Flowering and Fruit.: June-August	Common, associated with Oak-Rhododendron forest	Ripe fruits are sour-sweet in taste, much relished with salt and mustard and also make a refreshing drinks
52.	*Phoenix acaulis* Roxb. ex Buch-Ham/Palmae/ Khajur, Khajin, Wild date palm	Dwarf palm, fruit scaled on the spikelets at a narrow verticle angle, ovoid-oblong, 1.2-1.8 cm long, yellowing to reddish brown	Flowering and Fruit.: December-May	In open and dry places	Ripe fruits are eaten raw, sweet in taste. A sugary sap rich in vitamins is obtained from the trunk and is used as beverage
53.	*Phoenix humilis* Royle/ Palmae/Jhar-khajur; Dwarf date palm, Hill date palm	Tall erect palm, fruits ovoid, 1.2-1.8 cm long, generally mucronate, dull black whon ripo	Flowering and Fruit.: December-July	In open and rocky slopes	The petiole bases are removed from the young shoots and the tender white soft pith is eaten. Ripe fruits are also eaten raw, sweet in taste
54.	*Prunus armeniaca* L./ Rosaceae/Chulu, Khubani, Common apricot	Tree, flowers pink or white, fruits smooth, yellow, eaten raw, sweet-sour in taste	Flowering and Fruit.: April-August	Cultivated in margins of agricultural fields	Kernels extracted for fatty oil used for cooking

Contd...

Table 29.9–Contd...

Sl.No.	*Botanical Name/ Family/Vernacular*	*Brief Description*	*Season*	*Distribution*	*Uses/Method of Utilization*
55.	*Prunus cornuta* (Wall. ex Royle) Steud./Rosaceae/Jamani, Bagwaly, Himalayan bird cherry	Small deciduous tree, flowers white, drupe first red turning to purplish black when ripe, globose about 1.5 cm in diameter	Flowering and Fruit.: April-October	Common, associated with *Quercus flouribunda, Sorbus cuspidate, Acer caesium* forest	Ripe fruits are eaten raw, sour in taste
56.	*Prunus jacquemontii* Hook. f./Rosaceae/ Kursang, Targui	Deciduous shrub, flowers pink, fruits 2.5-3.0 cm long, ovoid, bright red	Flowering and Fruit.: May-August	Not common, near open stony places and screes	Fruits are eaten raw
57.	*Punica granatum* L./ Punicaceae/Anar, Darim, Pomegranate	Deciduous small tree, flowers orange and scarlet, berry globose, reddish-brown, 3.5-7.0 cm diam. thin, with coriaceous, woody epicarp and crowded with the persistent clayx lobes	Flowering and Fruit.: April-December	Wild on dry limestone soil, commonly cultivated	Fruits (seeds) are edible and are also used in beverages. Seed mixed with salt, green chillies and pudina (*Mentha sylvestris*) are ground to make a sauce. Also used as an acidulous spice in various food preparations
58.	*Pyracartha crenulata* (D. Don) Roem./ Rosaceae	Evergreen, glabrous, spiny shrub, flowers white, fruits orange-red, globose, about 6 cm in diameter	Flowering and Fruit.: April-November	In dry places, common along banks of streams	Ripe fruits are eaten raw
59.	*Pyrus lanata* Do. Don/ Rosaceae/Galao, Maul, Pahi, Thulmole, Pothi	Small deciduous tree, flowers white fragrant, fruits orange-yellow, spotted with red and speckled with small dots, about 2.5 cm in diameter, globose of pyriform	Flowering and Fruit.: April-August	Notcommon, associated with *Prunus cornata, Acer* and *Ribes* forests	Ripe fruit are eaten raw, sweet in taste, and also made into sauce and jams. This tree is dwindling due to indiscriminate exploitation (fodder, fuel and charcoal, house construction, agricultural implementation etc.) by the natives
60.	*Pyrus pashia* Buch. Ham. ex D. Don/Rosaceae/ Melu, Mole, Mol	Small deciduous tree, flowers white, fruits yellow-brown ultimately tuning black with raised spots, globose, about 2.0-3.5 cm in diameter	Flowering and Fruit.: January-December	In open places and associated with Oak-Rhododendron forests	Ripe fruits are edible

Contd...

Table 29.9–Contd...

Sl.No.	Botanical Name/ Family/Vernacular	Brief Description	Season	Distribution	Uses/Method of Utilization
61.	*Quercus leucotrichophora* A. Camus/Fagaceae/Banj, Liqual, Himalayan white oak	Evergreen tree, acorns single or in pairs at first covering the whole nut but when mature covering only half	Flowering and Fruit.: April-December	Common, often associated with *Rhododendron arboreum, Myrica esculenta, Benthamidia capitata* forests	Acorns are taken or ground with flour during famine
62.	*R. ellipticus* Smith/ Rosaceae/Hissalu, Hissar, Anchu, Yellow raspberry	Large evergreen shrub, flowers white, drupelets yellow, 1.2 cm in diameter, glabrous, succulent	Flowering and Fruit.: April-June	In open dry places	Ripe fruits have an agreeable flavour and sweetish taste and are relished with salt, also made into preserves, occasionally sold in the market
63.	*Randia uliginosa* DC./ Rubiaceae/Pindalu, Mainphal, Grey emetic nut	Small tree, flowers white or creamy coloured, berry ovoid, about 4.8-6.5 cm long, yellowish-brown, smooth	Flowering and Fruit.: April-May	Common, associated with *Shorea robusta* and *Anogeissus latifolia* forests	Fruits are eaten after thoroughly roasting or cooking
64.	*Rhus parvifloa* Roxb./ Anarcardiacecae/ Saunla, Tungla	Evergreen, shrub, flowers yellowish-green, drupe ovoid, red, brown, 4 cm in diameter, glabrous, shining. ripe fruits are edible, sweet-sour in taste	Flowering and Fruit.: May-November	In open places and associated with Oak-Rhododendron forests	Locally used to quench thirst
65.	*Rhus semialata* Murray/Anarcardiaceae/ Damila, Damphal, Dhamila, Titmulya	Small deciduous tree, flowers whitish or yellowish-green, fragrant, drupe greenish-yellow or dark reddish-brown, pubescent, globose, 4-6 mm diameter	Flowering and Fruit.: July-December	Common, associated with Oak-Rhododendron forest	Fruits are edible and sour in taste and are also ground to prepare sauce
66.	*Ribes alpestre* Wall. ex Decne/Grossulariaceae/Himoch, Chapchu, Lipchi, Imli, The Gooseberry, Asian Gooseberry	Stiff, small deciduous armed shrub, flowers white or greenish, berry globular or ovoid, about 1 cm long, dull red	Flowering and Fruit.: May-September	Common, dry ranges of the hills along with *Pinus excelsa* and *Cedrus deodra* forests	Ripe fruits are edible, sour in taste

Contd...

Table 29.9–Contd...

Sl.No.	Botanical Name/ Family/Vernacular	Brief Description	Season	Distribution	Uses/Method of Utilization
67.	*Ribes glaciale* Wall./ Grossulariaceae/Himoch, Kinlai, Kyanel, Gwaldum, Yellow current	Erect deciduous shrub, flowers green, berry globose or ovoid, 0.5-0.8 cm in diameter succulent, purplish red	Flowering and Fruit.: May-October	Common, associated with *Quercus semecarpifolia* and *Rhododendron arboreum* forest	Ripe fruits are edible, sour in taste. Fruits are made into pickles and sauce
68.	*Ribes himalense* Royle ex Decne/Grossulariaceae/Kim-kolia, Kaladani, Red current	Deciduous pubescent or glabrous, unarmed shrub, flowers greenish-yellow or brownish pink, fruits red or black	Flowering and Fruit.: May-October	Not common, associated with *Sorbus lanata, S. foliolosa* and *Prunus cornuta* forest	Ripe fruits are eaten raw and sour-sweet in taste, also made into pickles or preserves
69.	*Rubus biflorus* Buch. Ham ex Sp./Rosaceae/ Hissol, Achnoi, Himalayan yellow raspberry	Large spreading evergreen shrub, flowers white, fruits rounded, yellow or orange, 1.0-1.5 cm in diameter, ripe fruits sweet in taste	Flowering and Fruit.: April-July	In open and dry places	–
70.	*Rubus foliolosus* D. Don/Rosaceae/Hissol	Large rambling shrub, flowers light pink, fruits pink, globose, 0.8-2.0 cm in diameter	Flowering and Fruit.: April-August	Common, associated with Oak-Rhododendron forest	Ripe fruits are eaten raw, saur – sweet in taste
71.	*Rubus niveus*/ Rosaceae/Kala-hissar, Kayalcha, Katrana, Mysore raspberry, Mahabaleshwar raspberry	Large rambling shrub, flowers pink, fruits globose, orbicular or broadly ovate, 1.2-2.0 cm in diameter, black when ripe, fleshy, juicy	Flowering and Fruit.: March-October	Common	Ripe fruits are eaten raw, sour-sweet in taste and also made into sauce
72.	*Schleichera oleosa* (lour.) Oken./Sapindaceae/Kususm, Ceylon oak, Honey tree, Gum lac tree, Maeassar oil tree	Small tree, flowers greenish-yellow, fruits ovoid, 1.8-3.2 cm long with fleshy pulp, fleshy aril around the seeds is edible, pleasant, and acidic in taste	Flowering and Fruit.: March-September	Common in *Shorea robusta* and *Anogeissus latifolia* forests	Raw fruits are pickled

Contd...

Table 29.9–Contd...

Sl.No.	Botanical Name/ Family/Vernacular	Brief Description	Season	Distribution	Uses/Method of Utilization
73.	*Semecarpus anacardium* L./Anacardiaceae/Bhilao, Bhilawa, The marking nut tree, Oriental cashew-nut	Small deciduous tree, flowers greenish-white, drupe obliquely ovoid or oblong, shining black, seated in bright orange coloured fleshly receptacle	Flowering and Fruit.: June-February	Common, associated with *Shorea robusta* and *Anogeissus latifolia* forests	The orange coloured fleshy cup or hypocarp is eaten; it is acrid and astringent, roasted, it is some what like roasted apples and when dry somewhat like dates
74.	*Sorbus aucuparia* L./ Rosaceae/Moli, European mountain ash, Rowan	Small tree, flowers white, fruits red, globose, 0.8-1.2 cm in diameter	Flowering and Fruit.: June-October	Not common, associated with *Ribes* sp., *Sorbus* sp. and *Rosa* sp. forests	Ripe fruits are eaten raw, sour-sweet in taste, and also made into sauce
75.	*Sorbus cuspidata* (Spach.) Hedlund/ Rosaceae/Papamoli, Moli, Rikhmail, Himalayan white beam	Small deciduous tree, flowers white, fruits greenish-yellow often tinged with red and covered with small brown spots, 1.2-2.0 cm in diameter, globose	Flowering and Fruit.: May-December	Common, associated with Oak-Rhododendron forest	Ripe fruits are eaten raw, sweet in taste, and also made into jams and pickles
76.	*Spondias pinnata* L. Kurz./Anacardiaceae/ Amara, Ambara, Indian hog-plum, Wild mango	Small deciduous tree, flowers pinkish-green or greenish-white, fragrant, drupe yellow, 3.5-5.0 cm long, succulent, smooth containing a large fibrous stone surrounded by the soft juicy mesocarp	Flowering and Fruit.:March-October	Common, moist shady places	The aromatic flower buds used as agreeable curry. Fruits are made into sauce, pickles and jams. The yellow ripe fruits with sub-sweetish, flavoured juicy flesh are eaten raw and made into curry. Seed is also eaten and tasty
77.	*Stanvaesia nussia* (D. Don) Decne/ Rosaceae/Gadhmelu, Garimehal	Evergreen or sub-deciduous tree, flowers white, fruits orange-yellow, woolly tomentose, globose or ovoid, 0.6-0.8 cm in diameter	Flowering and Fruit.: April-December	Common, associated with *Myrica esculenta*, *Symplocos* sp., *Quercus* sp. Forest	Ripe fruits are eaten raw, sweet in taste
78.	*Syzgium cumini* (L.) Skeels/Myrtaceae/ Jamun, Phalenda, Black plum, Java plum	Large evergreen tree, flowers greenish-white, fragrant, berry globose or ovoid, dark purplish black with juicy flesh	Flowering and Fruit.: March-July	Wild as well cultivated	Ripe fruits are eaten raw, acrid, sour-sweet in taste or also taken with salt and mustard, fruits sold in the market

Contd...

Table 29.9–Contd...

Sl.No.	Botanical Name/ Family/Vernacular	Brief Description	Season	Distribution	Uses/Method of Utilization
79.	*Viburnum cotinifolium* D. Don/Sambucaceae/ Ghinwa, Ghenu, Bansura, Matyaana, Gabain, Cranberry bush	Deciduous shrub, flowers white, tinged with pink, drupe oblong, compressed, 0.8-1.2 cm long, first red ripening to black, shining, glabrous	Flowering and Fruit.: April-October	Common, associated with Oak-Rhododendron forest	Ripe fruits are eaten raw, sweet is taste, sweet in taste
80.	*Vitis parvifolia* Roxb./ Vitaceae/Baraem, Himalayan wild vine	Perennial, deciduous climber, flowers green, fragrant, fruits black, globose, 5.5-10.5 cm in diameter	Flowering and Fruit.: April-August	Common, mixed with *Pinus* sp. and Oak forest	Ripe fruits are edible, sweet and aromatic in taste
81.	*Xeromphis spinosa* (Thunb.) Keay/Rubiaceae/Rara, Maidul, Raadu; Emetic nut	Deciduous small tree, flowers yellow, berry globose or obovoid, yellow when ripe, aromatic, 2.4-3.2 cm in diameter	Flowering and Fruit.: February-April	Common, associated with *Anogeissus latifolia* and *Acacia catechu* forests	Ripe fruits are eaten after thoroughly roasted or cooked. Young leaves are eaten as vegetable
82.	*Zizyphus mauritiana* Lam./Rhamnaceae/Ber; Chinese date, Indian jujube, Indian plum, Indian cherry	Small tree, flowers green, drupe globose, first yellow, then orange and finally ripening to reddish-brown	Flowering and Fruit.: June-February	In open dry places	Ripe fruits are eaten

Sources: Negi *et al.*, 1985; Polunin and Stainton, 1985; Negi, 1988; Negi *et al.*, 1995; Gaur, 1999; Nautiyal *et al.*, 2009.

it is widely distributed in whole North-Western Himalaya including Himachal Pradesh, Uttarakhand and North-Eastern region. It is a shrub growing to 1–2.5 m tall, the stems whitish tomentose at first, becoming glabrous green to purple later. The flowers are with five dark pink to red petals. The fruit is 8–12 mm diameter, densely grey tomentose, dark red at first, ripening black. Average annual yields from a bush covering 2.5m^2 in the Himalayas are 657g. The fruit contains about 7.8 per cent sugars, 0.13 per cent protein, 0.77 per cent ash (Parmar and Kaushal, 1982). The fruits are enjoyed fresh alone or served with sugar and cream or ice cream. They are excellent for making pie, tarts, jam and jelly. The fresh fruits can be quick-frozen for future use.

Diospyros lotus

It is known as date plum in English and Kaki or Kaku in Hindi is a deciduous tree that is widely distributed in Middle East and South Asia, especially in China and India (Facciola, 1990). In India, it is widely distributed in whole North-Western Himalayan region including Kullu valley of Himachal Pradesh and lower and middle elevations of Uttarakhand hills.

Figure 29.6: Plants of *Diospyros lotus*

This tree can reach up to 30 m tall in warm areas, but will generally be smaller (10 m) in hills. Fruits are 1-2 cm diameter, yellowish and become bluish-black when completely ripe. The fruit contains about 1.9 per cent protein, 0.2 per cent fat, 47.7 per cent carbohydrate, 1 per cent ash. Fruits are eaten as raw or cooked. The fruit has an exquisitely rich flavour when it is fully ripe. The fruit can be dried, when it acquires a date-like flavour.

Pinus gerardiana

Chilgoza pine is a monoecious conifer tree of compact appearance with short lateral branches. It belongs to family Pinaceae. Chilgoza pine is very common in Afganistan, Tibet and Northern Baluchistan, where gregarious trees can be seen growing naturally. In India it covers an area of 2022 ha (Singh *et al.*, 1973) and grows naturally in inner Himalayan zone comprising Kinnaur and Pangi in Himachal Pradesh and Kistwar and Astur areas of Jammu and Kashmir. Tree at full maturity attains a height of about 16-20 m. The tree starts producing cones after the age 25-30 years. The nut is used raw, roasted or in confectionary products. It contains 51.3 per cent fat, 8.7 per cent water, 13.6 per cent protein, 22.5 per cent starch, 0.9 per cent fibre and 3.0 minerals and ashes. It posses tremendous wound healing properties.

Corylus jacquemontii

Corylus jacquemontii, the wild hazelnut tree, is also called as Himalayan filbert or Turkish Hazel. It is considered a variety of *Corylus colurna* (Kasaphigil, 1963). It is locally known as Bhotiya badam or Kabasi. In India, it is widely distributed in Northern temperate regions; the tree is found growing in forest as associated species with *Acer cappadocicum* from Kashmir to Uttarakhand at an altitude of 1600-3000 m (Pandey *et al.*, 2010).

Figure 29.7: Kernels and Fruit of *Corylus jacquemontii*

The seeds are nuts, 1–2.5 cm long and 1–2 cm in diameter, surrounded by an involucre (husk) which partly to fully enclose the nut (Rushforth, 1999). The kernels are rich source of fat and are edible. The oil content ranged from 45-70 per cent in different species of hazelnut.

Aesandra butyracea

The *Aesandra butyracea* (Roxb.) Baehni locally known as Chiura, is an oil seed bearing deciduous tree, belonging to the family Sapotaceae, restricted in the sub Himalayan tract and the outer Hamalayan ranges from Kumaon eastwards to Sikkim and Bhutan, ascending to 1500 m, chiefly along the sides of rivers and in shady valleys (Sundriyal and Sundriyal, 2003). The seeds of *Aesandra butyracea* yield an edible oil known as Phulwara butter, used mostly as a substitute for or and adulterant of ghee. The oil is also used as a substitute of coca butter in chocolate manufacture and in making of soap and candles. The yield of oil is 42-47 per cent of the weight of seeds or 60-70 per cent of the weight of the kernels and the average yield of seed per tree is 50 kg per year.

Palmitic acid content is reported to be very high (65 per cent) in the kernel fat, hence, this fat is one of the best source of palmitic acid, which has great demand in cosmetic and pharamaceutical industries. The other fatty acids are reported to steraic acid (3.3 per cent), myristic acid (0.03 per cent), oleic acid (27.6 per cent) and linoleic acid (3.7 per cent).

Berberis asiatica

It is an evergreen shrub 1.20 to 1.80 m high and stem up to 10 cm diameter. The fruit of this shrub is a berry, red or black in colour. It grows throughout the dry valleys of the Himalayas between 600 and 2400 m from Bhutan to Garhwal. It is common in open forests of Banj or Chir, and on gregarious waste land near cultivation. Fruits and flowers both are edible. They have a pleasant sour taste. Fruits are also made into jelly (Rawat *et al.*, 2011).

Refernces

Datta, S.K. (2007). Domestication of Underutilized Species for Ornamental Values. In: *Underutilized and Underexploited Horticultural Crops*, Vol. 1, (Ed.) K.V. Peter. New India Publishing Agency, New Delhi, pp. 105-114.

Dwivedi, S.K., Singh, R. and Ahmed, Z. (2006). *The Seabuckthorn*, FRL, DRDO, Leh, India, 57p.

Dwivedi, S.K., Singh, R. and Ahmed, Z. (2007). Identification and evaluation of biodiversity of Seabuckthorn in cold arid regions of Ladakh. *J. Agric. Sci.*, 77(7), 427-431.

Facciola, S. (1990). *Cornucopia: A Source Book of Edible Plants*. Kampong Publications, 677p.

Gaur, R.D. (1999). *Flora of the District Garhwal North West Himalaya (With Ethnobotanical Notes)*. Trans Media, Srinagar (Garhwal), India, 811 p.

Haq, N. and Hughes, A. (2002). *Fruits for Future in Asia*. Southamptun, UK, Sheikh Saeed Ahmad and Sumaira Javed.

Jalikop, S.H. (2006). Taxonomy, growth habit adaptability of underutilized fruits. In: *National Symposium on Underutilized Horticultural Crops*, held at IIHR, Hessarghatta, Bangalore, Abstract of leading papers, pp 4-5.

Kasaphigil, B. (1963). *Corylus colurna* and its varieties. *California Hort. Soc. J.*, 24: 95-104.

Nautiyal, M.C., Yadav, V.K. and Rao, V.K. (2009). Underexploited Fruits and Vegetables: Production Potential and Uses. In: *Agriculture in Uttarakhand Hills*, (Eds.) B.S. Bisht, and D.P. Singh. Directorate of Experiment Station, G.B.P.U.A.&T., Pantnagar, pp. 85-104.

Negi, K.S. (1986). Edible Wild Plants of Garhwal Himalayas: An Ethnobotanical Survey. *Ph.D. Thesis*, H.N.B. Garhwal University (Unpublished).

Negi, K.S. (1988). Some little known wild edible plants of U.P. Hills. *J. Econ. Tax. Bot.*, 12(2): 345-360.

Negi, K.S., Gaur, R.D. and Pant, K.C. (1995). Biodiversity: Untapped plant wealth in Uttarakhand. *Indian Hort.*, 40 (2): 32-40.

Negi, K.S., Tiwari, J.K. and Gaur, R.D. (1985). Economic importance of some common trees in Garhwal Himalaya: An ethnobotanical study. *Indian J. For.*, 8(4), 276-289.

Parmar, C. and Kaushal, M.K. (1982). *Wild Fruits of the Sub-Himalayan Region*. Kalyani Publishers. New Delhi.

Padulosi, S.E.P. and Hodgkin, T. (1999). Challenges and strategies in promoting, conservation and use of underutilized and neglected species. In: *Perspective on New Crops and New Uses*, (Ed.) J. Janick. *Proceedings of the Fourth National Symposium on New Crops and New Uses: Biodiversity and Agricultural Sustainability*, 8-11 November, 1998, Phoenix, Arizona, pp 140-145.

Pandey, G., Pandey, D. and Shukla, S.K. (2010). Underutilized nut fruits of hills. In: *Underutilized and Underexploited Horticultural Crops*, Vol. 5, (Ed.) K.V. Peter. New India Publishing Agency, New Delhi, pp. 123-162.

Patel, R.K., De, L.C., Singh, Akhat and Deka, Bidyut C. 2010. Lesser known edible fruits of North Eastern India. In: *Underutilized and Underexploited Horticultural Crops*, Vol. 5, (Ed.) K.V. Peter. New India Publishing Agency, New Delhi, pp. 163-174.

Polunin, O and Stainton, A. (1985). *Flowers of the Himalaya*. Oxford University Press, Delhi, India.

Ponnuswami, V., Senthamizh Selvi, B., Ramesh Kumar, A. and Prabhu, M. (2009). Value addition of underutilized fruits and its wastes. In: *National Workshop cum Seminar on Status and Future Strategies of Horticulture Development in A&N Islands*, held from 23-25 January, p. 45.

Rai, N., Asati, B.S., Patel, R.K., Patel, K.K. and Yadav, D.S. (2005). Undeutilized Horticultural crops in North-Eastern region. *Envis Bulletin, Himalayan Ecology*, 13(1): 3-26.

Rao, V.K., Yadav, V.K., Sharma, S.K., Sah, V.K and Singh, V. (2006. Seabuckthorn (*Hippophae* spp): The Wonder plant for higher Western Himalaya. In: *National Symposium Production, Utilization and Export of Underutilized Fruits with Commercial Potential*, held at BCKV, Kalyani, West Bengal w.e.f. 22-24, November, Book of Abstracts, pp. 11-12.

Ravindran, C., Sharma, M., Kohli, A. and Kher, R. (2002). Exploitation of Underutilized Fruit Species in Western Himalaya for Food Security and Enhanced Rural Livelihood. Regional Agricultural Research Station, Rajouri, Sher-e-Kashmir University of Agricultural Sciences and Technology-Jammu (Jammu-Kashmir), India.

Rawat, A., Mishra, N.K., Mishra, D.S. and Kumar, P. (2011). Wild fruits in Uttarakhand waiting their turn. *Indian Hort.*, 56 (5), Cover II, III, 36.

Rushforth, K. (1999). *Trees of Britain and Europe*. Collins.

Sah, V.K. and Tripathi, Shailesh (2009). Medicinal plants in North-West Himalaya: Perspectives and issues. In: (Eds.) *Agriculture in Uttarakhand Hills*, B.S. Bisht and D.P. Singh. Directorate of Experiment Station, G.B.P.U.A.&T., Pantnagar, pp. 174-183.

Singh, R.V., Khanduri, D.C. and Lal, K. (1973). Chilgoza pine (*Pinus gerardiana*) regeneration in Himachal Pradesh. *Indian Forester*, 99: 126-133.

Sundriyal, Manju and Sundriyal, R.C. (2003). Underutilized edible plants of Sikkim Himalaya: Need for domestication, *Current Sci.*, 85(6), 25, 731-736.

Tripathi, Shailesh and Sah, V.K. (2010). Underutilized medicinal plants of Uttarakhand hills. In: *Underutilized and Underexploited Horticultural Crops*, Vol. 5, (Ed.) K.V. Peter. New India Publishing Agency, New Delhi, pp. 425-448.

Yadav, V.K., Rao, V.K., Sharma, S.K. and Sah, V.K. (2011). Tapping potential of seabuckthorn an underutilized fruit. *Indian Hort.*, 56(5): 16-18.

Yadav, V.K., Sah, V.K., Singh, A.K. and Sharma, S.K. (2006). Variations in morphological and biochemical characters of Seabuckthorn (*Hippophae salicifolia* D. Don) populations growing in Harsil area of Garhwal Himalaya in India. *Tropical Research and Extension*, 9: 1-7.

Yadav, V.K., Singh, A.K. and Tripathi, S. (2005). Nutritional and medicinal importance of Seabuckthorn. Book of Abstract, In: *National Consultative Workshop on Medicinal and Aromatic Plants*, w.e.f. June 25-27, held at GBPUA&T, Pantnagar Uttarakhand.

Chapter 30

Small Millets: Livelihood and Nutritional Security for Mountain Agriculture

Birendra Prasad[1], *Abhishek Bahuguna*[2], *Rashmi Yadav*[3] *and Sunil Kumar*[4]

[1]*Senior Research Officer,* [2]*Senior Research Fellow and* [4]*Assistant Professor*
GBPUA&T Hill Campus, Ranichauri,
Tehri Garhwal – 249 199, Uttarakhand
[3]*Senior Scientist, NBPGR, New Delhi*

ABSTRACT

Millets are small grained cereals, the smallest of them include barnyard millets, finger millet, kodo millet, foxtail millet, proso and little millet. They are the staple food of the millions inhabitating the mountain farming, arid and semiarid tropics of the world. They are distributed in most of the Asian and African countries and parts of Europe. The grains of small millets, being nutritionally Superior to rice and wheat, provide cheap proteins, minerals and vitamins to poorest of the poor where the need for such ingredients is the maximum. The untapped grain yield potential coupled with nutritional superiority makes the small millets potential future food crops particularly in the more difficult rainfed areas under adverse soil and climatic condition. Even though the area under small millets is slowly decreasing, there will be still millions of hectares under small millets in India where other crops have less chance of adaptation. Because of their comparative photo insensitive nature, short growing season and low moisture demand, millets can be very well fitted into multiple cropping systems both under irrigation as well as dry farming conditions. During scarcity years they can provide nutritious grain as well as valuable fodder in a short span of time. Their long storability under ordinary conditions has made them 'famine reserves'. This aspect is perhaps the most important for Indian agriculture where the crop production suffers due to the vagaries of the monsoon. There are types to suit a wide range of rainfall situations which can be used for mid-season corrections when rains are delayed. Systematic improvement in small millets has not been attempted until very recently in India. Obviously, the varieties cultivated are local land races, which are the result of indirect human and natural selection over a long period of time. The success of any crop improvement programme depends on the availability of diverse germplasm. Therefore collection of germplasm is necessary, which then needs to be conserved, evaluated and distributed. Selective utilization of crops and varieties in recent times have threatened agro biodiversity leading to rapid erosion of natural resources and consequently affecting the nutritional security of people. One of the possible pathways for conservation of such neglected agro biodiversity resources is to bring them into use thereby making them viable crops within the contemporary social and economic context. However, in spite of a rich inter/intra-species diversity and wider climatic adaptability cultivation of diverse millet species/varieties is gradually narrowing in the recent past. In a way, a lack of institutional support for millet crops in contrast to the institutional promotion of rice and wheat continue to shrink the millet-growing region. Therefore, urgent attention must be given to the productivity enhancement of the rainfed mountain lands where millets are grown. Institutional finance and insurance which is offered generously to farmers who cultivate preferred grains must be extended to millet farmers also. A new thrust must be given by research institutions on millet farming issues and research initiatives must be led by farmers since they offer exciting perspectives which has to be people centered and people directed. There is also need to create a symbiotic relationship between farming and pastoralism and increased incomes for poorest to poor small millet farmers.

Introduction

Small millets may be defined as millets cultivated for their small grains which are borne on short, slender grassy plants. Small millets, sometimes referred to as minor millets, are a group of small seeded food grains, other than the major millets such as the sorghum and peal millet, that belong to the grass family Poaceae. Six small millets *viz.* finger millet, barnyard millet, foxtail millet, proso millet, little millet and kodo millet are important ones in various parts of the country. These small millets have their origin mainly in Asia and Africa. The most important domestication areas are East Asia, India sub-continent and regions from Southern margin of Sahara to the Ethiopian high lands of Africa. Some small millets are grown only in specific regions though they have long history of cultivation *i.e.* Kodo millet and little millet which are cultivated in India. On the other hand finger millet, foxtail millet, prosomillet and barnyard millet have much wider distribution in Asia, Africa and Europe (Seetharam and Gowda, 2007). India is considered as hub for the small millets, as a wide number of millets are grown in different parts of the country, starting from sea level in south to up to 3000 m altitude in North particularly in the Himalayas and North Eastern Hills.

There are many small millets in the world. Some researchers consider that they were developed from the corresponding wild grasses as the result of continued harvesting for food. Others believe that the prior spread of the idea of agriculture was needed in most cases before domestication could begin. The millets considered here are those being cultivated in the subtropical and tropical areas of the old world. The list contains Finger millet (*Eleusine coracana*), Proso millet (*Panicum miiliaceum*), Foxtail millet (*Setaria italica*), Little millet (*Panicum sumatrense*), (*formerly P. miliare*), Barnyard (Sawa) millet (*Echinochloa colona*) [formerly *E. frumentaceal*, Kodo millet (*Paspalum scrobiculatum*), Teff (*Eragrostis tep*, Fonio millet (*Digitaria exilis and D. iburua*).

Finger millet is an old tropical cereal still widely grown in eastern Africa and south Asia. Kodo and little millets continue to be important in Asia in times of famine or difficulty. The small millets are often grown in difficult conditions, and it is scarcely surprising that they involve high production risks (Joshi and Agnihotri, 1984). They have always been crops for situations where there is a risk of famine, as well as offering a low but more reliable harvest-relative to other crops-in low rainfall areas. Small millets have a wide adaptation. They can withstand a certain degree of soil acidity and alkalinity, stress due to moisture and temperature, and variations in soils from heavy to sandy infertile soils. Small millets are grown from the extreme southern tip of India at sea level to the temperate north Himalayan areas up to an altitude of 3000 metres with consequent variation in photoperiod from short to long days. Kodo millet was traditionally stored in the temples, so that seed would be available in times of crisis. The potential of these millets deserves more study. Food and seed reserves in the village are important, and should not be overlooked. They have good potential for livestock feed in the dry zones. The small millets should be developed both for their potential as good grain producers with modest water needs, and also as producers of forage. They can make good use of any irrigation water available after the main crops have been harvested, and so may be fitted in to more productive cropping patterns.

Millets need very little water for their production. Compared to irrigated commodity crops currently promoted by policy measures, millets need no irrigation and require just around 25 per cent of the rainfall regime demanded by crops such as sugarcane and banana. Thus, they do not burden the state with demands for irrigation or power. Millets are adapted to a wide range of ecological conditions often growing on skeletal soils that are less than 15 cm deep. It does not demand rich soils for their survival and growth. Hence, for the vast dryland area, they are a boon. Millet production is not dependent on the use of chemical fertilizers. Most millet farmers therefore use farmyard manures and in recent times, household produced biofertilisers. Therefore, they can significantly reduce the huge burden of fertilizer subsidy borne by the government. Grown under traditional methods, no millet attracts any pest. They can be termed as crops. A majority of them are not affected by storage pests either. Therefore, their need for pesticides is close to nil. Thus, they are a great boon to the agricultural environment. Millets are amazing in their nutrition content. Each of the millets is three to five times nutritionally superior to the widely promoted rice and wheat in terms of proteins, minerals and vitamins.

In spite of all these extraordinary qualities and capacities of millet farming systems, the area under millet production has been shrinking over the last five decades and rapidly, since the Green Revolution period. Between 1966 and 2006, 44 per cent of millet cultivation areas were occupied by other crops signifying an extraordinary loss to India's food and farming systems. Declining state support in terms of crop loans and crop insurance has significantly contributed to this decline and fall of millets in Indian agriculture. Unless this is halted urgently through a slew of policy

Table 30.1: Nutrient Composition of Small Millets and Other Cereals (per 100 g)

Crop	Protein (per cent)	Fat (g)	Ash (g)	Crude Fiber (g)	Carbohydrate (g)	Energy (kcal)	Ca (mg)	Fe (mg)	Thiamin (g)	Riboflavin (mg)	Niacin (mg)
Finger millet	7.7	1.5	2.6	3.6	72.6	336	350	3.9	0.42	0.19	1.1
Foxtail millet	11.2	4.0	3.3	6.7	63.2	351	31	2.8	0.59	0.11	3.2
Proso millet	12.5	3.5	3.1	5.2	63.8	354	8	2.9	0.41	0.28	4.5
Little millet	9.7	5.2	5.4	7.6	60.9	329	17	9.3	0.30	0.09	3.2
Barnyard millet	11.0	3.9	4.5	13.6	55.0	300	22	18.6	0.33	0.10	4.2
Kodo millet	9.8	3.6	3.3	5.2	66.6	353	35	1.7	0.15	0.09	2.0
Rice	7.9	2.7	1.3	1.0	76.0	362	33	1.8	0.41	0.04	4.3
Wheat	11.6	2.0	1.6	2.0	71.0	348	30	3.5	0.41	0.10	5.1

and financial incentives, millets might disappear from the agrarian landscape of India over the next fifty years. This will not only be a loss to India's food and farming systems, but will also prove to be a civilisational and ecological disaster. Therefore, there is an urgent need for Indian policy makers to refocus their attention towards millet farming systems and enact policies that create an enabling environment for millet farmers.

Features of Small Millets Farming

Climate Change Compliant Crops

Millet farming system makes them the climate change compliant crops. Climate change portends less rain, more heat, reduced water availability and increased malnutrition. If there is any cropping system that can withstand these challenges, survive and flourish, it is the millet system. It is important to note that with the projected 2 degree celsius temperature rise, wheat might disappear from our midst, since it is an extremely thermal sensitive crop. Similarly, the way rice is grown under standing water makes it a dangerous crop under climate change conditions. Methane emanating from water-drenched rice fields, is a green house gas, that severely threatens our environment. Millets are all-season crops cultivated round the year whereas wheat is season specific. While wheat and rice might provide only food security, millets produce multiple securities (food, fodder, health, nutrition, livelihood and ecological) making them the crops of agricultural security.

Low Water Consuming Crops

Small millets are astonishingly low water consuming crops. The rainfall needed for millet is less than 25 per cent of sugarcane and 30 per cent that of rice. We use 4000 litres of water to grow one kg of rice while all millets grow without irrigation. This can turn out to be a tremendous national gain especially in the ensuing decades of climate crisis. In a future, where water and food crisis stares us in the face, millets can become the food of security.

Warehouse of Nutrition

By any nutritional parameter, millets are miles ahead of rice and wheat In terms of their mineral content, compared to rice and wheat. Each one of the millets has more fibre than rice and wheat. Some as much as fifty times that of rice. Finger millet has thirty times more Calcium than rice while every other millet has at least twice the amount of Calcium compared to rice. In their Iron content, foxtail and little millet are so rich that rice is nowhere in the race. While most of us seek a micronutrient such as Beta Carotene in pharmaceutical pills and capsules, millets offer it in abundant quantities. The much privileged rice, ironically, has zero quantity of this precious micronutrient. In this fashion, nutrient to nutrient, every single millet is extraordinarily superior to rice and wheat and therefore is the solution for the malnutrition that affects a vast majority of the Indian population.

Absence of Chemical Fertilizers

Mountain agriculture is known for organic farming. Millets do not demand chemical fertilizers. In fact, under dry land conditions, millets grow better in the absence of chemical fertilizers. Therefore, most millet farmers grow them using farmyard manure under purely eco-friendly conditions. In recent years farmers have also started using bio-fertilizers such as vermin-compost produced in their backyard and growth promoters such as Panchagavya, Amritpani etc. These practices make millet production not only eco-friendly but stays under the control of farmers.

Thrives in the Poorest and Problematic Soils

Most millets can be grown on low fertility soils. Some in acidic soils, some on saline soils. In fact, finger millet grows well in saline soils. Barnyard millet too thrives in problem soils, where other crops, struggle to grow in such soils. Many of them are also grown to reclaim soils. Poor farmers especially in dryland India are owners of very poor lands. The only crops that sustain agriculture and food security on these lands are millets. In fact, the capacity of millets to grow on poor soils can be gauged from the fact that they grow in Sahelian soil conditions in West Africa which produces 74 per cent of all the millets grown in Africa and 28 per cent of the world production. If they flourish in such ecological zones where average rainfall can be less than 500 mm using soils that are sandy and slightly acid, it is a testimony for their, hardiness and extraordinary capacity to survive very harsh conditions. That is why millets can withstand drought like conditions in the Deccan and Rajasthan and produce food and fodder for people and livestock, respectively.

Insect Free

Growing traditional local landraces and under ecological conditions, most millets such as foxtail are totally pest free and hence do not need any pesticides. Even in storage conditions, most millets such as foxtail not only not need any fumigants, but act as anti pest agents to store delicate pulses such as green gram.

Ecological Crops

Millets are also called Ecological crops as they can bear large Environmental stress. Thery are grown in diverse soil, varying Rainfall regimes and in areas where thermo Photo periods vary. These can be grown in different ecological niches and whenever crop substitution is difficult one can grow millets. These group of plants secure food and nutritional security. Of course these millets are grown for human Consumption but their Straw is also an important fodder. The diseases like pest and other rarely occur in them. All mountain systems are fragile, vulnerable and marginal, only eco friendly crops are important for such ecosystem. Encouraging such crops meaning by sustaining the ecological functioning. Nutritionally these millets are not less than any crop but are more rich in certain aspects.

Nutra Cereals

Small millets are highly nutritious and even superior to rice wheat in certain constituents. Finger millet is the richest source of calcium (300-350 mg/100 gm grain) and other. The Small millets are good source of phosphorous and iron too. The protein content range from 7-12 per cent and fat content range from 1.12 to 5.0 per cent. The millet protein has well balanced amino acid profile and good source of methionine, cystine and lysine (Table 30.2). These essential amino acids are of special benefit to those who look upon plant food for their protein nourishmen specially vegetarians.

Table 30.2: Essential Amino Acids Contents of Millets and Cereals (g/100 g of protein)

Amino Acids	*Finger Millet*	*Kodo Millet*	*Proso Millet*	*Foxtail Millet*	*Barnyard Millet*	*Wheat*	*Rice*
Alanine	3.6	3.5	4.0	2.2	2.4	3.0	3.9
Cystine	2.2	2.6	1.0	1.6	2.8	2.2	1.4
Isoleucin	4.4	3.0	8.1	7.6	8.8	3.3	3.8
Leucine	9.5	6.7	12.2	16.7	16.6	6.7	8.2
Lysine	2.9	3.0	3.0	2.2	2.9	2.8	3.8
Methionine	3.1	1.5	2.6	2.8	1.9	1.5	2.3
Phenyl	5.2	6.0	4.9	6.7	2.2	4.5	5.2
Threonine	1.6	0.8	0.8	1.0	1.0	1.5	1.4
Tryptophan	6.6	3.8	6.5	6.9	6.4	4.4	5.5
Tyrosine	3.8	3.2	3.2	2.7	2.2	2.8	4.1
Valine	2.2	1.5	1.9	2.1	1.9	2.3	2.4

Nutritious Fodders

It has been emphasized that the animal production industry in the country can only be developed to yield results when sufficient feed input is available. In the absence of sufficient feed, there is considerable risk in improving animal production especially in mountain region. The development and spread of dwarf varieties of food crops has played a significant role in increasing the grain production of the country in recent years. But, this has reduced the bulk and fibrous portions which otherwise would have been available for meeting the needs of large animals. In the absence of sufficient land for pasture or forage production, the only alternative, is to augment the feed resources by developing small millets as crops for livestock feeding. There is also a need for obtaining this feed in increased quantities and with higher nutrient content. Therefore, an urgent need to make an attempt to examine the utility of small millets forage and crop residues as feed for small and large animals. The Chemical composition and digestibility coefficient of different small millets (per cent on dry basis) given in Table 30.3.

Multiple Securities

While single crops such as rice and wheat can succeed in producing food security for India millets produce multiple securities. They include securities of food, nutrition, fodder, fibre, health, livelihood and ecology. Most millets have edible stalks which are the most favoured fodder for cattle. Many a time, crops such as sorghum and pearl millet are grown only for their fodder value. Besides fodder, millets are storehouses of nutrition and hence produce nutrition security. Being hosts to diverse crops such as red gram, millet fields also produce fuelwood and fibre. The legume

Table 30.3: Chemical Composition and Digestibility Coefficient of Different Small Millets (Per cent on dry basis)

Stage/Type of Forage	*Composition*				*Digestibility*				*Reference*
	Crude Protein	*Ether Extract*	*Crude Fibre*	*Nitrogen Free Extract*	*Crude Protein*	*Ether Extract*	*Crude Fibre*	*Nitrogen Free Extract*	
Finger millet (*Eleusine coracona*)									
Fresh late vegetative	7.6	1.1	33.6	42.6	–	–	–	–	
Fresh dough stage	7.1	1.7	28.8	49.9	–	–	–	–	
Silage	3.6	1.5	38.8	46.2	8	44	69	52	Mudgal and Sampath (1970)
Straw	3.4	1.3	37.2	50	12	47	70	61	Mudgal and Sampath (1970)
Kodo millet (*Paspalum scrobiculatum*)									
Fresh early vegetative	11.4	1.4	28.8	44.1	–	–	–	–	Sen and Ray (1971)
Fresh dough stage	5.7	1.5	31.6	49.1	–	–	–	–	Sen and Ray (1971)
Straw	3.5	1.5	34.3	48.4	–	–	–	–	Sen and Ray (1971)
Foxtail millet (*Setaria italica*)									
Green	9.7	2.7	31.4	47.8	61	59	67	67	Morrison (1948)
Hay	8.2	2.7	25.3	44.7	60	64	62	57	
Setaria intermedia(Hay)	8.6	1	37.1	39.9	–	–	–	–	Amrith Kumar and Sampath (1973)
Setaria italica (straw)	8.2	1.2	32. 9	42.1	–	–	–	–	Mudgal and Sampath (1970)
Proso millet (*Panicum miliaceum*)									
Green	8.1	2.4	29.9	52.2	–	–	–	–	Morrison (1948)
Hay	12.5	2.5	33.9	44.5	56	41	60	61	Morrison (1948)
Straw	4.0	1.2	35.5	49.6	–	–	–	–	Morrison (1948)

crops that are companion crops for millets are also prolific leaf shedders. This leaf fall acts as natural manure and maintains soil fertility. Thus, millet farms not just use soil fertility for their growth but also return this fertility to the soil.

Finger Millet (Eleusine coracana)

Finger millet was developed in Africa from E. coracana subsp. africana, probably in the Ethiopian region. It was introduced to India perhaps more than 3,000 years ago. It is a tropical crop, grown from sea-level to 3,000 m asl. This is the most widely grown small millet in India and Africa, and can be very productive. de Wet *et al.* (1984) recognized five races. The millet is most rich in calcium and said to be poor's milk. It is rich in protein and carbohydrates and contains high phosphorous content. High content of leucine and other amino acids makes species important above others. The species can produce 15-20 q/ha with in better inputs or least 6-10 q/ha in poor conditions. This millet is grown from Andhra Pradesh to Uttarakhand from sea level to 8000 ft. above msl. The Finger millet contains important amino acids, vitamins and mineral making it nutritious pseudocereals. Biotechnology has been able to offer

Figure 30.1: Finger Millet (*Eleusine coracana*)

choice on species. There are various high yielding variety available for finger millet now for different agroclimatic region.

This particular crop is also well respected because of its ability to grow in variety of agroclimatic zones also in areas where soil management and other environmental constraints are limiting factor.

Finger millet is an erect tufted annual plant, growing 60 to 125 cm in height, Root system of the plant consists of a large number of adventitious and fibrous roots which are able to absorb moisture very efficiently from the soil. The stem is compressed but with round nodes. The plant tillers profusely terminating with branched finger likes inflorescence ranging from open to incurved fingers. Finger millet accounts for 8 per cent of the area and 11 per cent of production of all millets cultivation in the world. Finger millet is cultivated in more than 25 countries in Africa (eastern and southern) and Asia (from Near East to Far East), predominantly as a staple food grain. The major producers are Uganda, India, Nepal, and China. Finger millet has high yield potential (>10 t/ha under optimum irrigated conditions) and grain stores very well.

In India it is widely grown in Karnataka, Tamil Nadu, Andhra Pradesh, Orissa, Bihar, Gujarat and Maharashtra and in the hilly regions of Uttarakhand and Himachal Pradesh. In India, finger millet accounts for 60 per cent area (2.65 million ha) and three-fourths production (2.95 million tonnes) of the total small millets in the country. In Uttarakhand finger millet has the largest area (1.28 lakh ha) and production (1.78 lakh metric tonnes). However, the productivity is low as compared that of other parts of the country.

Origin

Finger millet (*Eleusine coracana* L. Gaertn) sub species *coracana* belongs to family poaceae. *Elusine*, the generic name, is after the "Eleusine" the Greek Goddess of cereals. The native home is presumed to be high lands of Central Africa (Hilu and de Wet, 1976) and was introduced into India approximately 4000 years ago. The long history of cultivation in India, accompanied by human selection has resulted in generation of large variability giving India the status of secondary centre of diversity. Finger millet is a tetraploid with genome composition AABB and basic chromosome number of 9 (2n=4x=36). The A genome donor is believed *to E. indica*, a ubiquitous weed of tropical and sub-tropical region, but there are conflicting reports on the B genome donor. Both *Eleusine tristachya* and *E. floccifolia* have been considered B genome donors to *E. coracana* based on DNA restriction patterns (Hilu and Johnson, 1992).

Uses

About 80-90 per cent of the ragi grown in India is consumed in the form of roti in northern India and mudde in south India. In addition to these several traditional dishes like gulgule, Ragi also considered second best for malt production. Malt based foods possess good nutrition and also high calorie density (Malleshi, 2007). The malting process involves soaking, germination, drying, de-rooting and kilning. The ragi malt is a good source of nutrients and serves as a source of amylases. By using ragi malt CFTRI, Mysore has developed techniques to produce Malted Weaning Food

(MWF) for children. Popping or puffing is another important use of ragi to prepare ready to use products. It has also developed techniques for decortication of finger millet, which traditionally was not practiced. The decorticated grain could be used as rice. Besides that, technology is also available for preparation of noodles, papad, biscuit and candies etc. from ragi flour/malt.

Cropping Systems

- ☆ Finger millet- Wheat/Barely/Mustard
- ☆ Finger millet-Lentil
- ☆ Finger millet-Potato-Maize
- ☆ Finger millet-Peas-Maize
- ☆ Finger millet + Soybean-Wheat/Barley/Oat
- ☆ Finger millet - Fallow

Small millets + soybean intercropping has been found to be a promising cropping pattern.

Genetic Resource and Breeding

At the ICRISAT in world gene bank, 5949 finger millet germplasm accessions from 24 countries are conserved for use in research and development. Similarly at Project Coordinating Unit of All India Çoordinated Small Millets, Bangalore 6000 accessions of finger millet collected from various parts of Indian subcontinent and several African and Eurasian countries are being maintained. Besides that all 13 centres involved in AICRP on Small millets are marinating their own working collections. On the basis of inflorescence morphology the species E. *coracana* is classified into two subspecies (*africana* and *coracana*). The subspecies *africana* is classified into races: *Africana*, and *Spontanea*, and subspecies *coracana* classified into races: Elongata, Plana, Compacta, and Vulgaris. The race Elongata is divided into subraces: Laxa, Reclusa, Sparsa, race Plana is divided into subraces: Seriata, Confundere, Grandigluma, race Compacta has no subraces and race Vulgaris is divided into subraces: Liliacea, Stellata, Incurvata, and Digitata.

In finger millet male sterile line INFM 95001 that is African origin and having white seed colour induced through mutation (Gupta *et al.*, 1997) has been utilized for development of composite populations at Ranichauri (Yadav *et al.*, 2008). The objective of the breeding programme was to develop early maturing with white or golden grain colour having resistance to blast. From multiple crosses attempted in F_2 and F_3 using genetic male sterile plants 130 white 299 golden 467 brown lines have been developed and advanced to F_5 generation. The line PRM 701 developed by crossing INFM 95001 x Gaja Local showed higher protein than varieties developed by brown x brown cross or pure line selection of brown land races (Table 30.4). Researched exhibited that crosses between white x brown to improve the nutritional composition.

Using RFLP, AFLP, ESTs and SSR were use to generate the genetic maps of finger millet, covering the nine homeologous chromosome groups. Genetic map was constructed through inter-sub specific F_2 population, from a cross between *E.coracana*

Table 30.4: Protein and Other Nutritional Content of Finger Millet Breeding Line as Compare to Checks

Genotype	Grain Colour	Crude Protein (per cent)	Crude Fat (per cent)	Crude Fiber (per cent)	Total Ash (per cent)	Calcium (mg/100g)	Iron (mg/100g)	In vitro Protein Digestibility (per cent)
PRM 701	Golden	11.26	1.80	4.64	3.65	421.14	9.17	72.09
PRM-1 (check)	Brown	8.59	1.45	3.87	2.38	398.41	4.07	66.42
VL -315 (check)	Brown	8.57	1.49	3.58	2.98	390.68	4.96	67.20

subsp. *coracana* and *E.coracana* subsp. *africana*. The maps span 721 cM on the A genome 787 cM on the B genome was mapped covering all 18 finger millet chromosomes partially (Dida *et al.*, 2007)

Barnyard Millet (Echinochloa spp.)

Japanese barnyard millet *Echinochloa crusgalli* was domesticated in Japan some 4,000 years ago. It belongs essentially to the temperate zone. Barnyard (Sawa) millet, *E. colona*, was domesticated in India, where it remains an important cereal in some areas. It has also been recorded from the Central African Republic, Tanzania, and Malawi. The two species have different chromosome numbers (2n = 54 and 2n = 36 respectively), and hybrids between them are sterile. The wild form is widespread as a tropical weed. de Wet *et al.* (1983) distinguished four races. Barnayard millet is the fastest growing among all the cultivated minor millets. The genus is widely distributed and includes some 20 species several of which are aggressive weeds. It is a tall plant having extremely loose to compact large ear head.

Figure 30.2: Barnyard Millet (*Echinochloa* spp.)

Origin

Barnayard millet is a common weed of both the temperate and the tropical worlds. Japanese millet (*E. crusgalli* (L.) P Beauv) is widely cultivated in China, Japan and Korea. Barnyard millet was grown in Japan dating back to 5000 years. It is widely cultivated in the Himalayan region and its cultivation extends up to the tropical plains. It is commonly known as Sawan or Madira. The commonly cultivated species in India is *E. colona* (L) Link (De Wet *et al.*, 1983). Sawan differs from the Japanese millet primarily because it is a tropical grass and lacks the beak to the spikelet which is present in *E. crusgalli*. The species cultivated in the Indian Himalayas has been recognized as *E. frumentacea*. Mid altitude Himalayan mountains show the maximum concentration of barnyard millet and it ranks second in area and production among the small millets after finger millet. Cultivated types are extremely variable having whitish green to purple inflorescence. Though *E. crusgalli* and *E. colona* have the same chromosome number (2n = 54) hybrids between them are sterile. Cytological studies point to the hexaploid *E. crusgalli* and *E. colona* as possible progenitors of *E. utilis* and *E. frumentacea*.

Uses

The grains of barnyard millet are extensively used as human food. Barnyard millet protein is highly deficient in lysine and compares well with little millet in its nutritional value. The dehusked grains are boiled in water and served as rice. Thin and thick porridge is prepared from the grains. Several food products have been developed from the flakes made from the flattened grains. Bread could be prepared from the composite flour made by co-milling with wheat. Several weaning foods have been developed from millet grains. These are planned to be integrated in the existing food based programme in Uttarakhand state. It is grown as a forage crop in the United States and can produce up to eight harvests per year. The green fodder and dry hay of the plant is fed to livestock in India and in terms of quality it has been found to be the best among all the millets.

Genetic Resource and Breeding

The world germplasm collections are being maintained at ICRISAT, India in which Uttar Pradesh (411) and Bihar (194) have largest contributions. However, many areas have not yet been explored adequately for collection of germplasm. Systematic evaluation has been undertaken at the AICRP at Bangalore in which 816 germplasm accessions have been evaluated.

Table 30.5: Performance of PRJ 1 in Tehri Garhwal and Uttarkashi Districts Under Frontline Demonstration

Year	*No. of Villages*	*No. of Farmers*	*Per cent Increase Over Farmers' Practice*
2001-02	06	27	44.00
2003-04	10	22	101.8
2004-05	04	21	53.34
2005-06	06	26	57.30
2006-07	09	46	57.30
2007-08	09	95	66.70
2008-09	17	144	57.86
2009-10	17	135	59.86

Barring finger millet, barnyard millet has good grain yield potential. Among the varieties released all but one belong to the species *E. frumentacea*. Recently, a high yielding dual purpose variety, PRJ 1 has been released for the mountain areas of Uttarakhand by GBPUA'T, Hill Campus, Ranichauri that belongs to the species *E. crusgalli*. It has very high grain and fodder yields, double to that of standard varieties, absolute resistance to grain smut disease, which is a serious problem in the hills. The variety has been popularized in many hill districts of the state and is in high demand because of its higher yields under farmers fields conditions (Table 30.5). In the germplasm collections an extra early maturing type suitable for contingency cropping has been isolated but it has very low yield potential. A dwarfing line has also been selected in the *E. crusgalli* type that may be useful in future breeding programmes. RAPD marker studies have established that *E. utilis* and *E. frumentacea* are very distinct but grouped with *E. crusgalli* and *E. colona* respectively (Hilu, 1994).

Foxtail Millet [Setaria italica (L.) P. Beauv.]

Foxtail millet is yet another ancient crop, probably domesticated in eastern Asia, and known to the Chinese as early as 2,700 BC (Purseglove, 1985). It is essentially a crop of the sub-tropical and temperate zones. The main production areas are in Japan, China, India and eastern Europe. Cobley (1976) noted that more than 12 rather variable groups of cultivars have been recognized. Foxtail millet, also known as Italian millet, German millet or Siberian millet, is an important crop of the Eurasian region and is grown widely throughout India from the temperate region to the semi-arid tropics. The common name of foxtail millet is Kauni or Kangni. It is widely preferred because of the short duration nature of the crop. The plant attains a height of about I-2m and the colour of the grains varies widely.

Figure 30.3: Foxtail Millet [*Setaria italica* (L.) P. Beauv.]

Origin

Foxtail millet is considered to have been domesticated in eastern Asia where it has been cultivated since ancient times. The domestication and cultivation of foxtail millet has been estimated at over 4000 years ago while, another study suggests that it has been grown in China for at least 5000 years (Ho, 1975). China has the largest area under foxtail millet and produces more than 90 per cent ofthe world's foxtail millet output. It is the most important millet in Japan and is widely cultivated in throughout India. The closest wild relative of foxtail millet is the weedy green foxtail, *Setaria italica* subsp. *viridis* (L.) Thellung, native to Eurasia. The wild species crosses naturally with the cultivated species to produce fertile hybrids but derivatives of such hybrids are generally obnoxious weeds. Isozymic studies on 831 land races of foxtail millet have revealed a close relationship between *S. italica* and *S. viridis*. The eastern Himalayan region exhibits more variability with respect to plant morphological characters and grain colour. Although foxtail millet is grown in many states its acreage in Andhra Pradesh, Karnataka and Tamil Nadu accounts for 95 per cent of the total area under this crop.

Uses

Foxtail millet is primarily grown as a food crop in India. The flour is mixed with wheat flour in various proportions to prepare bread, and baked products. Decorticated grains are boiled in water and served like rice. Flaking is done to prepare flat products. Thin or thick porridge is prepared from the grains and is sometimes fermented to increase its shelf life. A number of desserts are also made from the flour mixed with milk. Although the protein quality of the dehusked grains was found to be the lowest among the millets the ascorbic acid content and thiamin content have been found to increase in germinated seeds. The green fodder and hay of the plants are fed to animals while in developed countries the grains are preferred as caged bird feed.

Genetic Resource and Breeding

Foxtail millet is generally classified into a European (race moharia) and a Far Eastern complex (maxima). Race moharia includes cultivars with small, erect ear heads (inflorescences) while, race maxima is characterized by large and pendulous inflorescences. Cultivars from India are morphologically distinct from those of Europe and the Far East, and are recognized as race Indica. They are typically robust with panicles bearing branches that are loosely arranged along the primary axis. Collections from northern India resemble race maxima and are probably introductions from China. Branched panicle collections similar to that found in Meghalaya have been observed by us from the Himalayan collections. Systematic genetic evaluation work has been carried out at the AICRP Bangalore and catalogue has been published recently based on evaluation of 1,951 germplasm accessions for several descriptors (Seetharam, 1982). ANGRAU, Agricultural Regional Station, Nandyal, Andhra Pradesh is a lead centre for breeding of foxtail millet. The exotic collections are largely from China and the USA. Genetic divergence study has revealed no relation to geographical diversity.

In recent years, through concerted efforts under the AICRP on Small Millets, a total of 28 varieties have been evolved for various regions of the country (Seetharam and Halaswamy, 2003). Among the varieties released for the Himalayan region, the variety, PRK 1 developed by GBPUAT, Hill Campus was the first variety having earliest maturity period (65-75 days) while, the variety, PS 4 has been developed through mutation breeding using chemical mutagen, EMS. Sources of dwarfing gene in foxtail millet have been identified that may help in breeding ideal plant types. Regeneration of plants through tissue culture and protoplast culture from embryogenic callus has been reported and somaclonal variation in plantlets regenerated from adventitious buds of immature inflorescences and calluses has been observed. Comparative genetic mapping of foxtail millet and rice has revealed a high degree of conserved colinearity between the two genomes (Devos *et al.*, 1998).

Proso Millet (Panicum miliaceum L.)

Proso millet was well known to the Romans, and became the 'common millet'. This is essentially a crop of the temperate regions, but is also grown in the sub-tropics, and on high ground in tropical winters. Proso millet is also known as common millet, hog millet, broom-corn millet or Russian millet and is cultivated from ancient times. In the Himalayan region it is commonly known as Cheena. The plant attains a height of 0.3-1.5 m and has a shallow rooting system. It has very low water requirements and can adapt to a wide range of climatic and soil variations. Th e seeds of proso millet widely vary in colour.

Figure 30.4: Proso Millet (*Panicum miliaceum* L.)

Origin

Proso millet is an ancient crop and is believed to have been domesticated in Central and Eastern Asia (Purseglove, 1985). The progenitor of proso millet is native to Manchuria. The species was probably introduced to Europe as a cereal at least 3000 years ago. The cultivated proso millet is commonly subdivided into five subspecies, recognized as races in many areas without proper taxonomic validity. Among the small millets, proso millet has better success under temperate climates. It has been found up to an altitude of 3500 m in the Indian Himalayas. The Western Himalayan region harbours more variability in proso millet. Proso millet is cultivated in many central and southern states as well.

Uses

Like other cereals the protein of common millet is also deficient in lysine and threonine. The grains contain a comparatively high percentage of indigestible fibre as the seeds are enclosed in the hulls which are difficult to remove. The extremely early maturing nature of the crop gives it an added advantage to be grown successfully as a catch crop, a practice widely prevalent in Bihar where it is cultivated in the spring season before the Kharif crop or after the harvest of Rabi crop under irrigated condition. In the low altitude Himalayan region the same practice is also followed, while in the high hill areas it is grown as a rain-fed summer- rainy season crop. It is chiefly used as a food grain crop in the developing countries, with a variety of food preparations made from the roasted, popped or flattened grains. The grains are dehusked and boiled in water and served like rice. Porridge is prepared from the boiled grain or flour. Indian breads and baked products can be prepared from the composite flour made by co-milling with wheat. Noodle and pasta are prepared from proso millet flour. Malted grains are used to make traditional beverages.

Genetic Resource and Breeding

Germplasm collection and evaluation work in proso millet have not received due attention so far. Genetic evaluation is being under taken up by the AICRP on Small Millets at Bangalore, where 577 accessions are being maintained. Andhra Pradesh (135), Bihar (28) and Himachal Pradesh (20) are the states which have contributed most to the germplasm augmentation but a significant number of collections (376) remain unidentified (Seetharam, 1982).

Proso millet is a self-pollinated crop but natural cross-pollination may at times exceed 10 per cent. Several high yielding varieties have been released in the United States in recent years. In India 14 improved varieties have been recommended for different regions of the country. The Dholi centre of RAU, Bihar is the lead centre in breeding proso millet. Recently a variety, PRC 1 has been released by Hill Campus for the hill region of Utrarakhand. Tissue culture from immature embryos and inflorescences has been employed to derive successful regeneration in proso millet.

Conclusion

Small millet have rightly been termed as the crops for the future of food and farming at the global level signifying their importance for the future. Small millets are grown in areas with adverse soil and climatic conditions and produce reasonable grain and forage yield for the largely resource-poor farmers. The superior nutritional quality of the small millet grains coupled with low input requirements, have attracted attention in recent years. Small millets are important for the livelihood and nutritional security in many parts of the world, particularly in the drier regions and mountain agriculture. The world collections of small millets are being maintained at ICRISAT, India while national collections are being maintained at the AICRP on Small Millets centre at UAS, Bangalore. With the expanding market and a growing demand for organically produced food grains, small millets grown in the hill areas has great potential for export because of limited use of chemical fertilizer and pesticides in these crops. Post green revolution, there has been a systematic decline in the production of millets. This can be understood from the production trends of millets vis a vis other crops such as rice and wheat that were relentlessly promoted for intensive farming in select few resource rich areas under irrigated conditions. Much of the decline in

Table 30.6: Recommended Improved Varieties of Small Nillets for NWH Region

Sl.No	Crop/ Variety	Year of Release	Average Yield (q/ha)	Characteristics	Recommended Areas
A. FINGER MILLET					
1.	PES 110	1985	25-30	Ear head top incurved, resistant to blast, protein content 9.0 per cent.	All ragi growing areas of the country
2..	PES-176	1985	20-25	Moderately resistant to blast, suitable to both irrigated and rainfed cultivation.	All ragi growing areas of the country
3.	Pant Madua 3 (PES 400)	1986	20-22	Early maturing, ear head semi compact, resistant to blast.	All ragi growing areas of the country including Uttarakhand hills
4.	VL-149	1991	25-30	Resistant to blast, high grain and fodder yield	Rainfed areas except for Tamil Nadu and Andhra Pradesh
5.	VL-146	1997	20-25	Early maturing and resistant to blast	Rain-fed areas of Uttarakhand and Karnataka
6.	VL-315	2004	20-25	Moderate in maturity duration, resistant to blast.	Rain-fed Areas of Uttarakhand hills.
7.	VL-324	2005	19-22	Resistant to finger and neck blast and high yielding.	Rain-fed Areas of Uttarakhand hills
8.	PRM-1	2006	25-30	Dual purpose variety for grain and high fodder, moderately resistant to blast.	All ragi growing areas of Uttarkhand
9.	PRM-2	2010	20-25	Early maturing, moderately resistant to blast and *Cercospora* leaf spot.	Mid and high hills of Uttarakhand
B. BARNYARD MILLET					
1.	VL 29	1988	20-25	Tolerant to smut and leaf spot	Rainfed areas except Tamil Nadu and Andhra Pradesh
2.	VL Madira 172	2000	20-23	Tolerant to grain smut and leaf spot disease	Rain-fed areas of Uttar Pradesh, Uttarakhand, Gujarat and Karnataka
3.	PRJ-1	2003	25-30	Dual purpose type with high grain and fodder yield, field resistant to most diseases.	Rain-fed areas of mid and high hills of Uttarakhand
C. FOXTAIL MILLET					
1.	PS-4	1999	17-18	High protein (13-15 per cent), tolerant to blast, high grain: husk ratio (80:20) and early maturity.	All foxtail growing areas of the country
2.	PRK-1	1995	18-20	Early maturity, compact ear head, yellowish brown grains.	Mid and high hills of Uttarakhand
D. PROSO MILLET					
1.	PRC-1	2008	10-12	Compact and droopy ear head, high protein (13.3 per cent).	Hilly areas of Uttarakhand

production can be attributed to the way millet cultivation areas have shrunk over the last fifty-five years. The urgent and immediate need is to put millets into the Public Distribution System (PDS). Different parts of India grow diverse kinds of millets. Southern Andhra Pradesh, Tamil Nadu, Orissa and Southern Karnataka are the home of Finger millet. Uttarakhand and other hill and tribal areas cultivate a range of small millets such as Foxtail, Proso, Kodo and Barnyard. The Indian PDS system will be enriched with the high nutritive quality of these millets if they are included in it. Government must urgently provide space for millet based foods in the school meals and welfare hostel programmes. This will overcome the problem of malnutrition of young children a problem where India fares worse than the sub Saharan region, the poorest in the world. Urgent attention must be given to the productivity enhancement of the rainfed lands where millets are grown. This could be achieved through special watersheds on millet lands and dovetailing government's employment programmes such as NREGA to support millet cultivation from sowing to harvesting.

References

Arnrith Kumar, M.N. and Sampath, S.R. (1973). *Seminar on recent advances in Milk Production Technology*. Feb. 27-28, U.P. Inst. of Agric. Sci., Kanpur

Cobley, L.S. (1976). *Introduction to the Botany of Tropical Crops*, 2nd edition. Longman, London.

CSE (2007) "Rainfed Areas of India-Center for Science and environment", downloaded on 10th Dec. 2011 from http://www.cseindia.org/programme/nrml/rainfed_specials.htm

De Wet, J.M.J., Rao, K.E. Prasada, Brink, D.E. and Mengesha, M.H. (1984). Systematics and Evolution of *Eleusine coracana*. *Am. J. Bot*. 71: 550.

De Wet, J.M.J., Rao, K.E. Prasada, Mengesha, M.H. and Brink, D.E. (1983). Domestication of Sawa millet (*Echinochloa colons*). *Econ. Bot*. 37: 283.

De Wet, J.M.J. (1989). Origin, evolution and systematics of minor cereals. In: *Small Millets in Global Agriculture*, (Eds.) A. Seetharam, K.W. Riley and G. Harinarayana. Oxford and IBH Pub. Co. Pvt. Ltd., New Delhi, India. pp. 19-30.

Devos, K.M., Wang, Z.M., Beales, J., Sasaki, T. and Gale, M.D. (1998). Comparative genetic maps of foxtail millet (*Setaria itlaica*) and rice (*Oryza sativa*). *Theoretical and Applied Geneics*, 96(1): 63-68

Dida, M.M., Srinivasachary, Ramakrishna, S., Bennetzen, J.L., Gale, M.D. and Devos, M.K. (2007). The genetic map of finger millet, *Eleusine coracana*. *Theoretical and Applied Geneics*, 114: 321-332

Doggett, H. (1986). Small millets-a selective overview. In: *Small Millets in Global Agriculture*, (Eds.) A. Seetharam, K.W. Riley and G. Harinarayana. Oxford and IBH Publishing Co. Pvt. Ltd., New Delhi, pp. 3-17.

DoMD, (2011). *Area and Production Analysis of Millets in India*. Directorate of Millets Development (DoMD), Ministry of Agriculture and Cooperation, Government of India, downloaded on 10th Dec., from http://dacnet.nic.in/millets/all_india_apy_trend.htm

Gupta, S.C., Muza, F.R. and Andrews, D.J. (1997). Registration of INFM 95001 finger millet genetic male sterile line. *Crop Science*, 37: 1409.

Hegde, B.R. and Linge Gowda, B.K. (1986). Cropping systems and production technology for small millets in india. In: *Small Millets in Global Agriculture*, (Eds.) A. Seetharam, K.W. Riley and G. Harinarayana. Oxford and IBH Pub. Co. Pvt. Ltd., New Delhi, India. pp. 209-235.

Hilu K.W. and De Wet J.M.J. (1976). Domestication *of Eleusine coracana* (L.) Gaertn. *Economic Botany*, 30: 199-200.

Hilu, K.W. (1994). Evidence from RAPD markers in the evolution of *Echinochloa* millets (Poaceae). *Plant Systematics and Evolution*, 189(3-4): 247-257.

Hilu, K.W. and Johnson, J.L. (1992). Ribosomal DNA variation in finger millet and wild species of *Eleusine* (Poaceae). *Theor. Appl. Genet.*, 83: 895-902.

Ho, P. (1975). *The Cradle of the East*. Chicago University Press, Chicago, USA, 440p.

Joshi, P.K. and Agnihotri, A.K. (1984). Millet Production in India. Problems and prospects. *Agricultural Situation in India*, 39: 329.

Malleshi, N.G. (2007). Nutritional and technological features of ragi (finger millet) and processing for value addition. In: *Food Uses of Small Millets and Avenues for Further Processing and Value Addition,* (Eds.) K.T. Krishne Gowda and A. Seetharam. Project Coordination Cell, AICSMIP, ICAR, UAS, Bangalore, pp. 9-19.

Morrison, F.B. (1948). *Feeds and Feeding*, 20th Edn. Morrison Public Co., Ithaca, New York.

Mudgal, V.D. and Sampath, S.R. (1970). Review of research work carried out in Animal Nutrition, Southern Regional Station, National Dairy Research Institute, Bangalore.

Purseglove, J.W. (1985). *Tropical Crops: Monocotyledons*. Longman, London, U.K.

Seetharam, A. (1986). Genetic resources of small millets in india. In: *Small Millets in Global Agriculture*, (Eds.) A. Seetharam, K.W. Riley and G. Harinarayana. Oxford and IBH Pub. Co. Pvt. Ltd., New Delhi, India. pp. 45-57.

Seetharam, A. (1982). Conservation of minor millets germplasm in India. *MILWAI Newsletter* 1: 14.

Seetharam, A. and Gowda, K.T.K. (2007). Production and utilization of small millets in India. In: *Food Uses of Small Millets and Avenues for Further Processing and Value Addition,* (Eds.) K.T. Krishne Gowda and A. Seetharam. Project Coordination Cell, AICSMIP, ICAR, UAS, Bangalore, pp. 1-8

Seetharam, A. and Halaswamy, B.H. (2003). *Handbook of Small Millets Varieties*. Project Coordinating Cell, All India Coordinated Small Millets Improvement Project (ICAR), University of Agricultural Sciences, Bangalore, India, 183p.

Sen, K.C. and Ray, S.N. (1971). *Nutritive Value of Indian Cattle Feeds and the Feeding of Animals*, 6th Edn. Indian Council of Agricultural Research, New Delhi.

Sharma, J. (2007). *National Conference on Agriculture for Rabi Campaign 2007*. Presentation by Dr, J.S. Sharma, CEO, National Rainfed Agriculture Authority.

Shivaramaiah, M.T. and Sampath, S.R. (1981). *Mysore J. Agric. Sci.*, 15: 549.

Yadav, V.K., Kumar, B., Malik, N., Yadav, R. and Kumar, J. (2008). *Annual Progress Report of AICRP on Small Millets*, Ranichauri Centre. Hill Campus, GBPUA&T, Ranichauri, 76 p.

Chapter 31

Farming Practices for Promotion of Sustainable Agriculture in Dryland Areas

M.K. Tiwari

Agriculture Extension Officer,
Zila Parishad, Collectorate Campus, Bhilwara – 311 001, Rajasthan

Agriculture has changed dramatically, especially since the end of World War II. Food and fiber productivity soared due to new technologies, mechanization, increased chemical use, specialization and government policies that favored maximizing production. These changes allowed fewer farmers with reduced labor demands to produce the majority of the food and fiber.

Although these changes have had many positive effects and reduced many risks in farming, there have also been significant costs. Prominent among these are topsoil depletion, groundwater depletion and contamination, rise to problems of accumulation of toxic pesticides in the environment, the decline of family farms, continued neglect of the living and working conditions for farm laborers, increasing need for chemical fertilizers and pesticides resulting higher cost of production, and the disintegration of economic and social conditions in rural communities.

Agriculture in the desert region is naturally risky because of the large variation in annual rainfall. Farmer recognizes that maximum crop productivity cannot be achieved every year. Year of drought are managed by coping strategies like, using the food grains for consumption only, preference to drinking water needs, sale of livestock and migration of one or more family members in search of employment etc.

Sustainable agriculture is a broad concept that covers a variety of approaches to production agriculture, but rather a range of practices – a system – including: planting practices, integrated nutrient management, integrated pest management, organic farming, crop rotations, and crop selection that reduce soil erosion. Sustainable agricultural practices are also applied to the production of livestock. All approaches try to achieve an environmentally sound, economically viable, ethically acceptable and socially responsible form of agriculture.

Climate change is manifesting itself in many ways across the country. Among the indicators, while long term rainfall data analysis shows no clear trend of change, regional variations as well as increased rainfall during summer and reduced number of rainy days can be noticed. In the case of temperature, there is a 0.6°C rise in the last 100 years and it is projected to rise by 3.5-5° C by 2100. The carbon dioxide concentration is increasing by 1.9 ppm each year and is expected to reach 550 ppm by 2050 and 700 ppm by 2100. Extreme events like frequency of heat and cold waves, droughts and floods have been observed in the last decade. The sea level has risen by 2.5mm every year since 1950 while the Himalayan glaciers are retreating. These are all symptomatic of climate change.

Farming in India is affected by numerous disasters including droughts, floods, cyclones, hailstorms, attacks of pests and diseases. Drought has become a familiar feature of life in Rajasthan and has become the most frequent difficulty faced by the rural people; it is generally accepted that every third year will be a year of drought.When disasters strike, the worst affected are the landless agricultural labourers, for whom relief work is the only hope. Starvation of the most deprived people has been one of the devastating results. As disasters prevail in Rajasthan and people struggle to survive, many have been forced to migrate to the cities and other states.

A growing movement has emerged during the past two decades to question the role of the agricultural establishment in promoting practices that contribute to these social problems. Today this movement for sustainable agriculture is garnering increasing support and acceptance within mainstream agriculture. Not only does sustainable agriculture address many environmental and social concerns, but it offers innovative and economically viable opportunities for growers, laborers, consumers, policymakers and many others in the entire food system.

India is a large country with 15 agro-climatic zones, with diverse seasons, crops and farming systems. Rajasthan is predominately an agrarian state with about 70 per cent of the population depending on agriculture. Water resources being scarce, agriculture is essentially rain fed and continues to be susceptible to the vagaries of monsoon. Out of the total cultivable area of 256 lacs hectare, the area under assured irrigation is only 43.43 lac hectare, which is 16.91 per cent, resulting in the state's, low rate of food grain production. The problem is aggravated by frequent visits of drought every other year.

The sandy loam soils of this region are characterised by poor water retention capacity, low organic matter content and poor fertility. Soil erosion through wind and water is very severe and common phenomenon leading to the formation of gullies. Thus, fertile topsoil has been washed out in the area and adversely affects the crop production. At the same time frequent droughts making the condition very critical.

Recurring droughts with subsequent production failures leading to high order effects on migration of both like human and livestock. Farmers of the state prefer varieties that fulfill multiple objectives of grain and fodder yield instead of focusing only on grain yield. M.A.Altieri (2008) shows that many small farmers cope and even prepare for climate change, minimizing crop failure through increased use of drought tolerant local varieties, water harvesting mixed cropping, opportunistic weeding, agroforestry and a series of traditional techniques.

Watershed approach is the way for rain fed regions like Rajasthan which envisages ecological regeneration and optimum utilization of soil, water, plant and human resources in a planned and scientific way for prosperity of rural areas. The main objective of watershed approach to optimize land use pattern, to conserve the soil and water resources, through controlling erosion to manage the land and biological resources so as to control degradation, to recycle runoff water to boost up the production of food and non-food crops, fuel, fodder, timber and to improve the condition of the resource user as well as village community and enhance their livelihoods through the following interventions.

State of Rajasthan has high dependence on agriculture and allied activities for its economic growth where cereals account for 60 per cent of the total agriculture production, oilseeds account for 44 per cent. In spices state is the largest producer of fenugreek, cumin and coriander seeds. It is the third largest producer of milk (9 per cent) and largest producer of wool. Livestock products (meat/poultry) contribute to over 13 per cent to the NSDF.

The scenario right now marks more of agro based industries in Rajasthan so as to commensurate with the increasing demand of ready to eat organic foods and to generate more employment. This analysis clearly indicates a convergence between strength and opportunities mentioned above for growth and better quality of life for its citizens.

Sustainable agriculture integrates three main goals–environmental health, economic profitability, and social and economic equity. A variety of philosophies, policies and practices have contributed to these goals. People in many different capacities, from farmers to consumers, have shared this vision and contributed to it. Despite the diversity of people and perspectives, the following themes commonly weave through definitions of sustainable agriculture.

Sustainability rests on the principle that we must meet the needs of the present without compromising the ability of future generations to meet their own needs. Therefore, *stewardship of both natural and human resources* is of prime importance. Stewardship of human resources includes consideration of social responsibilities such as working and living conditions of laborers, the needs of rural communities, and consumer health and safety both in the present and the future. Stewardship of land and natural resources involves maintaining or enhancing this vital resource base for the long term.

A *systems perspective* is essential to understanding sustainability. The system is envisioned in its broadest sense, from the individual farm, to the local ecosystem, *and* to communities affected by this farming system both locally and globally. An emphasis on the system allows a larger and more thorough view of the consequences of farming practices on both human communities and the environment. A systems approach gives us the tools to explore the interconnections between farming and other aspects of our environment.

A systems approach also implies *interdisciplinary efforts in research and education.* This requires not only the input of researchers from various disciplines, but also farmers, farm-workers, consumers, policymakers and others.

Making the transition to sustainable agriculture is a process. For farmers, the transition to sustainable agriculture normally requires a series of small, realistic steps. Family economics and personal goals influence how fast or how far participants can go in the transition. It is important to realize that each small decision can make a difference and contribute to advancing the entire system further on the "sustainable agriculture continuum." The key to moving forward is the will to take the next step.

Finally, it is important to point out that *reaching toward the goal of sustainable agriculture is the responsibility of all participants in the system,* including farmers, laborers, policymakers, researchers, retailers, and consumers. Each group has its own part to play, its own unique contribution to make to strengthen the sustainable agriculture community. Farming and natural resources and Plant and animal production practices.

Natural Resources Management and Farming

Natural resources are national assets. These are the symbol to great economic well being. In our country faulty economic policies and lack of environment sensitive planning have led to a vastly eroded resource base and a rapidly degrading natural environment. Recently focus on natural resource management has been built up into MGNREGA through the priority on works related to land development water conservation, and micro irrigation works. This programme is promoting sustainable cultivation practices and livelihood opportunities while leading to a more sustainable environment. Sustainability has been described (Whyte, 1994) as "managing the use, development, and protection of natural and physical resources in a way or at a rate which enables people and communities to provide for their social, economic and cultural wellbeing, and for their health and safety while: Sustaining the potential of natural and physical resources (excluding minerals) to meet the reasonable foreseeable needs of future generations; Safeguarding the life supporting capacity of air, water, soil and ecosystems; and Avoiding, remedying or mitigating any adverse effects of activities on the environment.

Water

Water has become the most vulnerable and endangered commodity due to imbalance caused in its demand and supply. This is the principal resource that has helped agriculture and society to prosper, and it has been a major limiting factor when mismanaged. In Rajasthan state there is remarkable variation in the availability of water on account of the regional rainfall and geography. Rain water harvesting and checking of soil erosion is essential at farmer's field by constructing farm pond, check dam, gully plug, village ponds (*Johad*), farm field bunding, Nadi, *Khadine*, etc.

Figure 31.1: Farm Pond

Figure 31.2: Check Dam

Energy

Modern agriculture is heavily dependent on non-renewable energy sources, especially petroleum. The continued use of these energy sources cannot be sustained indefinitely, yet to abruptly abandon our reliance on them would be economically catastrophic. However, a sudden cutoff in energy supply would be equally disruptive. In sustainable agricultural systems, there is reduced reliance on non-renewable energy sources and a substitution of renewable sources or labour to the extent that is economically feasible.

Air

Many agricultural activities affect air quality. These include smoke from agricultural burning; dust from tillage, traffic and harvest; pesticide drift from spraying; and nitrous oxide emissions from the use of nitrogen fertilizer. Options to improve air quality include incorporating crop residue into the soil, using appropriate levels of tillage, and planting wind breaks, cover crops or strips of native perennial grasses to reduce dust.

Soil

Soil erosion continues to be a serious threat to our continued ability to produce adequate food. Numerous practices have been developed to keep soil in place, which include reducing or eliminating tillage, managing irrigation to reduce runoff, and keeping the soil covered with plants or mulch. According to CAZRI livestock grazing after harvest of field crops is a common practice which helps in regaining soil fertility.

Crop Production Practices

Sustainable production practices involve a variety of approaches. Specific strategies must take into account topography, soil characteristics, climate, pests, local availability of inputs and the individual grower's goals. Despite the site-specific and individual nature of sustainable agriculture, several general principles can be applied to help growers select appropriate management practices:

- ☆ Selection of species and varieties that are well suited to the site and to conditions on the farm;
- ☆ Diversification of crops (including livestock) and cultural practices to enhance the biological and economic stability of the farm;
- ☆ Management of the soil to enhance and protect soil quality;
- ☆ Efficient and humane use of inputs; and
- ☆ Consideration of farmers' goals and lifestyle choices.

Selection of Site, Species and Variety

Preventive strategies, adopted early, can reduce inputs and help establish a sustainable production system. When possible, pest-resistant crops should be selected which are tolerant of existing soil or site conditions.

Agro-Biodiversity

By growing a variety of crops, farmers spread economic risk and are less susceptible to the radical price fluctuations associated with changes in supply and demand. Research shows that small farms are much more productive than large farms if total output is considered rather than yield from a single crop. There are so many issues related to agro diversity at grass root level:

- ✰ Loss of climatically adapted gene pool requiring low inputs
- ✰ Loss of seed due to repeated droughts, seed sector meets 40 per cent of demand
- ✰ Extension/food security policies aimed at improving productivity only (farmers' viewpoint, risk reduction, steady production, local preferences)
- ✰ Monsoon forecasting at District level, of no use at lower level, not reliable
- ✰ Local sustainable agriculture practices exist, but not known widely and not supported
- ✰ Compartmentalisation of water use – irrigation, drinking, sanitation

Mixed crop and livestock operations have several advantages. First, growing row crops only on more level land and pasture or forages on steeper slopes will reduce soil erosion. Second, pasture and forage crops in rotation enhance soil quality and reduce erosion; livestock manure, in turn, contributes to soil fertility. Third,

Figure 31.3: Mixed Cropping

livestock can buffer the negative impacts of low rainfall periods by consuming crop residue that in "plant only" systems would have been considered crop failures. Finally, feeding and marketing are flexible in animal production systems. This can help cushion farmers against trade and price fluctuations and, in conjunction with cropping operations, make more efficient use of farm labor.

Soil Management

Soil fertility and soil health are managed through diversified sources of organic matter in organic farming. Erosion, desertification, salinization and alkalinization make the soil infertile and non productive. This may be the results of inappropriate cultivation and faulty management practices. A healthy soil will produce healthy crop plants that have optimum vigor and are less susceptible to pests. While many crops have key pests that attack even the healthiest of plants proper soil management can help prevent some pest problems brought on by crop stress or nutrient imbalance. In sustainable systems, the soil is viewed as a fragile and living medium that must be protected and nurtured to ensure its long-term productivity and stability. According to the CAZRI study farmers follow certain management practices which help to maintain the soil fertility to some extent. Allowing herd of Sheep in the field for one or two days and "Khejri" trees to grow in the field are some of the practices followed to increase soil fertility.

Efficient Use of Inputs

Dry lands are most potential areas for organic farming as it a traditionl farming and climatic conditions support it. This is primarily meant for sustainable and efficient resource utilization.

In contemporary agriculture, however millions of farmers across the world use chemical pesticides. These chemicals are sold with the promise of efficiency seemingly taking away the drudgery of farming. Pesticides are poisonous substances intended to harm or kill living organisms considered as pests. However, there are also huge quantities of pesticides being used in other arenas of life: mostly in public healthcare and in agriculture.

The bottom line is that pesticides are poisons. In addition to innumerable health problems created by pesticides (as residues and as drift) to urban consumers, pesticides are proving to be fatal to farming community members too.

Growers frequently ask if synthetic chemicals are appropriate in a sustainable farming system. Sustainable approaches are those that are the least toxic and least energy intensive, and yet maintain productivity and profitability. Preventive strategies and other alternatives should be employed before using chemical inputs from any source. However, there may be situations where the use of synthetic chemicals would be more "sustainable" than a strictly nonchemical approach or an approach using toxic "organic" chemicals. This approach may use less energy and may compact the soil less than numerous passes with a cultivator or mower.

Land Use

The emerging challenge of feeding the climate refugees by increasing the areas under agriculture and enhancing its productivity. This issue can be addressed by utilizing wastelands. According to Integrated Waste Land Development Programme approximately 68.35 million hectares land is lying as wastelands in India. Conversion of agricultural land to urban uses is a particular concern in Existing farmland conversion patterns often discourage farmers from adopting sustainable practices and a long-term perspective on the value of land. At the same time, the close proximity of newly developed residential areas to farms is increasing the public demand for environmentally safe farming practices. By helping farmers to adopt practices that reduce chemical use and conserve scarce resources, sustainable agriculture research and education can play a key role in building public support for agricultural land preservation. Educating land use planners and decision-makers about sustainable agriculture is an important priority.

Consideration of Farmer Goals and Lifestyle Choices

Management decisions should reflect not only environmental and broad social considerations, but also individual goals and lifestyle choices. For example, adoption of some technologies or practices that promise profitability may also require such intensive management that one's lifestyle actually deteriorates. Management decisions that promote sustainability, nourish the environment, the community *and* the individual.

Agriculture Policy

Existing federal, state and local government policies often impede the goals of sustainable agriculture. New policies are needed to simultaneously promote environmental health, economic profitability, and social and economic equity. For example, commodity and price support programs could be restructured to allow farmers to realize the full benefits of the productivity gains made possible through alternative practices. Tax and credit policies could be modified to encourage a diverse and decentralized system of family farms rather than corporate concentration and absentee ownership. Government and land grant university research policies could be modified to emphasize the development of sustainable alternatives. Marketing orders and cosmetic standards could be amended to encourage reduced pesticide use. Coalitions must be created to address these policy concerns at the local, regional, and national level. On the other side free or cheap electricity enables farmers to draw excessive water that results in a lowering of water tables soil salinity and over production of food grains.

The "food system" extends far beyond the farm and involves the interaction of individuals and institutions with contrasting and often competing goals including farmers, researchers, input suppliers, farm-workers, unions, farm advisors, processors, retailers, consumers, and policymakers. Relationships among these actors shift over time as new technologies spawn economic, social and political changes.

Various Strategies for Sustainable Agriculture in Dryland Areas

1. Support for biomass and livestock development should be the foundation of the policy which would ensure that the inputs for our agriculture are organized at the local level and not provided by giant synthetic chemical fertilizer companies as it is happening today. So it may be better to provide

direct input subsidies to the farmers than other players. Sustainable agriculture can contribute significantly to increased food production, as well as make a significant impact on rural people's welfare and livelihoods. But without appropriate policy support at a range of levels, these improvements will remain at best localised in extent or, at worst, wither away.

2. Local agro diversity needs to conserve and promote best practices through community partnership supported by a policy. This strategy will be helpful to address the problem of food security especially for marginalised community. There are following actions can be taken -
   - ✰ Increase farmers' control over seed and technologies
   - ✰ Promotion of indigenous varieties and local knowledge of pest management – documentation, training, demonstration
   - ✰ Promotion of seed saving and exchange, seed bank
   - ✰ Documentation of traditional early warning systems
   - ✰ Documentation and dissemination of best practices for inclusion into development planning for up-scaling
   - ✰ Survey of local gene pool with farmers
   - ✰ Initiate Participatory Plant Breeding that is farmer led, with farmers' objectives – farmers' control over genetic resources and on-site conservation
3. New policy must include direct support for the farmer for taking up organic-ecological agricultural practices that are more suited for a dry state like Rajasthan. Farmers face frequent droughts and sometimes incidents of intense rainfall leading to floods. Under the scenario of increasing frequency of such environmental hazards, there is need for participatory discussion among farmers and researchers on what can be the appropriate strategies in sustainable agriculture – growing drought resistant varieties, or a mix of different varieties to utilise all opportunities. Which types of livestock are better suited to the current situation of rainfall, fodder availability, and markets - for marginal and other types of farmers.
4. There is a need to strengthen adaptive research and technology assessment, refinement and transfer capabilities so that the existing wide technology transfer gaps are bridged. For this, an appropriate network of extension service needs to be created to stimulate and encourage both top-down and bottom-up flows of information between farmers, extension workers, and research scientists to promote the generation, adoption, and evaluation of location specific farm technologies.
5. Besides maintenance breeding, greater effort should be made towards developing hybrid varieties as well as varieties suitable for export purposes. Agronomic and soil researches in the region need to be intensified to address location specific problems as factor productivity growth is decelerating in major production regimes.

6. Generation and effective assessment and diffusion of packages of appropriate technologies involving system and programme based approach, participatory mechanisms, greater congruency between productivity and sustainability through integrated pest management and integrated soil-water-irrigation-nutrient management, should be aggressively promoted to bridge the yield gaps in most field crops.
7. It is necessary to enlarge the efforts for promoting available dry land technologies. Promoting efficient fertiliser practices, improving soil-testing services, strengthening distribution channel of critical inputs specially quality seed and development of physical and institutional infrastructure will help resource-poor farmers.
8. Vast untapped potential in the yield exists for all crops in most of the states accounting for more than three-fourths of crop area. Emphasis must be given to the states in which current yield levels are below the average yield.
9. Agriculture is the biggest user of water, accounting for about 80 per cent of the water withdrawals. There are pressures for diverting water from agriculture to other sectors. An integrated water use policy is formulated and judiciously implemented. Resource-poor farmers in the rainfed ecosystems practice less-intensive agriculture, and since their incomes depend on local agriculture, they benefit little from increased food production in irrigated areas.
10. Diversification towards the high value and labour intensive commodities can provide adequate income and employment to the farmers dependent on small size of farms. Due importance should be given to quality and nutritional aspects. High attention should be given to develop post-harvest handling and agro-processing and value addition technologies not only to reduce the heavy post-harvest losses and also improve quality through proper storage, packaging, handling and transport. The role of biotechnology in post-harvest management and value addition deserves to be enhanced.
11. To enhance the productivity of marginalized famer's field, integration of sustainable agriculture with MGNREGA may be a solution of food security.
12. Agricultural cooperatives and Gram Panchayats must play a leading role in this effort. In doing so, the needs of small farmers should be kept in mind. The agro-processing facilities should preferably be located close to the points of production in rural areas, which will greatly promote off-farm employment. Gender sensitization is needed among stakeholders.
13. Government may facilitate and support community level action through voluntary organizations, Bharat Nirman Volunteers (BNVs), and youth club members including farmers groups aimed at improving food security, reducing poverty, and assuring sustainability in the management of natural resources. For which adequate attention is required to create mass awareness amongst the farming communities in the selected villages.
14. Action may be initiated for effective convergence of the developmental activities through various governmental and non governmental agencies.

Sincere efforts for mass participation and voluntary services are to be encouraged and properly oriented.

15. Artificial recharge to ground water is essential to arrest the decline of water level and degradation of quality. When it comes to Risk Management, it should be acknowledged first that the existing risk management strategies and mechanisms have failed farmers badly. There is a need for complete recasting of the existing models and mechanisms. We need new mechanisms to assess damage and loss and better ways to deliver support - including weather insurance, livestock insurance and effective crop insurance.
16. Mobilization of the best of science and development efforts (including traditional knowledge and modern scientific approach) through partnerships involving national and international research institutions, NGOs, farmers' organizations and private sector in order to tackle the present and future problems of food security and production should be done.
17. Donors and Government must urgently increase funding for agricultural research targeted at the needs of the rural and urban poor, and every effort must be made to ensure the free flow of information, technology and germplasm so that a proper sustainable agriculture can be achieved.

Conclusion

A wide diversity of strategies and approaches are necessary to create a more sustainable food system while technology improvement has its limit and the productivity of natural resources cannot be enhanced indefinitely. These will range from specific and concentrated efforts to alter specific policies or practices, to the longer-term tasks of reforming key institutions, rethinking economic priorities, and challenging widely-held social values.

References

Altieri, M.A. (2008). *Small Farms as a Planetary Ecological Asset*. WASSAN, Secunderabad.

Despande, W.R. (2009). *Basics of Organic Farming*. All India Biodynamic and Organic Farming Association, Indore.

Roy, S. (2001). *Gender Sensitization in Agriculture. A training Manual*. MoA, Govt. of India, New Delhi.

Sharma, A.K. (2010). *A HandBook for Organic Farming*. Agrobios (India) Jodhpur.

Vidyadharan, Veena and Tiwari, M.K. (2009). Mr. Bairwa's Diverse Farm. *LEISA India*, Vol. 25.1

Vidyadharan, Veena and Tiwari, M.K. (2009). Tribal women attain food sovereignty through seed banks. *LEISA India*. Vol. 11.3

Whyte, A.G.D. (1994). Multiple Objective Modeling of the Sustainability of Forests. *Proceedings of IUFRO International Workshop on Sustainable Forest Managements*, Furano, Japan, October 17-21, pp. 31-32.

Chapter 32

Hematological Effects in Lead Induced Chick Embryos: An Experimental Study in Regards to Environmental Hazards

S.K. Pyne[1], R.C. Ghosh[2] and A.K. Aditya[3]

[1]Professor in Animal Science and Head of the Department, ASEPAN, PSB, Visva-Bharati

[2]Professor, Pathology Department, College of Veterinary Science and A.H., Durg – 491 001, Chhattisgarh

[3]Formality Professor, School of Life Science, Visva-Bharati, Santiniketan

ABSTRACT

Present study was made with 300 fertile White Leghorn eggs; these were equally divided into five groups. Group I was kept as control. Group II was injected with 1 mg lead nitrate/egg, likewise group III with 2 mg lead nitrate/egg, group IV with 1 mg lead acetate/egg and group V with 2 mg lead acetate/egg was given. In all the lead treated groups, only a single dose of lead salt in 0.05 ml distilled water was injected into the yolk sac on the tenth day of incubation. 50 per cent control eggs were injected with 0.05 ml distilled water and rest were kept intact. All the incubated eggs of 5 groups were opened on 19th day and pooled blood samples were taken for different haematological tests. Haemologibin (gm per cent), packed cell volume (per cent) and red blood cell counts were decreased ($P<0.05$), whereas leucocyte counts were increased ($P<0.01$) among the treated groups as compared to normal. In differential leucocyte counts, heterophils were increased ($P<0.01$) and monocytes decreased ($P<0.01$) but lymphocytes, eosinophils and basophils showed decreasing tendency in treated groups.

Introduction

Lead is widely used in a variety of modern industries because of its properties *viz.* low boiling point, early oxidised, mixes easily with other metals to form alloys and anticorrosive. But the importance of lead as an environmental pollutant cannot be overstressed. The soil lead enters the ground water and ultimately reaches drinking water to be consumed by animals. Another major source of lead is combustion of gasoline from automobiles. It has been found that the concentration of lead in air near heavy traffic may be 20 times more than that in rural areas. Uses of grass bordering the roads carrying great deal of traffic for livestock feeding may result in deleterious consequences from nutritional and veterinary point of view (Somlyay *et al.*, 1986). The lead concentration of street dust may range from 200 to 20,000 mcg/g (Swinyard *et al.*, 1983). After absorption into the body system, there is a constant exchange of lead between blood, soft tissue and bone.

Definitive experimental studies on the effect of lead on prenatal developing were carried out by different researchers using chick, hamster, rat and mouse (Goyer and Rhyne, 1973). But it has been found in chick embryos that responses of lead salt are different in different days of incubation and there are still doubts with regard to the actual mechanisms involved (Nerbaitz *et al.*, 1985).

Surprisingly little attention has been given for various biological effects of lead compounds in later stages of development in laboratory animals as well as in chick embryos. Therefore, the present investigation has been undertaken after injecting the lead compounds into the yolk sac of chicken eggs on the 10th day of incubation.

Materials and Methods

The present investigation has been conducted with 300 fertile White Leghorn eggs from a stock known to be nutritionally and genetically healthy. The eggs were incubated in a standard egg-hatching incubator with devices for adjustment of temperature and humidity with an arrangement for mechanical turning of eggs. Fresh selected eggs marked with date considered as 'Zero' day on the date of setting in incubator and the subsequent intervals of 24 hrs. were counted as day one, two, three and so on. The eggs were incubated at a temperature of 37 ± 1°C and in relative humidity of 65 to 85 per cent. The eggs were turned at least four times a day. The eggs are supposed to free from any exposure to any heavy metals.

Eggs were equally divided into five groups. Group I was kept as control. Group II was injected with 1 mg lead nitrate/egg, likewise group III with 2 mg lead nitrate/egg, group IV with 1 mg lead acetate/egg and group V with 2 mg lead acetate/egg was given. In all the lead treated groups, only a single dose of lead salt in 0.05 ml distilled water (fresh solution) was injected into the yolk sac with the help of sterile tuberculin syringe on the 10th day of incubation. 50 per cent control eggs were injected with 0.05 ml distilled water and rest were kept intact. A special device was applied to inject the solution aseptically. All the incubated eggs of five groups were opened on 19th day. Pooled blood samples from the chick embryos were collected and blood smears were made for different haematological studies. DLC and TLC both were determined as per method described by Natt and Herrick (1952). For PCV microhaematocrit method (Cohen, 1967) was employed. The Hb concentration was estimated by cyanmethaemoglobin method as described by Makarem (1974). For DLC freshly prepared blood smears were stained with wright's stain (Lucas *et al.*, 1974) and different leyucoytes were counted by Zig-Zag method in uniformly cell distributed areas. Length and breadth of the RBC were made in microns (m) with the help of calibrated oculometer.

Results

Table 32.1 presents the average value of haemologibin (g per cent) packed cell volume (per cent) total red blood cells and total leucocyte count (TLC) of control and treated groups exposed to different doses of lead compounds. The HB, PCV and total R.B.C. decreased significantly (P<0.01) whereas TLC increased significantly (P<0.01) in all treated groups as compared to normal value (Table 32.1). For Hb per cent, no significant variation was observed between the treated groups. PCV per cent of group III was significantly (P<0.01) decreased as compared to other treated groups. Red blood cell counts was significantly (P<0.01) decreased in group III as compared to group II, and IV. Difference between group III and V was also significant (P<0.01). TLC was highly increased (P<0.01) in group IV as compared to group II, III and V.

Table 32.1: Effects of Lead Compounds of the Hb (g per cent), P.C.V. (per cent), R.B.C and W.B.C of chick embryos (Mean ± S.E.)

Groups	*Haemoglobin (g per cent)*	*Packed Celll Volume (per cent)*	*Red Blood Cells X Million*	*White Blood Cells X Thousand*
I (Control)	6.70^{a}±0.15	17.48^{a}±0.39	2.05^{a}±0.01	40.40^{a}±0.42
II	5.43^{b}±0.20	15.52^{b}±0.31	1.67^{b}±0.16	44.00^{b}±0.50
III	5.12^{b}±0.18	13.66^{c}±0.27	1.26^{c}±0.10	45.90^{be}±0.55
IV	5.53^{b}±0.17	14.62^{de}±0.29	1.75^{b}±0.08	52.60^{c}±0.96
V	5.35^{b}±0.19	14.74^{be}±0.33	1.60^{b}±0.12	46.20^{de}±0.79

Figures with different superscripts vary significantly (P<0.05).

Table 32.2 summarise the average values of differential leucocyte count (DLC) in the control and after inoculation of different doses lead compounds within the embryonated eggs. Heterophils increased significantly (P<0.01) in all treated groups as compared to normal (Table 32.2). No significant variation was observed between the treatments.

Decreasing tendency for lymphocytes (Table 32.2) was noticed among the treated groups but the values were not statistically significant than that of normal.

Monocyte significantly (P<0.01) reduced in group II, III and IV only as compared to normal (Table 32.2). Among the treated groups, group III showed significant (P<0.01) differences (reduction) as compared to group IV. The values of group II and IV also differ (P<0.05) between themselves. Group II showed more or less same effect with that of group III.

Table 32.2: Effects of Lead Compounds on the Differential Leukocyte Counts (per cent) of Chick Embryos (Mean ± S.E.)

Groups	Heterophil	Lymphocyte	Monocyte	Eosinophil	Basophil
I (Control)	39.29^{a}±0.24	49.30^{a}±0.25	6.40^{a}±0.32	4.30^{a}±0.14	0.80^{a}±0.13
II	45.40^{b}±0.40	46.60^{a}±0.32	4.60^{b}±0.25	3.30^{ab}±0.32	0.30^{a}±0.14
III	46.90^{b}±0.67	45.50^{a}±0.51	4.10^{b}±0.22	2.60^{bc}±0.25	0.50^{a}±0.16
IV	46.00^{b}±0.37	45.60^{a}±0.53	5.50^{c}±0.16	2.30^{b}±0.14	0.60^{a}±0.15
V	45.50^{b}±2.25	44.20^{a}±2.80	6.50^{a}±0.45	3.50^{ac}±0.68	0.30^{a}±0.14

Figures with different superscripts vary significantly (P<0.05).

For eosinophils, decreasing tendency (Table 32.2) was noticed in the treated groups. High reduction was noticed only in group III and IV as compared to normal. The overall differences among the treated groups were not significant except between the group IV and V, where the values differ significantly (P<0.05).

Decreasing tendency for basophil (Table 32.2) was observed in the treated groups but the values were not statistically significant as compared to normal.

The average values of the length and breadth of the R.B.C. of control and of treated groups after exposure of lead compounds were measured. Length (11.17 m) increased significantly (P<0.01) in all treated groups as compared to normal values. Among the treated groups R.B.C. length of group V significantly (P<0.01) differs (increased) from other groups. No such variation was found in breadth of R.B.C., except group V where breadth significantly (P<0.01) increased as compared to other groups including control.

Discussion

The marked reduction (P<0.01) of haemoglobin, packed cell volume and red blood cells may be explained due to prolonged action of lead compounds upon the haemopoietic system. Among the treated groups, higher doses of lead nitrate affects more in haemopoietic system. Total leucocyte account (TLC) in control was 40.4 thousands which was increased (P<0.01) in treated groups in general (Table 32.1) among which more pronounced effects were noticed in lead acetate treated groups than that of lead nitrate groups. In differential leucocyte counts, heterophils increased (P<0.01) in all the treated groups as compared to normal. Whereas lymphocytes, eosinophils and basophils counts showed decreasing tendency among the treated groups. It this regard monocyte only decreased significantly as compared to normal. Marked leucopenia, decreased haemoglobin, increased heterophils and reduction of hymphocytes were also reported by Sharma and Seetharaman (1950) in RIR males infected with a virulent strain of NDV (New Castle disease virus). On the contrary, Chandrasekharan and Krishnan (1958) showed increased Hb concentration, reduced leucocyte count and reduced lymphocyte percentage with increased percentage of heterophils in clinical outbreak of New Castle disease among the 8 to 12 weeks old chicks.

Average size of the R.B.C. increased in all treated groups compared in normal values, and this may be due to the presence of more immature erythrocytes in circulation, Gilman *et al*. (1980) and Braunwald *et al*. (1987) have stated that nearly all (95 per cent) circulating inorganic lead remains associated with erythrocytes. Hatch (1982) has demonstrated in dog that lead compounds caused immature erythrocytes, anaemia from haemolysis, bone marrow hyperplasia, leucocytosis with neutrophilia and shift to left in neutrophils, eosinopenia or monopenia or neuttropenia.

This findings is in conformity with the present investigation. Immature red blood cells in circulation and resultant shortened erythrocyte lifespan contributes to anaemia due to lead (Jones and Hunt 1983). Inhibition of haem-biosynthesis and subsequent lowering the haematocrit values due to lead acetate have also been reported by Madej *et al*. (1988). Haemolytic effect o lead may be the direct effect on mature red blood cells, which occur independently due to the effect of lead on haem-biosythesis in erythroid precursors. Goyer and Rhyne (1973) have stated that impairment of haem-synthesis by lead might have occurred due to some biochemical reasons. On the other hand, Bischoff *et al*. (1928) have showed in their comparative study that inorganic lead caused 40 per cent reduction in haemoglobin and organic lead has caused no drop in haemoglobin against the same doses given to rabbit. Basophilic stippling, which is the most common effect of lead (Vermande and Meigs 1960), is not observed in this experiment but immature R.B.Cs are very common which has also been noted by Jones and Hunt, 1983. Laucranjan *et al*. (1975) have reported that the male workers in the lead industry who had a mean blood lead level of 74.5 mcg/gl. had a significantly higher incidence of treatospermia when compared to another group which had a mean blood lead level of 52.8 mcg/dl. They also reported that initial pathophysiologic effects of lead results from the interference with enzymatic pathways essential to the synthesis of haem. Lead interferes with transmitochondrial transport of iron and it also prevents iron from uniting with protoporphyrin and the latter accumulates in erythrocytes. This free erythrocytic protoporphyrin can be estimated and is a more sensitive indicator of the effects of lead.

Conclusion

In the present investigation, from the observations of different parameters it was noticed that both lead nitrate and lead acetate were showing deleterious effects in different systems of the body. Further, it was also noticed that lead nitrate was found to be more teratogenic than have lethal effect, whereas lead acetate showed more lethal effect than teratogenic effects.

References

Bischoff, F., Maxwell, L.C., Evans, R.D. and Nuzum, F.R. (1928). Studies on the toxicity of various lead compounds given intravenously. *J. Pharmacol. Exp. Therapeut.*, 34: 85-109.

Braunwald, E., Isselbacher, K.J. Petersdorf, R.G. Wilson, J.D., Martin, J.B. and Fauci, A.S. (1987). *Harrison's Prionciples of Internal Medicine*, 11th edn. Pp. 852-853. Mc Graw-Hill-Book Company, New York, Maxico, New Delhi, Tokyo, Toronto.

Chandrasekharan, K.P. and Krishnan, R. (1958). Haematology in Ranikhet disease. *Indian Vet. J.*, 35: 616-619.

Cochen, R.R. (1967). Anticoagulation, centrifugation time and sample replicate number in the microhaematocrit method for avain blood. *Poult. Sci.*, 46: 214-218.

Goyer, R.A. and Rhyne, B.E. (1973). Pathological effects of lead. *Int. Rev. Exp. Pathol.*, 12: 1-77.

Gilman, A.G., Goodman, L.S. and Gilman, A. (1980). *The Pharmacological Basis of Therapeutic* 6th edn. Pp. 1616-1622. Macmillan Publications. New York.

Hatch, R.C. (1982). *John's Veterinary Pharmacology and Therapeutic*, 5th edn. (Eds.) N.H. Booth and L.E. McDonald, pp. 1005-1010. Kalyani Publishers, Ludhiana. New Delhi.

Jones, T.C. and Hunt, R.D. 1983. *Veterinary Pathology*, 5th edn. Pp. 1000-1005. Lea and Febiger. Philadelphia.

Lucas, A.M. and Jamroz, C. (1974). *Atlas of Avian Hematology*, Oxford and IBH Publishing Co. New Delhi, pp. 228-229.

Laucranjan, L., Popescu, H.I. Gavanescu, O., Klepsch, L. and Serbanescu, M. (1975). Reproductive ability of workmer occupationally exposed to lead. *Arch. Environ. Health*, 3: 396-401.

Makerem, A.M. (1974). Haemoglobin, myoglobim and hepatoglobins. In: *Clinical Chemistry: Principles and Technics*, 2nd edn, (Eds.) R.J. Henry, D.C. Cannon and J.W. Winkelman, Maryland: Harper and Row Publishers Inc., pp. 1111-1214.

Madej, J.A., Radzanowska, G., Kalimentowski, S. and Zarka, M. (1988). The reactions of some haematological and biochemical indices and the histopathological changes in the organs of hens given lead, selenium of lead plus selenium by mouth. (Cited in *Vet. Bull.* 1989. 59(9) 5437 pp. 776).

Natt, M.P. and Herrick, C.A. (1952). A new blood diluent for counting the erythrocytes and leucocytes of the chicken. *Poult. Sci.*, 31: 735-738.

Narbaitz, R. Marino, I. and Sarkar, K. (1985). Lead induced early lesions in the brain of the chick embryo. *Teratology*, 32: 389-396.

Sharma, G.L. and Seetharaman, C. (1950). Blood picture in Ranikhet disease of fowls. *Indian J. Vet. Sci.*, 20: 203-210.

Swinyard, C.A. Sutton, D.B. and Saloum, L.M. (1983). Lead in the environment. Experimental studies of lead toxicity in animals and their relevance to marginal lead toxicity in children.*Cong. Anom.* 23: 29-60.

Somlyay, L. Laszio, V. and Marta (1986). Lead contamination of vegetation bordering the roads in the neighbouring *Allatorv Lapja.*, 41(7): 423-434. Abstracted in *Biol. Abst.* 82(11) 1010, 1986.

Vermande, V.E. G.J. and Meigs, J.W. (1960). Change in the ovary of the Rhesus monkey after chronic lead intoxication. *Fertil. Steril.*, 11: 223-234.

Chapter 33

Horticulture Production in West Garo Hills District and its Scope for Sustainability in Rural Areas

Akaibel D. Arengh

Research Scholar,
RDAP Department, NEHU, Tura Campus

ABSTRACT

The West Garo Hills district of Meghalaya has suitable agro-ecological regions with the humid climate.Diversification in topography, altitude, and a climatic condition offers a wide scope for subtropical fruits and tropical fruits. The West Garo Hills district occupies around 27.07percent of the total area of the state for fruit cultivation with the total production and yield of 59261 million tones and 35841kg/ha against the state average of 238977 million tonnes and 35948 kg/ha respectively. The district leads in banana, areca nut, cashewnut and tea leaf production occupying around 28.2 per cent, 35.3 per cent, 68.7 per cent and 20.5 per cent respectively. Land use planning is the basic tool for bringing sustainability in the developing countries so that the ecological carrying capacity of the land is maintained.The shifting cultivation can be brought under permanent cultivation by replacing it with scientific methods of cultivation and substituting by alternatives like horticulture crops and plantations which can help to wean people away from Jhum. Proper utilization of land for bringing sustainability depends on taking up of such alternatives. Garo Hills offer a fertile ground for Horticulture but the main bottleneck is of market linkages and non-existence of adequate processing facilities.

Keywords: Sustainable horticulture production, Scope, West Garo Hills District.

Introduction

Meghalaya state with a geographical area of 22,429 sq km, is situated in northeast India. Meghalaya is basically an agricultural state with 70 per cent of the population depending on agriculture and contributes about 22 per cent of the GDP. In Meghalaya climate is monsoonal with distinct warm wet and cold dry periods. The western and southern part of the state is warmer than the central upland where mean minimum temperature stands at 2°C. Average maximum and minimum temperature varies from 5°C to 32 °C. The torrential rainfall variation varies from place to place and from altitude to altitude. The highest amount of rainfall ranges from around 11400 mm in Cherrapunjee to 400mm in some part of the state. Due to variation in topography and climatic factor a unique array of vegetation can be found in the region ranging from tropical and sub tropical to temperate and near temperate climate.

Among the constituent units of the state of Meghalaya, the Garo Hills are of comparatively low elevation and are warmer than the districts of Khasi and Jaintia Hills. The Garo Hills lie between 25°9′ and 26°1′ north latitude and 89°49′to 91°2′ east longitude has vast potential and ecological diversification. The total geographical area of the region is 8,167sq kms (1981 census). The West Garo Hills district of Meghalaya has suitable agro-ecological regions with the humid climate. The landscape is medium to steep slope and the soil is red loamy, brown and red. Hence, there are ample opportunities to grow a variety of horticulture crops. An estimate reveals that more than 14373 hectares of land is covered by horticulture crops and the annual production is about 68279 metric tones (District Statististical Handbook 2008-09).

Table 33.1: Land Utilization in2008-09

Sl.No.	Particulars	Area in hectares
1.	Total geographical area	367700 hectares
2.	Reporting area	307700 hectares
3.	Forest area	165508 hectares
4.	Area under non agriculture uses	14292 hectares
5.	Barren and uncultivable land	7391 hectares
6.	Permanent pasture and other grazing land	858 hectares
7.	Land under misc. trees,crops and groves	23902 hectares
8.	Cultivable waste land	33220 hectares
9.	Fallow and other than current fallows	42690 hectares
10.	Current fallows	13958 hectares
11.	Net area sown	65881 hectares
12.	Area sown more than once	24896 hectares
13.	Total cropped area	90777 hectares

Source: District at a glance 2009-10, Government of Meghalaya Department of Economics and Statistics, West Garo Hills, Tura.

Diversification in topography, altitude, and climatic conditions offers a wide scope for subtropical fruits and tropical fruits. Similarly wide and diverse type of vegetables including indigenous one are cultivated in the region. A wide variation in the agro climatic situation in the region enables cultivation of different varieties of fruits, vegetables and plantation crops. The region has been exploring the cultivation of different plantation crops.

Several programmes had been implemented for transformation of horticulture in Garo Hills like Technology mission and minor irrigation projects. India is one of the leading countries in the world for areca nut production where Meghalaya make a significant contribution with a production of 4 per cent from an area of 3 per cent. Cashewnut plantation was started in recent years and is definitely making an impact in many part of the region. Spices like black pepper, ginger, turmeric etc., is also occupying an important part of horticulture production of the country. As per the department of Economic and Statistics (2009-10) around 90777 hectares of the district is under cropped area and additional cultivable wasteland of about 33220 hectares (provisional) and fallow land of 13958 hectares (provisional) is available which can be brought for cultivation and expansion of horticulture in the district.

Production Scenario

The West Garo Hills district occupies around 27.07 per cent of the total area of the state for fruit cultivation with the total production and yield of 59261 million tones and 35841kg/ha against the state average of 238977 million tonnes and 35948 kg/ha respectively. The West Garo Hills district grows diversity of crops and shares major part in the production status of the state and some part in the production status of the country. In pineapple production Ri bhoi district leads with 41.74 per cent of the total state production followed by West Garo Hills district with 22.09 per cent. In citrus and in papaya production West Garo Hills contributes around 7.27 per cent and 22.7 per cent respectively, to the state production. The district leads in banana, areca nut, cashewnut and tea leaf production occupying around 28.2 per cent, 35.3 per cent, 68.7 per cent and 20.5 per cent respectively. India is one of the leading countries in the world for areca nut production where Meghalaya make a second largest contributor in the production with the West Garo Hills making a contribution of around 35.3 per cent out of the total area of 27.1 per cent.

Spices like coriander, black pepper, tezpatta, ginger, turmeric, chilies and cinnamon etc., occupy an important part of horticulture production of the state with the West Garo Hills District as the leading contributor of the production of spices. With the production per cent of 39.6 in coriander from the area of 32.6 per cent, 27.3 per cent in black pepper from the area of 37.7 per cent, 83.4 per cent in garlic from the area of 59.1 per cent, 21.9 per cent in ginger from the area of 26.09 per cent, 16.9 per cent in turmeric from the area of 19.7 per cent and 44.1 per cent in chilly from the area of 48.09 per cent.

Major oilseeds grown in the state are rapeseed and mustard, sesamum and castor. The West Garo Hills district covering around 79.6 per cent area of the total spices grown area in the state hold the production of 86 per cent in rapeseed and mustard,69.1 per cent in sesamum and 28.2 in castor of the total state production.

Production and productivity of the region is low as compared to the national level which can be improved by harnessing the potential of the region.

Table 33.2: Plantation Production in 2006-07 in Meghalaya (Area in hectare, Production in M.T, Yield in kg per hectare)

Districts		*Pineapple*	*Citrus*	*Banana*	*Papaya*	*Areca Nut*	*Cashew Nut*	*Tea Leaf*
Ri bhoi	Area	3515	199	905	155	133	–	875
	Production	40415	1445	14802	1159	74	–	4340
	Yield	11498	7261	16356	7477	556	–	4960
East Khasi	Area	952	4557	765	89	4510	–	8
Hills	Production	6862	20532	8558	617	4754	–	–
	Yield	7208	4505	11187	6932	1054	–	–
West Khasi	Area	798	1053	745	44	1208	–	20
Hills	Production	4570	4636	4731	255	1112	–	–
	Yield	5727	4403	6351	5795	920	–	–
Jaintia	Area	175	1141	382	18	1635	–	17
Hills	Production	1311	6210	1457	40	3207	–	–
	Yield	7493	5442	3815	2222	1961	–	–
East Garo	Area	980	318	1570	149	1035	184	53
Hills	Production	15201	1430	18314	1323	1174	182	283
	Yield	15511	4497	11665	8879	1134	989	5340
West Garo	Area	2573	1271	1675	120	3276	3216	394
Hills	Production	21382	2703	20467	1090	5850	7770	1199
	Yield	8310	2127	12119	9083	1786	2416	3042
South Garo	Area	1119	306	380	34	284	3423	8
Hills	Production	7071	183	4038	315	390	3348	–
	Yield	6319	598	10627	9270	1373	978	–
Meghalaya	Area	10112	8845	6422	609	12081	6823	1375
	Production	96812	37139	72366	4799	16561	11300	5822
	Yield	9574	4199	11268	7880	1371	1656	4234

Source: State Level Crop Reports; Directorate of Economics and Statistical, Government of Meghalaya, Shillong.

The West Garo Hills District has a good potential for cultivation of tea with the production of 28.6 per cent from the area of 20.5 per cent and stands next to Ri Bhoi District (74.5 per cent from the area of 63.6 per cent). Rubber plantation has also been initiated although their processing through large scale plantation activities has not been taken up on an organized scale.

Table 33.3: Condiments and Spices Production in 2005-06 in Meghalaya (Area in hectare, Production in M.T, Yield in kg per hectare)

Districts		Coriander (P)	Black Pepper (P)	Garlic (P)	Tez-patta (P)	Ginger	Turmeric	Chillies	Cinnamon (P)
Ri bhoi	Area	–	147	14	–	994	78	89	–
	Production	–	89	16	–	7430	660	140	–
	Yield	–	605	1143	–	7475	8462	1573	–
East Khasi	Area	–	173	26	3010	515	89	94	–
Hills	Production	–	110	31	6968	3842	491	68	–
	Yield	–	636	1192	2315	7460	5517	723	–
Wesr Khasi	Area	–	103	12	52	390	83	48	–
Hills	Production	–	65	20	105	2553	313	34	–
	Yield	–	631	1667	2020	6546	3771	708	
Jaintia Hills	Area	5	34	18	21	251	1039	132	
	Production	21	25	12	40	2450	6362	102	–
	Yield	4200	735	667	1900	9761	6123	773	–
East Garo	Area	16	58	65	117	4738	103	332	–
Hills	Production	76	93	84	280	24486	568	197	–
	Yield	4750	1603	1292	2391	5168	5515	593	–
West Garo	Area	16	339	220	393	2512	358	895	17
Hills	Production	105	156	946	1072	11764	1779	559	15
	Yield	6562	460	4300	2729	4683	4969	625	882
South Garo	Area	12	43	17	26	225	67	271	–
Hills	Production	63	32	25	44	1084	335	167	–
	Yield	5250	744	1470	1680	4818	5000	616	–
Meghalaya	Area	49	897	372	3619	9625	1817	1861	17
	Production	265	570	1134	8509	53609	10508	1267	15
	Yield	5408	635	3048	2351	5570	5783	681	882

Source: State Level Crop Reports; Directorate of Economics and Statistical, Government of Meghalaya, Shillong.

The West Garo Hill District is one of the good reservoirs for the variety of horticultural crops; however, the practice of unregulated shifting cultivation poses a threat to sustainable diversity and sets as an obstacle to diversification. Non availability of quality planting material, including seeds can be major drawbacks for the production of horticultural crops. Availability of quality planting material both in terms of quantity and in time from within the region is a major setback for the success of horticulture industry. Numerous new technologies developed by various research centers/Institutes/Universities in different parts of the country and appropriate for the region is required to be identified and undertaken in the region for taping the available resources up to its maximum level. Proper marketing channel, prices including proper storage system for the perishable product is lacking in the region. Promotion of marketing channel and export coupled with price stabilization and reducing the risk through insurance are some important issues which needs to be addressed.

Table 33.4: Oilseed Production in 2006-07 in Meghalaya (Area in hectare, Production in M.T, Yield in kg per hectare)

Districts		*Rapeseed and Mustard*	*Sesamum*	*Castor*
Ri bhoi	Area	167	–	6
	Production	70	–	6
	Yield	419	–	1000
East Khasi Hills	Area	65	46	–
	Production	38	32	–
	Yield	585	696	–
Wesr Khasi Hills	Area	30	24	–
	Production	26	18	–
	Yield	867	750	–
Jaintia Hills	Area	65	34	–
	Production	62	16	–
	Yield	954	470	–
East Garo Hills	Area	719	248	8
	Production	350	124	4
	Yield	487	500	500
West Garo Hills	Area	6003	1128	19
	Production	4142	606	11
	Yield	690	537	579
South Garo Hills	Area	212	155	–
	Production	127	80	–
	Yield	599	516	–
Meghalaya	Area	7261	1635	82
	Production	4815	876	39
	Yield	663	536	476

Source: State Level Crop Reports; Directorate of Economics and Statistical, Government of Meghalaya, Shillong.

Sustainability through Horticulture

Sustainability is a grassroots movement, and it revolves round sustainable land use, culture and land use change has to be compatible, though in the developing countries land use planning primarily aims at increased agricultural output while the developed countries have environmental concerns uppermost in their mind

(Venkatesvarlu, 2003). Further Reijtjes *et al.* (1992) defines sustainable agriculture as 'management of resources for agriculture to satisfy changing human needs, while maintaining or enhancing the quality of the environment and conserving the natural resources. Land use planning is the basic tool for bringing sustainability in the developing countries so that the ecological carrying capacity of the land is maintained.In India Watershed Development Approach has been adapted to Soil Conservation and Water management. The same is followed in the West Garo Hills District for land development which is very important as sizable population still practice Jhum Cultivation. Jhum cultivation or shifting cultivation has been in practice since 7000 B.C and is the practiced basically in all the north eastern states including in all the parts of Meghalaya. The area in shifting cultivation in North Eastern states is around 2.70 m.ha and the per cent of area annually cultivated in the shifting cultivation is 16.9 (Venkatesvarlu). As per the Soil and Water Conservation Department Government of Meghalaya (2001) annual area under Jhum in Sq. Km is 155.45 and the per centage of annual Jhum area from the total geographical area. 4.19. Due to disturbance of the forest by shifting cultivation have not only resulted in the degradation of the forest but also have brought the shifting cultivation below the subsistence level causing further environmental degradation. This calls for other sustainable form of agriculture through fruit trees and agroforesrtry (Sen Raj Kumar, 2010). Some of the researchers have suggested that in agroforestry fruit trees and medicinal trees can be included along with the crops as an intercropping. Chundawat (1996) identified the period for intercropping in some of the fruit trees which are given as follows.

Table 33.5: Time Required to Occupy Allotted Space and Suggested Period of Intercropping to Various Fruit Species

Fruit	*Spacing*	*Time to Complete Coverage of space*	*Period of Intercropping*
Mango	10×10	7 years	5 years
Citrus	6×6	5 years	4 years
Banana	1.8×1.8 or 2.0×2.0	6 months	5 months
Guava	6×6	5 years	3 years
Apple	6 ×6	6 years	4 years
Stone fruits	6× 6	4 years	2 years
Litchi	10× 10	10 years	7 years
Sapota	10× 10	10 years	7 years
Grape	3× 3	3 years	1 years
Papaya	2.0 ×2.0 or 2.5× 2.0	9 months	5 months
Ber	6× 6	4 years	2 years
Pomegranate	6× 6	5 years	3 years
Date	8 ×8	8 years	5 years
Pineapple	90 ×90× 25	9 months	6 months

Source: Chundawat and Pareek (1996).

The shifting cultivation can be brought under permanent cultivation by replacing it with scientific methods of cultivation and substituting by alternatives like horticulture crops and plantations which can help to wean people away from Jhum. Proper utilization of land for bringing sustainability depends on taking up of such alternatives.

Annual crops like vegetables, tuber crops etc, can be cultivated with the perennial crops like fruit trees, spices, areca nut etc, as an inter crop or the companion crop without compromising the yield which could increase in the initial year returns per unit of area.

Horticulture-Based Units

The state has considerable scope for the development of horticulture. The hilly terrain and acidic soils are especially suitable both for citrus fruits and other plantation crops. Farmers grow fruits for home consumption. Large quantities of fruit that cannot be marketed locally are wasted during good harvests. Proper storage, processing and marketing facilities has to be encouraged from going in for commercial horticulture. West Garo Hills has a competitive advantage in horti-based industries like fruit processing preservation and canning, areca nuts processing and scented supari making, banana powder making, dehydration and canning of fruits and vegetables products,ginger dehydration, integrated Unit for the production of fruits juices, squashes, ketch up, Jams and Jelly, extraction of pine apples fibers, pickles and chutney making, starch from banana stems, spices grinding and packing in poly pouched etc., which will not only further strengthen the horticulture base of the district and the State but also enhance employment opportunities. The state can be benefitted with the appropriate linkages between the growers and the processors of horticulture produce. All these products have a good market not only in the state and rest of India but also in the neighboring countries.

A number of spices and aromatic plants grow in different parts of Meghalaya. This can be used for the production of high value export items such as spice oils, oleoresin and other essential oils. State Government encourages the setting up of such units in this sector for upgrading the enterprise.

Garo Hills offer a fertile ground for Horticulture but the main bottleneck is of market linkages and non-existence of adequate processing facilities. Major Corporate if invited with the involvement of CII, APEDA, Ministry of Food Processing and Govt. of Meghalaya to take up contract growing activities and putting up processing units will prove to be a great boost in benefitting the farmers as well as the government. Farmers will be encouraged to arrange adequate quantities of quality inputs such as planting materials, suitable fertilizers and pesticides etc.

Government are committed to safeguard the interest of tribal in respect of land and to maintain the population structure by the operation of some of the provisions of the Meghalaya Transfer of Land (Regulation) Act.1971 which encourages setting up of industrial units by Public Sector, Joint Sector, Private Sector and also in the Small Scale Sector of proprietary nature.

Conclusions

Horticultural crops can play a unique role in improving the economy of the region and the country. North Eastern region is gifted with ideal agro climatic conditions for the production of large number of horticultural crops. The resources for the horticultural production can be tapped up to the full extend by inculcating right technology and cultural practices along with the effective support from the concern departments which in turn can bring sustainability. The potential of horticulture crops and plantation development is rated very high on account of the soil and climate suitability. Horticulture and plantation crops are high value, labour intensive and yield more per unit area, and citrus fruits are more suited to the soil. Horticulture has been able to generate additional employment opportunities but also providing sustainable income to small, marginal and tribal farmers.

References

Area Production and Yield of principal crops in Meghalaya from 1999-2000 to 2006-07(include land use statistics and irrigation statistics) volume III. Directorate of Economics and Statistics Government of Meghalaya, Shillong.

Chundawat, B.S. (1996). Intercropping in orchards. In: *Advances in Horticulture*, (Ed.) K.L. Chadha. Malhotra Publising House, New Delhi, p. 763-776

District At A Glance (2009-10).Govrnment of Meghalaya, district statistical officer, West Garo Hills, Tura.

District Statistical Handbook West Garo Hills 2007-08.Govrnment of Meghalaya, district statistical officer, West Garo Hills, Tura.

Sinha, Komol (2010). Rural Development in India Retrospect and Prospect. Concept Publishing Company Private Limited. New Delhi-110059

Venkateshwarlu, J. (2003). *Rainfed Agriculture in India: Research and Development Scenario*. Directorate of Information and Publications of Agriculture. Indian Council of Agricultural Research. Krishi Anusandhan Bhavan, Pusa, New Delhi.

Chapter 34

Rural Entrepreneurial Innovation and its Role in Rural Development

G Singaiah, Nirdosh Kumar Srivastava** and V. Purohit****

Department of Management,
North Eastern Hill University, Tura Campus, Tura, Meghalaya

ABSTRACT

This paper highlights that entrepreneurs who have created innovative organizations and service provisions platform contributes to sustainable development. Entrepreneurial innovations are considered to be able to contribute to the economy by creating a new system with their sustainable innovation. Innovation enhances economic growth and development. However, innovation in rural areas is more likely to arise through small- scale ventures and entrepreneurs than industrial research and development. The processes, structure and outcomes of their initiatives are constructed with more traditional effort. This research paper aims to address this issue by exploring the linkages between rural development and entrepreneurship. In this paper much attention has been given as to how entrepreneurial innovation, can stimulate change in the rural areas leading to rural development.

Keywords: Rural entrepreneurship, Rural development, Entrepreneurial innovation.

Introduction

Entrepreneurship in itself has a wide range of conceptualisation. On the one end, anyone who wants to work for himself or herself is considered to be an

---

E-mail: *gsingaiah@yahoo.co.in; **nirdoshsri@gmail.com; ***vandanapurohit11@gmail.com

entrepreneur. While on the other end, entrepreneur is a person having a very high aptitude and who pioneers change, encompasses characteristics found in only a few of the masses. Entrepreneur is an innovator, co-ordinator and an arbitrager who increases the 'dynamism' of the economy (Karayiannis, 1990), entrepreneurship and innovation systems theory both share a common origin (Schumpeter, 1994; Schumpeter, 1934; Schumpeter and Opie, 1961), innovation is indispensable for economic development (Edquist, 1997, Freeman, 1987, Hall *et al.*, 2003, Lundvall 1992, Spielman, 2005). In developing countries rural innovation is more essential as it is in rural areas that most of the poor live (World Bank, 2008). Furthermore, rural entrepreneurship and small business ventures have been considered as a strategy for rural development (Hoy 1996). Rural entrepreneurship as defined by Wortman (1990) is the creation of a new organization that introduce a new product, serves or creates a new market or utilizes a new technology in rural environment (Yu and Artz, 2009). Although traditional strategies of recruiting plants and relocating business to rural areas are inadequate to enhance rural economic growth, opening and growing new business in rural areas can motivate local economies by creating job opportunities, providing an extended variety of products and services as a result expanding the quality of life in rural areas.

Moreover, entrepreneurship is fundamental for the maintenance of healthy economy and for effective economic development new business in low- income areas must be started through local initiatives (Acs and Malecki, 2003, Lichtenstein and Lyons 2001, Smilor 1997, Winters 1997). Although micro enterprise and small business programs are hugely supported by the government, relatively less priority has been given to entrepreneurship and small business development as the basis for widespread rural development of low- income rural communities (Robinson and Christy, 2002). The present paper is an attempt to examine the linkages between rural development and entrepreneurial innovation and small business development. The goal of the paper is to understand the conceptual underpinnings of entrepreneurship as well as its potential as a factor of rural development on the perspective of rural entrepreneurship.

Entrepreneurship and Rural Development

Entrepreneurship has been considered as one of the important strategy of development in all progressive nations. The reason behind its popularity amongst policy makers is its potential as a source of job creation and economic growth. Entrepreneurship and innovation could be fostered in a systematic way to transfer traditional urban or rural economies into entrepreneurial ones. Entrepreneurship is 'the process of uncovering or developing an opportunity to create value through innovation and seizing that opportunity without regard to either resources or the location of the entrepreneur in a new existing company' (Slaughter 1996). Whereas, rural development can be defined as the economic, social and environmental development and the development of governance of territories characterised by low population density, heavy reliance on land as a source of livelihood and a non-urbanised settlement structure (Fieldsend and Katona, 2007). The above definition has been considered to successfully distinguish rural development from the closely associated activities of regional development and agriculture development.

There are various factors which determine the prospects of entrepreneurship in rural communities. Nolan (2003) points out in *Main Street of Tomorrow*, a Federal Research Bank conference report on rural entrepreneurship that determinants of enterprise creation include, demographics, unemployment, wealth, educational and occupational profile of the workforce, prevalence of small firms, the extent of owner-occupied housing, infrastructure endowment and regional history. Porter (1995) emphasize that sustainable economic growth strategies require 'improving the environment for business'. Fundamental principles underlying an entrepreneurship-led- rural development policy are- an economic versus a social focus, highlighting the creation rather than the redistribution of wealth within economically disadvantaged rural communities, an emphasis on the private sector rather than the government and social service sector but highlights the supporting role of government and social services, an emphasis on having trained and experienced entrepreneurs. (Robinson, Lyson, and Christy 2002).

Characteristics of Rural Entrepreneurship

Rural entrepreneurship signifies that type of entrepreneurship which is able to offer added value to all type of rural resources (natural, material, financial, human) in rural areas. (North and Smallbone 2000). Yu and Artz (2009) explain the characteristics of rural entrepreneurship in their study on the basis of four dimensions of business as business nature, human capital and expertise, family background and social capital. As is commonly known business located in rural areas will suffer from less local demand, infrastructure or supportive assistance than in the urban areas, but will benefit from lower land rent and wages. Shields (2005) elucidates that rural settings are characterised by distinctive socio -cultural values and preferences that can affect small business development, through its influence on gender roles, co-operation, communications and network composition. Starting a business in rural or urban area is an important but complex decision to make. In rural entrepreneurship the business has to be consumer oriented and have a clear market. Another distinctive feature of rural entrepreneurship is that the business is identified by the owner (Jack and Anderson, 2002).

Hoy *et al.* (1980) in a study found that a rural entrepreneur appears to be "independent, capable of taking chances, results-oriented, optimistic, confident in their powers, hardworking and innovative". It was also noted that rural entrepreneurship should be clearly focused on creating new workplaces by implementing investments in rural communities. The rural residency experience matters in the effects of rural entrepreneurship role model and in attitudes toward geographically- specific economic and social change and toward rural entrepreneurship. It acts as part of the rural entrepreneurship related resources endowment (Westlund and Bolton, 2003). Under the social capital connected to entrepreneurship both as resources for action and as preference for like an entrepreneurial lifestyle has been integrated into economic model for decades (Westlund and Bolton, 2003). As Becker (1996) state that utility function at any moment depends not only on the different goods consumed but also on the stock of personal and social capital at that moment (Yu and Artz, 2009).

David J. Lewis (1996) in a study on rural entrepreneurship in Bangladesh villages recognised the nature of rural entrepreneurship in the light of green revolution and asserts that the processes of technological change under privatisation and structural adjustment have created new prospects leading to the creation of new form of technology- based enterprises in rural areas. He further elucidates that these enterprises need not always be captured by local elites. For NGOs and other development agencies there may be new openings for innovative support to enterprise efforts by the poor.

Rural Entrepreneurship Perspectives

Due to its strong relationship with economic growth, entrepreneurship is a key determinant of economic development at different economic levels. As rural communities search for new engines of growth, entrepreneurs will be a mechanism that transforms opportunity into prosperity. To accomplish this goal, the functional role of entrepreneur in economic theory is linked to a set of strategic initiatives based upon various economic development strategies and community models designed to foster entrepreneurship. On the basis of the previous literature, the following perspectives of rural entrepreneurship are considered for the present study as strategic initiatives to promote rural development.

Rural Institution

The World Bank (2002a) offers a working definition of institutions as rules, enforcement mechanisms and organizations that promote market transactions. This definition points out that institutions provide multiple functions to markets; they pass on information, mediate transactions, assists the transfer and enforcement of property rights and contracts, and handle the level of competition. Along with these concepts, we define market institutions as rules of the game, enforcement mechanisms and organizations that facilitate market interaction, coordination, contract formation and enforcement. Kraybill and Weber recognize the dominance of political institutions over market and civic institutions. The federal government is the moving force in creating institutions and organization that raised living standards of low income rural residents (Christy and Dassie, 2000).

Voslee (1994) states that while there has been a gradual accumulate of knowledge and insight as thinking on development has evolved over the past fifty years, the question of how to generate entrepreneurship and sustainable economic growth remain unanswered. He argues that evidence from developing and developed countries supports that government should not manage development in detail. He state that:

> "Many forms of intervention have proved counterproductive and the importance of openness and competition has been acknowledged. Theory and practical experience indicate that interventions are more likely to hinder than help development unless they are market friendly" (Voslee, 1994).

This view asserts the argument that federal interventions are most successful in augmenting growth if they provide a favourable environment for business. Porter (1995) emphasize that this can be accomplished if government uses public funds in ways that do not alter business incentives, but rather to focus on providing infrastructure to support genuinely profitable business. Various private and public sector market-supporting institutions and institutional arrangements have been projected to bridge market imperfections, reduce transaction costs, increase opportunities for the poor in markets and to make the market systems more inclusive and integrated (World Bank 2002a). Among the potential market-supporting institutions that can improve market functions in rural areas are farmer organizations such as Producer Marketing Groups (PMGs). Their prospects in this process lie in enabling contractual links to input and output markets (Coulter *et al.*, 1999) encouraging economic harmonization in liberalized markets (Rondot and Collion 1999) and in leveraging market functions for smallholder farmers.

However, their success in this process depends on their capability in transmission of market information; coordinating marketing functions; defining and implementing property rights and contracts; assisting smallholder competitiveness in markets (World Bank 2002a), and more crucially in mobilizing their members to engage in markets. For cooperatives and rural organizations to be successful in helping a broad set of socio-political and economic objectives of small producers, new policies and institutional reforms are required to facilitate their transformation from public sector service providers to private sector enterprises with clear business plans. Therefore, farmer organizations tend to succeed only when farmers can manage them autonomously with minimal government interference; farmers participate actively in decision-making at every stage of the process; and their cooperative activities are profitable (World Bank 1995).

Training Approach

Training is a specialized and very practical form of education, which prepares people to do their job well. It increases the skills that build intelligent actions and attitudes so as to bring willing cooperation with fellow employees and management (Planty and Macord, 1948). Training acts as a means of transferring the new knowledge and skills in scientific agriculture of an external origin to a local farming system (Dwarakinath and Padmasini, 1977). Regarding farmer training into rural entrepreneurship, Rao (1969) defined it as an intensive learning activity for a group of selected farmers, supported by competent trainers to understand and practice the skills required in implementation of new technology at a place where appropriate facilities exist and at a time and duration considered suitable by the farmer.

The various studies have revealed that the training programmes have been successful in equipping the rural masses with self confidence and high self esteem. Udayakumar and Shreedhara (2002) in an analysis of entrepreneurship development programmes assured that training is an integral part of economic development programmes and they are focussed on developing entrepreneurship with the objectives of increasing the number of entrepreneurs who start new business units. The training programmes have helped in generating more employment and income to the public.

Women Empowerment

Women constitutes half of human resources and have been acknowledged as the key agents of sustainable development and women's equality is essential to a more holistic approach towards establishing new patterns and process of development that are sustainable (Jha, 2009). The contribution of women and their role in the family as well as in the economic development and social transformation is crucial. Women comprise 90 per cent of total marginal workers of the country. 78 per cent of all regular working women are engaged in agriculture (Kumar, 2009). The rural women are endowed with basic indigenous knowledge, skill, potential and resources to establish and manage enterprise. But they lack in knowledge regarding accessibility to loans, various funding agencies procedure regarding certification, awareness on government welfare programmes, motivation, technical skill and support from family, government and other organization. Furthermore, formation and strengthening of rural women entrepreneurs network must be encouraged (Sathiabama, 2010).

However, empowerment is a process of awareness and capacity building, leading to greater participation and greater decision making power. The process of empowerment strengthens their innate ability through acquiring knowledge, power and experience (Murugan and Dharmalingam, 2000). To raise the status of women, they must be empowered socially, economically and politically. Empowerment can serve as a powerful instrument for women to achieve upward social and economic mobility and to achieve power and status in the society (Sushma, 2007). Presently, rural economic development is one of the factors that have changed the whole scenario of social and cultural environment of the country especially for the women. The rural women are occupied in small-scale entrepreneurship programme with the help of Self Help Groups. Now they are economically empowered and attaining status in family and community. The micro entrepreneurships are strengthening the women empowerment and remove the gender inequalities.

Sustainable Entrepreneurship

Sustainable entrepreneurship can be defined in a broad sense as innovative behaviour of actors in the perspective of sustainability, comprising of actors from governmental and non-governmental, profit and non-profit organisations. In a narrow sense, sustainable entrepreneurship is defined as innovative behaviour of single actor or organisation operating in the private business sector that are seeing environmental or social issues as a core objective and competitive advantage. It has recently become of interest to various researchers with a focus on environmental entrepreneurship (Gerlach, 2003; Dijkema, 2006; Cohen and Winn, 2007; Lepoutre, 2008). This new notion trying to establish new theory on sustainable entrepreneurship has mainly emphasised on entrepreneurial action through opportunity recognition in relation to market failures (Cohen and Winn, 2007; Dean and McMullen, 2007; Hall and Lobina, 2007). While on the other hand, a pro-active strategy is pursued by entrepreneurs that are constantly improving the sustainability of their business by looking forward to the future regulations, social trends and building resources, interacting with social conditions and creating value, beyond what is legally required or accepted as standard practice (Aragon-Correa and Sharma, 2003; Sharma and Henriques 2005, Lepoutre, 2008).

Sustainable entrepreneurship is a follow-on concept from sustainable development that envelops many new evolutions in business like corporate social responsibility, ethical entrepreneurship, ecological care, stakeholder participation and so on and so forth. It can also be defined as the regular assurance by business to behave ethically and contribute to economic development while improving the quality of life of the workforce, their families, local communities, the society and the world at large as well as future generations (Crals and Vereeck, 2004). The material and immaterial contributions of dynamic corporate behaviour to nature and to global and local communities which shape our natural and social environments are subject to changing view on people and planet (Bos, 2002).

Janssen (2001) provide a list of ten ground rules for becoming a sustainable entrepreneur:

1. The corporation should start reducing the environment damage, respecting human rights and treating its employees with great care.
2. Sustainable entrepreneurship has to be a self- initiated process and should not simply be a response to external pressure.
3. If corporation wants to practice sustainable entrepreneurship, it should identify clear aim and targets.
4. The aim should be closely related to the corporation's practice and should match the corporate values and its primary activities.
5. The aims have to be closely related to the consumer's needs.
6. The corporation has to be capable of explaining the relationship between sustainability and its activities and production process.
7. The corporation should adhere to these aims on a long term basis.
8. Consumer and pressure groups should have a transparent overview of investments made by the corporation related to sustainable entrepreneurship.
9. Sustainable entrepreneurship practiced by the corporation should not shift to the consumers via a price increase.
10. A corporation should not attempt to over emphasise its effort.

Bos (2002) added an additional rule to the list.

11. A corporation should make sure that its practices are shared by the corporation as a whole and those they are not solely effort of the management. (Adapted from Crals and Vereeck, 2004).

From the above literature, it can be alleged that sustainable entrepreneurship is beneficial to the rural economy not only from the business point of view but also overall development by improving the quality of life of the workforce.

Measuring Rural Entrepreneurship

Entrepreneurship is gradually being recognized as a major engine of economic growth. By merging existing resources with innovative ideas, entrepreneurs add value through the setting up of new firms, commercialization of new products and creation of new jobs. In short, entrepreneurs are the link between new ideas and economic growth. Different types of entrepreneurs provide different benefits to their community but high-growth entrepreneurial businesses are considered as providing the largest economic value to communities (Henderson, 2002).

Johnson (2006) proposed a conceptual basis for measuring the welfare outcome of economic development programme and practices in general and entrepreneurship oriented programs in particular. The theoretical association between entrepreneurship and economic growth is usually accredited to the early work of Joseph Schumpeter (1934). Schumpeter's concept of creative destruction offered us with a hypothetical relationship between entrepreneurship and economic growth and development.

In an attempt to formalise the more dynamic and disequilibrium conceptualisation of entrepreneurship Wennekar and Thurik (1999) merged many of the key ideas into an amalgam theory of entrepreneurship which portrays a dynamic model of entrepreneurship. Figure 34.1 demonstrates the dynamic model of entrepreneurship which classify three levels at which entrepreneurship may be viewed- the individual within which entrepreneurship resides, the firm which serves to internally organize factors of production and the region which serves as the external organizer of factors of production. It also recognizes three dimension of entrepreneurship- the condition which leads to entrepreneurship, the attributes of entrepreneurial individuals, firm and regions, and impact or consequences of entrepreneurship.

The combined theory of entrepreneurship above put forward at least two important observations about entrepreneurship. Firstly, entrepreneurship is not just a part of economic development; it is the root of long- term economic development. Secondly, entrepreneurship is highly place- based. Entrepreneurs function through membership in networks and by exploiting their place-based human capital. Not only is entrepreneurship shaped by spatial attributes and not only does it transfer the socio-economic of places but more fundamentally economic development is less likely to occur in the aggregate if space is not effectively organized into regional entrepreneurial system.

Conclusion

This study explores two facets that many consider to be crucial for rural entrepreneurship in general, namely entrepreneurial innovation and rural development. The results of this paper indicate that to increase the understanding of rural economic growth, researchers could extend their analysis of rural entrepreneurship to analyzing the factors supporting entrepreneurship development in rural places. Increasingly, entrepreneurs that innovate are thought to be able to contribute to sustainability by creating new system with their sustainable innovation. On this basis we have tried to create the relationship between the entrepreneurial

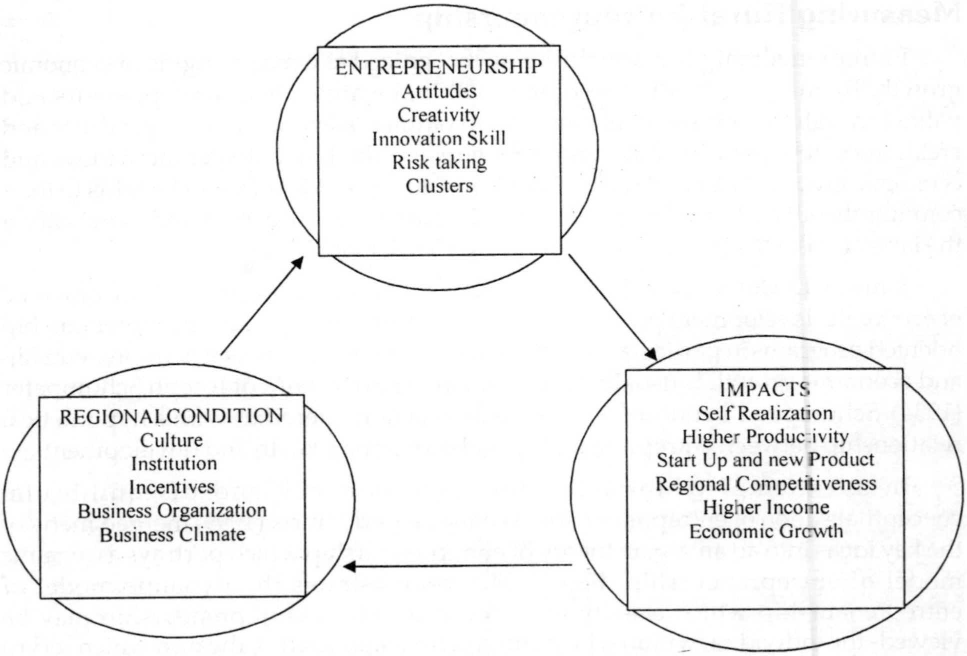

Level	Conditions for (Entrepreneurship)	Element of (Entrepreneurship)	Impact of (Entrepreneurship)
Individual	Culture Incentives	Attitudes Skill Creativity	Self Realization Income
Firm	Business Culture Size Distribution Incentives Organization	Start-up New Market Firm Size R&D	Firm Productivity Growth New Sectors Size Distributions
Region	Culture Institutions Regulations	Variety Competition Cluster	Competitiveness Economic Growth

Source: Wenneker and Thurik (1999).

Figure 34.1

innovation and rural development. These insights adds to further understanding the interactions between rural entrepreneurship and rural development in order to more effectively stimulate and support innovating entrepreneurs to contribute in the transition toward sustainability.

References

Acs, Z. J. and E. Malecki. (2003). "Entrepreneurship in Rural America: The Big Picture" Main Streets of Tomorrow: Growing and Financing Rural Entrepreneurs. Kansas City: Federal Reserve Bank.

AragonCorrea, J. A., and Sharma, S. (2003). A Contingent ResourceBased View Of Proactive Corporate Environmental Strategy. Academy Of Management Review, 28(1): 7188.

Audretsch, David B. (1995). Innovation and Industry Evolution. Cambridge, MA: MIT Press.

Bank, W (1999) World Development Report 1998-1999: Knowledge for Development. World Bank.

Becker, G. (1996). Accounting for Tastes, Cambridge. MA: Harvard University Press.

Becker, G. (1993). Nobel Lecture: The Economic Way of Looking At Behaviour. *The Journal of Political Economy* 101: 385-409.

Bos, A. (2002), 'Sustainable Entrepreneurship in a Changing Europe: Pedagogy of Ethics for Corporate Organizations in Transformation', in: G.F. Simons, D. Min *et al.* (eds.) *EuroDiversity: A Business Guide to Managing Differences*, Oxford: UK and Woburn, USA: Butterworth-Heinemann.

Cohen, B., and Winn, M.I. (2007) Market Imperfections, Opportunity And Sustainable Entrepreneurship. *Journal of Business Venturing* 22 p.2949.

Crals, E. and Vereeck L. (2004). Sustainable Entrepreneurship in SMEs. Theory and Practice. Limburgs Universitair Centrum, Diepenbeek, Belgium.

David, J. Lewis (1996). "Understanding Rural Entrepreneurship in Bangladesh Village Individuals, Roles or Structures" Small Enterprise Development, Vol. 7, No. 4.

Dean, T.J., and McMullen, J.S. (2007). Toward A Theory Of Sustainable Entrepreneurship: Reducing Environmental Degradation Through Entrepreneurial Action. *Journal of Business Venturing* 22 p.5076.

Dijkema, G.P.J., Ferrao, P., Herder, P.M., and Heitor, M. (2006). Trends And Opportunities Framing Innovation For Sustainability In The Learning Society. Technological Forecasting and Social Change 73 p.215227.

Dwarakinath, R. and Padmasini, (1977). Strategy of Training Farmers, Karnataka State Department Of Agriculture, Bangalore.

Edquist, C (1997). Systems of Innovation: Technologies, Institutions And Organizations. London: Pinter Publishers.

Fannin, J. Matthew, and Thomas G. Johnson. (2004). The Spillover Effects of Knowledge Production on Missouri Counties. Working paper. University of Missouri - Columbia.

Fieldsend, A.F. and Katona Kovacs, J. (2007). HVTK- giving Local relevance to European- wide rural development practice. Rural Areas and development series 5, Values and Challenge in Designing the Europian Rural Structure, Research Network Experience.

Freeman, C (1987). Technology Policy And Economic Performance: Lessons From Japan. London, New York: Pinter Publishers.

Gerlach, A. (2003). Sustainable Entrepreneurship And Innovation. Centre For Sustainability Management (CSM), Univeristy Of Lueneburg.

Hall, D., and Lobina, E. (2007). Profitability And The Poor: Corporate Strategies, Innovation And Sustainability. Geoforum 38: 772–785.

Hall, A, Clark, N, Sulaiman, R, Sivamohan, MVK and Yoganand, B (2003). From Measuring

Impact To Learning Institutional Lessons: An Innovation Systems Perspective On Improving The Management Of International Agricultural Research. *Agricultural Systems* 78: 213-241.

Henderson, J. (2002). Building The Rural Economy With High-Growth Entrepreneurs. *Economic Review- Federal Reserve Bank Of Kansas City*: Third Quarter.

Hoy, F. (1996). Entrepreneurship: A Strategy for Rural Development. Rural Development Research, 29-46.

Hoy F., Vaught B. C., 1980 The rural entrepreneur: A study in frustration. Journal of Small Business Management, 7(3):19-24. INSSE, 2010 http://www insse.ro (last view: 13.04.2010).

Jack S. L., Anderson A. R. (2002) The effects of embeddedness on the entrepreneurial process. Journal of Business Venturing **17**(5):467-487.

Janssen Groesbeek, M. (2001), Sustainable Entrepreneurship – Theory, Practice, Instruments. Amsterdam: Business Contact.

Johnson T.G. (2006). Measuring the Returns to Rural Entrepreneurship Development, Paper Presented at the Exploring Rural Entrepreneurship: Imperatives and Opportunities for Research Conference, Washington.DC.

Kalantaridis C. (2006) A study into the localization of rural businesses in five European countries. European Planning Studies 14(1):61-78.

Karayiannis, A.D. (1990). The Entrepreneurial function in Economic Literature: A Synoptic Review, Rivista Internazionale di Scienz Economiche and Commerciali, Vol. 37, (2), 245-268.

Kraybill, Davis S. and Weber, Bruce A. (1995). Institutional Change and Economic Development in Rural America. *America Journal of Agricultural Economic*, December, 1265-1270.

Lepoutre, J. (2008), Proactive Environmental Strategies In Small Businesses: Resources, Institutions And Dynamic Capabilities, Dissertation submitted to the Faculty of Economics and Business Administration Department of Management, Innovation and Entrepreneurship, Universiteit Gent.

Lichtenstein, G. A. and T. A. Lyons. (2001). "The Entrepreneurial Development System: Transforming Business Talent and Community Economies," *Economic Development* Quarterly, 15(1): 3-20.

Lundvall, B-e (1992). National Systems of Innovation: Towards a Theory of Innovation and Interactive Learning. Pinter Publishers.

Matthews, C. and Human, S. (2004). Family Background, Chapter 8. Handbook of Entrepreneurial Dynamics, the Profess of Business Creation. Sage Publications. London.

Murugan, K.R. and Dharmalingam, B. (2000). Self-Help Groups-New Women's Movement In Tamil Nadu. Social Welfare,47(80):9-12.

Nolan, A. (2003). "Entrepreneurship and Local Economic Development: Policy Innovations in Industrialized Countries," Pp. 77-90, edited by M.Drabenstott, N. Novack, and B. Abraham. Kansas City, MO: Federal Reserve Bank of Kansas City, The Center for the Study of Rural America.

North D., Smallbone D., 2000 Innovative activity in SMEs and rural economic development: Some evidence from England. European Planning Studies 8(1):87-106.

Planty, E.G. and Macord, C.A. (1948). Training Employees And Managers For Production Work, Ronald Press Company, New York.

Porter, M. E. (1995)."The Competitive Advantage of the Inner City." *Harvard Business Review*,May- Jun. 55-71.

Rao, M.K.S.(1969). A Critical Analysis Of Farmers Training In IADP And IAAP Idstricts In Relation To High Yielding Varieties Programme. *Ph.D. Thesis*, IARI, New Delhi.

Robinson, K. L., T. A. Lyson, and R. D. Christy. (2002). "Civic Community Approaches to Rural Development in the South: Economic Growth with Prosperity," *Journal of Agricultural and Applied Economics* 34(2): 327-338.

Sathiabama, K (2010) Rural Women Empowerment and Entrepreneurship Development retrived from http://www.microfinancegateway.org/p/site/m/template.rc/1.9.43775/ dated 08/12/2011.

Schumpeter, J.A. (1934). The Theory Of Economic Development An Inquiry Into Profits, Capital, Credit, Interest And The Business Cycle (English Translation Of 1912 German Language Volume), Harvard University Press.

Schumpeter, JA (1944) *Capitalism, socialism, and democracy*. London: Allen and Unwin.

Schumpeter, JA and Opie, R (1961) *The theory of economic development: an inquiry into profits,capital, credit, interest, and the business cycle.* New York: Oxford University Press

Sharma, S. and Henriques, I. (2005). Stakeholder influences on sustainability practices in the Canadian forest products industry. *Strategic Management Journal*, 26(2): 159180.

Shields, J. F. (2005), "Does Rural Location Matter? The `Significance of a Rural Setting for Small Businesses", Journal of Developmental Entrepreneurship, Vol.10, No. 1, pp. 49-63.

Slaughter, M. P. (1996). Entrepreneurship: Economic Impact and Public Policy Implications: An Overview of the Field. Kansas City, MO: Kauffman Center for Entrepreneurial Leadership.

Smilor, R. W. (1997). "Entrepreneurship and Community Development," A Paper Presented At A Kauffman Center For Entrepreneurial Leadership Conference On Entrepreneurship As a Community Development Strategy.

Spielman, D.J. (2005). Innovation Systems Perspectives on Developing-Country Agriculture: A Critical Review. *ISNAR Discussion Paper*. International Food Policy Research Institute.

Sushma, K.C. (2007). An Analysis of Entrepreneurship Development In Women Through EDP Trainings, Thesis, Degree Of M.Sc. in Agriculture Extension Education, College of Agriculture, University of Agricultural Science, Dharwad.

Udaykumar, M.A. and Sreehdhara, T.N. (2002), Entrepreneurship development programmes:An analysis. *Southern Economist*, pp.10-12.12/2011.

Voslee, W.B. (1994). Entrepreneurship and Economic Growth. Pretoria: HSRC Publisher.

Westlund, H. and Bolton, R.(2003). Local Social Capital And Entrepreneurship. *Small Business Economics* 21: 77-13.

Winders, R.M. (1997). "Small Business Creation and Economic Wellbeing of Nonmetropolitan Counties: The Case of Georgia." Tennessee Valley Authority Rural Studies Program.

Wennekers, A.R.M. and A.R. Thurik, (1999). "Linking entrepreneurship and economic growth," *Small Business Economics*, Vol. 13, 27-55.

Yu, Li and Artz, Georgeanne M. (2009). Migration and Rural Entrepreneurship, Working Paper No. 09017, IOWA state University.

Chapter 35

Underutilized Crops: Sustainable Development for Hill Agriculture

Birendra Prasad[1], *Abhishek Bahuguna*[1] *and V.K. Yadav*[2]

[1]*Senior Research Officer and Senior Research Fellow, GBPUA&T Hill Campus, Ranichauri, Tehri Garhwal – 249 199, Uttarakhand*
[2]*Principal Scientist, IGFRI, Jhanshi, Uttar Pradesh*

ABSTRACT

Underutilized crops are usually characterized by having local importance in consumption and production systems, requiring relatively low inputs, adapting to specific agro-ecological niches, receiving scarce attention by national agricultural and biodiversity conservation efforts, mainly consisting of local types or landraces, and being cultivated with indigenous knowledge. There is obviously an imperative need for diversifying our present day agriculture to not only meet our food requirements but for future posterity as well. The predicted climatic change and other factors leading to eco-system degradation are likely to change the face of agriculture completely. An increasingly key issue in improving the use of underutilized species is the globalization of the agricultural market. With specific reference to agriculture, globalization is bringing about a further specialization in favour of a few crops that might best serve "Global Village". Obviously, this trend will lead to much narrower agricultural basket, estimated to hold only some 150 widely commercialized crops, in favour of some commodities which will enter newer areas and displacing local ones. Diversification in agricultural systems is indeed an important asset for those fragile social groups who may never be able to afford certain commodities and to whom the more diversified is the portfolio the greater is their self-sustainability and self-reliance in difficult areas where they live, often characterized by conditions not suitable for the cultivation of improved varieties of commodity crops. Therefore, the cultivation of underutilized crops for the benefits in terms of more balanced diets, diversified income to farmers as well as related sectors of the society, better maintenance of agroecosystems and greater use of marginal lands along with enhanced preservation of cultural identity can be shared by all the humankind. Efforts so far have been directed to raise the awareness on the need to work on underutilized species and to start redressing their status of neglect. We want to make sure our efforts will lead ultimately to the "full promotion stage" beyond which the crop will no longer be considered underutilized and promote a specific crop have the desired effects with respect to the interests and needs of poor hill farmers, the maintenance of diversity and strengthening sustainability of production.

Keywords: Underutilized crops, Sustainability, Hill-agriculture, India.

Introduction

Underutilized crops are often considered 'minor crops' and were once grown more widely or intensively, but are falling into disuse for a variety of agronomic, genetic, economic and cultural reasons. Farmers and consumers are using these crops less because they are in some way not competitive with other species in the same agricultural environment (IPGRI 2002). Consequently, these species have been neglected and genetic erosion of their genepools has become severe. Underutilized crops can be found in many different agricultural ecosystems, but they are mainly grown by small landholders in the marginal areas especially in hill agriculture. They are usually characterized by having local importance in consumption and production systems, requiring relatively low inputs, adapting to specific agro-ecological niches, receiving scarce attention by national agricultural and biodiversity conservation efforts, mainly consisting of local types or landraces, and being cultivated with indigenous knowledge. Many underutilized species are adapted to low-input agriculture and depended on by a large number of people in marginal areas in hill agriculture of developing countries.

The hill region is characterized by terraced slope farming, small holder – resource poor farmers and is extremely fragile from geo- morphological point of view. It harbours rich bio-diversity and is part of the Hindustan centre of origin. The highly heterogeneous and rainfed nature of the mountain agriculture has given rise to Baranaja (literally 12 types of grains or crops) system, a form of diversified agriculture mainly as a risk avoidance measure and in situ conservation. Small millets (finger millet, barnyard millet, chenopods) and grain legumes (rice bean, horse gram, faba bean and adzukibean) are the traditional crops of the mountain region. There are multi- utility, nutritionally rich crops providing livelihood support and nutritional and ecological security to the mountain peasants living in remote areas far away from the public food distribution system. These crops were receding in the recent past due to the encouragement of elite food crop (rice and wheat) and off- season vegetables that require high external inputs.

Underutilized is commonly applied to refer to species whose potential has not been fully realized. The term itself does not provide any information as to geographical (*underutilized where*?), social (*underutilized by whom*?) and economic (*underutilized to what degree*?) implications. It is thus not surprising that whenever underutilized species are being addressed in national or international fora there is inevitably a call for a clarification over the exact meaning of such a term. The following are some examples to explain the source of confusion on this term. With regard to the *geographical distribution*, often a species could be underutilized in some regions but not in others.

Underutilized crops are being often presented as *'new crops'* (Vietmeyer, 1990) for the fact that commercial companies/researchers are only recently working on them. In reality, local populations over generations have used these species. Yet the loss of local knowledge and thus the increasing ignorance of new generations on the traditional uses of these crops also contribute to portraying such a misleading image. Of course, a crop can be completely new to an area simply because it has been introduced there recently from a distant country.

Neglected crops are those grown primarily in their centers of origin or centers of diversity by traditional farmers, where they are still important for the subsistence of local communities. Some species may be globally distributed, but tend to occupy special niches in the local ecology and in production and consumption systems. While these crops continue to be maintained by socio-cultural preferences and use practices, they remain inadequately characterized and neglected by research and conservation. The understanding of the causes behind the low level of use and/or neglect of a crop is however what is ultimately needed in order to design an appropriate strategy to address its improvement.

Dependence of mankind on plants for survival has been indispensable ever since the dawn of civilization. Transformation from food gathering to settled agriculture necessitated domestication of plant species useful for mankind leading to co-evolution of plants. It has been estimated that out of a total 80,000 plants identified for possible economic use, about 30,000 plants have been found edible in nature and approximately 7,000 plants have been cultivated by mankind at one time or the other, of which only 158 plants are widely used for food. Among these, 30 crops provide 90 per cent of the world's food and only 3 crops *viz.* rice, wheat and maize provide 60 per cent of the world's total food requirement (Wilson, 1992).

This dangerously narrow level of food base may imperil the existence of mankind during the time of impending crisis in unforeseen times. It hardly needs to be emphasized that there is an immediate necessity to enlarge our food basket with alternative food crops to safeguard against such a catastrophe. In recent years there has been a steady increase in problem soils and a rapid diversion of fertile lands suitable for elite crop cultivation to alternative uses such as industrialization and urbanization. The high input agriculture once the harbinger of green revolution has caused large scale ecosystem degradation and environmental pollution. As a result, throughout the globe there has been a rekindling of interest to collect, introduce, evaluate and utilize the vast array of neglected and under-utilized alternative crop plants that can adapt themselves to the harsh environmental and degraded soil conditions, produce economic yield and ensure livelihood, nutritional and ecological security to the impoverished people residing in remote inhospitable areas.

Neglected and under-utilized plant species are those plant species whose potential has not been exploited to the fullest extent. These are variously known as underutilized and under exploited species, neglected species, orphan crops, future food crops, crops for the future, nutrition crops, etc. These crops could be the key to sustainable agriculture in most developing countries facing a resource crunch as well as rapid depletion of natural resources. With the implementation Intellectual Property Rights through WTO it has become imperative to search for novel genes and products to retain competitive advantage at the global arena. Therefore, these crops are gaining importance world wide. The term "neglected and under-utilized crops" has a different connotation depending on global, regional and national contexts as well as socio-cultural point of view. Among the vast array of plants deserving attention, only a few could be prioritized for scientific exploitation in a phased manner based on their economic potential and niche advantages. At the national level AICRPs have been initiated to include some such neglected crops to provide a necessary thrust on their collection, conservation, systematic evaluation and utilization in crop improvement programmes (Dutta *et al.*, 2007). Value addition of their products is also a prerequisite to make them commercially acceptable among the nontraditional consumers and elite strata of people. In this chapter we deal with a few representative food crops such as small millets, pseudo cereals and some important grain legumes, that have special significance in Himalayan agriculture leaving aside the vast group of other crops.

Himalayan Agriculture

Biodiversity

Among the centres of diversity of crop the Hindustan Gene Centre is one of the major ones. Within India there are eight sub-centres of which three come under the Himalayan belt. These are Western Himalaya, Eastern Himalaya and the Noth-Eastern Region. The important crop plants, the origin of which could be traced back to the Himalayan gene centre are *Oryza, Hordeum, Avena, Cicer, Vigna, Brassicae, Solanum* etc. Altogether about 164 species of crop plants are supposed to have originated in the Indian sub-continent. Mixed cropping is the norm which has given rise to the system called *Baranaja*, literally twelve types of grains. However, this enormous crop diversity has not been harnessed effectively for the enhancement of agricultural productivity and for the enrichment of the lives of the people living in the Himalayan region. As the gene rich countries are generally considered poor so also the gene rich regions within the country itself.

There is obviously an imperative need for diversifying our present day agriculture to not only meet our food requirements but for future posterity as well. The predicted climatic change and other factors leading to eco-system degradation are likely to change the face of agriculture completely. The development of new virulent biotypes and pathotypes calls for alternative crops and or their resistant varieties for keeping a continuous supply of food grains for our ever increasing population. Not knowing the exact nature of changes that are expected to occur and the genetic changes that will be favored most under these circumstances it is of paramount importance to conserve as wide a genetic base as possible for the future

Physiographic Information

The Himalayan ranges are about 2500 km long extending from 73°–97' East longitude to 27°–37' North latitude, and covers an area of approximately 2,36,000 sq km. In the Himalayas four factors- latitude, longitude, altitude and aspect play a pivotal role in determining the range of agro-climates and add to the multiplicity of habitats. This in turn provides appropriate ecological niches not only for plants and animals but also for different ethnic groups of human beings to prosper in their natural habitat. There are three ranges, the outer or lesser, central or middle and inner or greater Himalaya running almost parallel from east to west. The altitude of the ranges varies from less than a few hundred meters to more than 8000 m above mean sea level. Places in the outer and inner Himalaya having similar altitude will have distinctly different agro-climates owing to the proximity of the latter to the snowline. Himalaya is the youngest and tallest mountain system in the world but geo-morphologically it is the most fragile among all the mountain systems. The continuous rise of the Himalayas and consequent tectonic effect gives rise to many devastating earthquakes and natural calamities to which the area is specifically prone.

The Indian Himalaya has two distinct geographical regions: the Western Himalaya and the Eastern Himalaya. The Western Himalayan ranges are much wider and have colder and drier climate, in contrast, the Eastern Himalaya is among the wettest regions of the world with high concentration of biodiversity. The three major river systems- the Indus, Ganges and Brahmaputra have their origin either in the Himalaya or in the highlands of Tibetan plateau and flows through the Himalayas along with their tributaries. The watersheds of these river systems harbour the richest biodiversity. Shifting cultivation is the norm in the Eastern Himalaya while terraced farming is predominant in the Western Himalaya.

Pseudo-cereals

Pseudo-cereals are a group of food grain crops that do not belong to the family Poaceae. Predominant among these are grain amaranth (Amaranthaceae), buckwheat (Polygonaceae) and chenopodium (Chenopodiaceae). Among these, the former two have attained a reasonable status as food grain crops, particularly in the high altitude Himalayan region. Grain amaranth has larger coverage in Uttarakhand while buckwheat is grown in large areas in Himachal Pradesh and Jammu and Kashmir. Chenopodium too is widely grown in Himachal Pradesh and its adjoining areas.

Table 35.1: Nutritional Composition of Pseudocereals as Compared to Wheat

Crop	*Protein (%)*	*Fat (%)*	*Carbohy-drate (%)*	*Mineral Content (%)*	*Crude Fibre (%)*	*Calcium (mg)*	*Phos-phorus (mg)*	*Iron (mg)*	*Energy (K cal)*
Amaranth	15.6	6.3	72.7	2.93	2.43	222.0	548.0	13.9	410.0
Buckwheat	12.0	2.4	72.9	2.9	10.3	114.0	282.0	13.2	355.0
Chenopod	16.0	–	66.0	3.0	–	139.0	452.0	6.5	395.0
Wheat	13.3	2.0	71.0	–	–	41.0	372.0	10.5	333.0

Table 35.2: Amino Acid Composition of Pseudo-cereals as Compared to Wheat

Amino Acids	*Amaranth*	*Buckwheat*	*Chenopods*	*Wheat*
Leucine	5.7	6.2	5.7	5.8
Isoleucine	3.9	3.7	3.3	3.3
Lysine	5.5	6.2	6.0	2.2
Arginine	–	10.1	6.9	3.6
Histidine	2.5	2.2	1.8	1.7
Phenylalanine	4.0	4.2	4.1	4.2
Threonine	3.6	3.7	4.0	2.7
Valine	4.5	5.1	4.0	3.7
Tyrosine	3.3	1.5	3.2	–
Cysteine	2.1	1.6	1.2	–
Methionine	2.6	1.6	2.2	2.1

Grain-Amaranth (Amaranthus spp.)

Amaranths are fast growing annual, semi-hardy plants with broad leaves and long inflorescences of different attractive colors. Grain amaranths occupy a unique position in mountain agriculture because of their high cash earning value and richly nutritious grains and leaves that provide livelihood and nutritional security to resource-poor farmers of the Himalayan region.

Figure 35.1

Origin

The origin of cultivated grain amaranth species is difficult to trace because of the cosmopolitan weedy nature of the wild ancestors. Based on archaeological, historical and folkloric evidence, the New World has been suggested as the centre of origin. The major cultivated grain producing species are: *A. hypochondriacus*, *A caudatus*, *A. cruentus* and *A. edulis*. They were domesticated prehistorically in the highlands of tropical and subtropical Central America and were the staple food of the Mexicans during the

Figure 35.2

Aztec civilization. Subsequent analysis revealed that *A. edulis* was in fact a variety of *A. caudatus. A. caudatus* may have been domesticated from an early introduction of *A. cruentus* that then crossed with a wild Amaranthus species. Isozyme and RAPD marker studies have revealed that *A. hybridus* is a common ancestor to all cultivated grain amaranth species. Archeological evidence suggests that *A cruentus* predated *A. hypochondriacus*. The major vegetable types are *A. dubius, A. blitum, A. viridis,* and *A. tricolor* and are grown under largely tropical conditions. Among the grain types *A. hypochondriacus* and *A. caudatus* are better adapted to the cold conditions of the high altitude areas.

Grain amaranth is cultivated as a minor crop throughout the world. In India, grain amaranth is widely grown in the mountainous regions as well as in the plains extending upto Kerala in the south to Gujarat in the west. Amaranths are well adapted to both temperate and tropical conditions. However, its cultivation in the high altitude Himalayan region beyond 1500 m is widespread and extends up to an altitude of 3000m. Grain amaranth is extremely drought-hardy and can survive in areas receiving a precipitation of only 200 mm while vegetable amaranth can tolerate acidic soil conditions. This C_4 plant species is considered to be highly energy efficient because of its superior photosynthetic pathway.

Uses

Amranth is a crop of multiple utility. Its grains are superior to cereals in terms of protein, amino acid composition, mineral contents, vitamins and energy. Amaranth grains contain twice the amount of lysine available in wheat and thrice that of rice and maize. The grain is used in the preparation of bread, biscuits, flakes, cake, pastry, crackers, ice cream and lysine rich baby-foods. High forage yield, high protein content coupled with low oxalates and nitrates make grain amaranth a promising forage crop. Ethnobotanical uses include its utility against measles, snake bites, kidney stones, for meat preservation, preparation of beverages and also as a blood purifier. The tocopherol fraction of amaranth oil contains cholesterol-lowering properties and could be useful in treating cardiovascular diseases.

Cropping System

Grain amaranth is cultivated under both pure and mixed cropping systems. Pure cropping is predominant in higher hills while mixed cropping is practiced in the low altitude areas. The following cropping systems are important:

- ☆ Grain amaranth - Wheat/ Barley/Lentil/ Mustard
- ☆ Grain amaranth - Peas (February sown)
- ☆ Grain amaranth + Pulses (Rajma/Rice bean/Soybean) – Wheat
- ☆ Grain amaranth + Small millets – Wheat
- ☆ Grain amaranth + Maize –Wheat
- ☆ Grain amaranth - Fallow

Among these Grain amaranth + Rajma in 2:1 ratio has been found to be the most promising inter cropping pattern.

Genetic Resource and Breeding

In India, initial efforts were made in the 60's by the Indian Agricultural Research Institute, New Delhi that held a collection of less than 50 germplasm lines. Germplasm collection efforts were geared up only from 1978 onwards. Presently, more than 5000 collections are being conserved in India at various institutes including National Bureau of Plant Genetic Resources (NBPGR) Regional Stations, Shimla and Akola,; National Botanical Research Institute, Lucknow and others covering more than 40 species. The hill collections are being maintained at NBPGR Regional Station, Shimla and the collections from the plains are kept at NBPGR Regional Station, Akola.

About 800 accessions were evaluated for 29 descriptors at NBPGR Regional Station, Shimla earlier while the evaluation of an additional 2215 accessions from different regions including 298 exotic collections was taken up later for 40 descriptors and significant variations have been reported for grain yield and its components. Prasad *et al*. (2011) also reported genetic variability of some selected grain amaranth (*Amaranthus spp.*) genotypes for morphological traits under mid hills of Uttarakhand. Under the AICRP 100 germplasm lines are screened every year under hill and plain locations to identify desirable genetic donors.

The breeding system in grain amaranths could be classified into three types: (a) obligate out-crossing in dioecious species, (b) relatively greater out-crossing in monoecious members of section Amaranthus, and (c) relatively greater self-pollination in members of Section Blitopsis. Breeding system has been reported to vary from a high rate of selfing (over 90 per cent) to mixed mating with as much as 30 per cent cross pollination. Although barriers exist between the cultivated species and the wild forms, introgression does take place. Successful interspecific hybridization between *A. spinosus* and *A viridis*, the two wild species, has been achieved paving the way for further such studies between the cultivated and wild forms for successful gene transfer.

Table 35.3: Extent of F_1 Heterosis (Per cent) for Various Characters in Inter-Varietal Crosses of Grain Amaranth

Character	*Heterosis*	
	Mid Parent	*Better Parent*
Heading time (days)	8.89	8.37
Maturity period (days)	5.14	4.87
Plant height (cm)	15.14	6.13
Inflorescence length (cm)	29.03	7.00
Number of fingers /ear head	34.49	34.46
Number of leaves/plant	38.94	5.47
Length of finger (cm)	15.89	4.79
Basal stem girth (cm)	25.71	12.82
Test weight (g)	4.15	2.73
Seed yield (g/plant)	26.76	22.75

Table 35.4: Range of Yield and Frequency of Transgressive Segregants in the F_2 Generation of Six Grain Amaranth Crosses

Cross/Parent	*Average Seed Yield /Plant*	*Range of Yield*	*Frequency of Transgressive Segregants*
Suvarna x PRA 8801	16.49	8.0–25.0	40.0
PRA 8801 x Suvarna	13.65	5.0–45.0	15.75
Annapurna x VL 33	10.37	5.0–25.0	5.25
VL 33 x Annapurna	17.05	5.0–45.0	40.0
PRA 8801 x VL 33	16.89	5.0–35.0	46.50
VL 33 x PRA 8801	16.00	5.0–30.0	45.35
Suvarna	15.34	5.0–35.0	–
Annarpurna	18.00	10.0–35.0	–
PRA 8801	17.34	15.0–35.0	–
VL 33	15.00	5.0–25.0	–

Male sterility in *a hypochondriacus* has been detected in the USA and Nepal. It is reported to be controlled by a single recessive gene. This opens up the possibility of hybrid seed production through proper exploitation of heterosis. In six intervarietal crosses of grain amaranth F_1 heterosis, to an extent of 22.75 per cent over the better parent, has been observed although heterosis in general, was low (Table 35.3). Substantial transgressive segregation in F_2 generation in some of the crosses (Table 35.4) was also observed (Dutta, 1998). Mutation breeding has been extensively applied for seed colour and plant morphological improvement in India, but development of varieties through mutation breeding has not been successful so far. Since the first variety, Annapurna was released during 1980s, a number of varieties have been developed and released both at the state and national levels (Table 35.4) for the Himalayan region.

Restriction enzymes, cloning, and nucleotide sequencing have been applied to genomic DNA studies of grain amaranth. The molecular cloning of a gene encoding seed specific protein from grain amaranth has been patented by National Research Centre for Plant Genomic Research (NRCPGR), New Delhi. A non- allergenic gene from grain amaranth has been successfully transferred to potato to improve its protein content substantially. Genes isolated from amaranthus that have antimicrobial properties, have also been patented (Datta *et al.*, 1998).

Table 35.5: Performance of Frontline Demonstration for PRA 3 in Tehri Garhwal Districts Under AICRP ON UUC

Year	*No. of Villages*	*No. of Farmers*	*Per cent Increase Over Farmers' Practice*
2005-06	05	08	55.15
2006-07	03	12	65.23
2007-08	01	06	34.21
2008-09	04	08	30.43
2009-10	06	65	20.42
2010-11	02	09	78.28

Buckwheat (*Fagopyrum* spp.)

Common buckwheat is a herbaceous erect annual belonging to the family polygonaceae. It is a multipurpose, fast growing, annual crop cultivated throughout the world. In India, buckwheat is widely grown in the north-western Himalayan region mostly in pure stands.

Origin

Buckwheat is supposed to have originated in temperate central Asia from where it has migrated to other countries and regions. The perennial wild species *Fagopyrum cymosum*, native to China and India, was considered to be the progenitor of the two commonly cultivated species, *F. esculentum* (common or Japanese buckwheat) and *F. tatricum* (Tatary buckwheat). Recently, a newly discovered subspecies.*F. esculentum* ssp. *ancestralis* Ohnishi has been suggested to be the wild ancestor of common buckwheat, previously known as wild tatary buckwheat and *F. tataricum* ssp.*potanini* Batalin is the wild ancestor of tatary buckwheat. The place of origin of common buckwheat was also traced to the northwestern region of Yunnan province and that of tatary buckwheat to the northwest part of Sichuan province in China. Diploid populations of *F. cymosum* from eastern Tibet are considered as progenitors of tetraploid buckwheat found in the Tibet-Himalayan region.

Figure 35.3: Buckwheat (*Fagopyrum* spp.)

Buckwheat is cultivated throughout Europe, East Asian countries, the Himalayan countries, North America and even in Australia, Brazil and South Africa. In India buckwheat crop is widely cultivated in the states of Jammu and Kashmir, Himachal Pradesh, Manipur and the mountainous region of West Bengal. In southern India it is sporadically cultivated in the Nilgiris and Palni hills. Buckwheat is cultivated between 500 m to more than 4200 m altitude mainly in pure stands. *F. esculentum* is more prevalent in the low altitude areas while *F. tatarícum* is better adapted to high altitude regions.

Uses

Buckwheat is a crop of multifarious uses. The tender shoots are used as a leafy vegetable. The grain is used in various culinary preparations, particularly during fasts observed on the occasion of religious festivals. The amino acid composition of buckwheat grains is much superior as compared to that of the cereals. Its grains though possess comparatively less protein than that of wheat or barley but are an excellent source of lysine for which cereals have been found to be deficient. The grains are also used as livestock and poultry feed. Buckwheat crop is used as fodder, green manure and as a cover crop. Fresh and dried stems, leaves and flowers are a good source of *rutin* (rutoside), a flavonol glycoside (quercetin-3-0-rutinoside), which is used in the treatment of fragile blood capillaries and protects them from rupturing due to high blood pressure. *Rutin* is supposed to provide protection against the harmful effects of radioactive rays. Buckwheat has higher anti-oxidant properties superior to most cereals and buckwheat protein product suppresses gall-stone formation and reduces the cholesterol level.

Cropping System

The cropping system involving buckwheat varies according to altitude and the maturity duration of the crop. With short duration varieties even three crops of buckwheat can be taken in a year. However, the major crop rotations involving buckwheat are as under:

- ☆ Buckwheat - Toria
- ☆ Buckwheat – Wheat/Barely/Mustard
- ☆ Buckwheat – Peas
- ☆ Buckwheat - Fallow

Unlike grain amaranth which is more prevalent in Uttarakhand, buckwheat crop has extensive coverage in Himachal Pradesh.

Genetic Resource and Breeding

Collection of buckwheat has been mainly made through multicrop explorations by the NBPGR, Regional Station, Shimla in association with other agencies. The centre has undertaken systematic evaluation studies in which 408 accessions, both indigenous as well as exotic, belonging to 6 species, have been characterized for 31 descriptors (Joshi and Paroda, 1991). Currently, 250 accessions are being conserved in the National Gene Bank under long term storage out of the total 561 active collections maintained at NBPGR, Regional Station, Shimla. Under the AICRP a set of 25 germplasm lines are regularly evaluated every year at multi-locations. Germplasm lines collected from Sangla region in Himachal Pradesh have shown to be superior in terms of grain yield. Physiological and biochemical studies have also been undertaken for evaluation of genotypes.

F. esculentum is a heteromorphic self-incompatible species with dimorphic white, pink and red flowers. *F. tataricum* is self fertile with inconspicuous flowers having light green sepals. Homomorphic, highly self-compatible diploid lines have been isolated but they suffer from severe inbreeding depression. A study of heteromorphic sporophytic type of incompatibility in *F. esculentum* suggested that inheritance of style length is controlled by a single gene, the allele for short style (S) being dominant over long style (s). Attempts to transfer self-compatibility of *F. tataricum* to *F. esculentum* have proved unsuccessful, but inter-specific crosses between *F esculentum* and *F. cymosum* have proved successful through ovule culture. Concerted multilocation testing has resulted in the release of several varieties of various maturity groups (Table 35.6). In a recent paper, buckwheat breeding work in India has been reviewed and strategies for buckwheat improvement have been suggested (Dutta, 2004).

Chenopods (*Chenopodium* spp.)

Grain chenopods are the ancient food crops belonging to the family Chenopodiaceae. These are annual, fast growing, multi-utility crop plants grown throughout the world.

Table 35.6: Recommended Varieties of Underutilized Crops for NWH Region

Sl.No	*Crop/ Variety*	*Year of Release*	*Average Yield (q/ha)*	*Special Features*	*Recommended Areas*
A.	**GRAIN AMARANTH**				
1.	Annapurna	1984	22.50	High yield potential, high protein (15 per cent), wider adaptability.	Mid and higher Himalayan region
2..	PRA-1	1997	14.50	High grain yield	Uttarakhand hills
3.	PRA-2	2001	14.50	High grain yield	Entire NWH region except Jammu and Kashmir
4.	PRA-3	2003	16.50	High grain yield and early maturity.	Whole NWH region except Jammu and Kashmir
5.	Durga	2006	21.00	High grain yield and early maturity	Whole NWH region except Jammu and Kashmir
6.	VL Chua 44	2006	13.20	Early maturing (110-120 days).	Mid and higher hills of Uttarakhand
B.	**BUCK WHEAT**				
1.	Himpriya	1991	12.00	*F. tataricum*, high grain yield, medim maturing.	High altitude areas of Himachal Pradesh and Uttarakhand
2.	VL Ugal 7	1991	8.00	*F. esculentum*, early maturing, moderate yield.	Mid hills of Uttarakhand
3.	PRB 1	1998	12.00	*F. esculentum*, high yielding, medium maturity.	Hill region of UP, HP and North Eastern states
4.	Himgiri	2006	11.50	Early maturing (81-95 days)	Dry temperate region of Himachal Pradesh and Jammu and Kashmir
5.	Sangla B-1	2006	12.70	Medium in maturity (104-108 days) and high yielding	Mid and high hills of Himachal Pradesh and Uttarakhand
C.	**CHENOPODIUM**				
1.	Pusa Bathua 1	-	300 q/ha green leaf	Vegetable type	Northern plains
	RICE BEAN				
1.	PRR 1	1995	15.00	High yielding	Uttarakhand hills
2.	PRR 2	1997	12.00	High seed yield, light yellow seed.	Uttarakhand, HP and North Eastern states.
3.	BRS 1	2003	14.50	Early maturing and high seed yield	NWH hills (Identified)
	HORSE GRAM				
1.	VL G 8	2005	7.0-8.0	Cream colour grain, 125-130 days maturity	Hill areas of Uttarakhand
2.	VL G 10	2006	9.0-12.0	Light yellow grain, 113-117 days maturity	Hill areas of Uttarakhand under organic conditions
3.	VLG 15	2007	9.0-12.0	Light yellow grain, 113-118 days maturity	Hill areas of Uttarakhand Under organic conditions

Origin

Chenopodium has been identified as a promising future crop. Four species are known to be cultivated in different parts of the world: *Chenopodium album* L., *C. quinoa* Willd., *C. nuttaliae* Safford and *C. pallidicaule* Allen. In India, cultivation of grain chenopods is extensive in the North-western Himalayan region and extends up to the Khasi and Garo hills in the North-East. The cultivation areas in the North—Western Himalayan region having been delineated (*Pratap et al., 1998*). Cultivation of a new species, *C. giganteum* was reported from the Himalayan region but this has been identified as a variety of *C. album* (L.) Steward. The other three species originated in tropical America and are referred to as grain chenopods. The most important of Andean chenopods is *C. quinoa* which was considered as a sacred plant by the Incas.

Figure 35.4: Chenopods (*Chenopodium* spp.)

Uses

Chenopods are multi-utility crop plants grown for various uses. Chenopod grains are highly nutritious. The protein quality of grain chenopods is comparable to that of milk protein and are rich in essential amino acids particularly lysine. The mineral content of leaves is comparable to those of spinach. The grain is sometimes used as a staple food and consumed in the form of porridge, pudding and also cooked with rice. Like other pseudo-cereals it is also consumed during religious fasts. The grain is used in the preparation of local fermented beverages. Occasionally the seeds are made into flour to make Indian bread. The Himalayan chenopod has been used for making leaf protein concentrates. Powdered chenopod seeds are given to children to treat intestinal worms. The plant may serve as a field indicator for magnesium.

Chenopodium starch has a high swelling and water-binding capacity, similar to those of potato starch. Quinoa starch is more highly branched and has lower amylose content than cereal starches. As a result, a patent application has been filed in Europe for the manufacture of a carbohydrate-based cream substitute from quinoa starch.

Genetic Resource and Breeding

NBPGR Regional Station, Shimla is currently maintaining 116 accessions of chenopods. Out of these, 99 are indigenous and 17 exotic quinoa obtained from Peru, Bolivia, Ecuador and the USA. Eighteen accessions belonging to nine species are being maintained at NBRI, Lucknow. Systematic effort has not been made so far for germplasm collecting and conservation in the Himalayan region. Germplasm evaluation work was undertaken for 116 lines at NBPGR Regional Station, Shimla and a great deal of variability was reported for most of the morphological characters. A great morphological character along with yield and maturity variation were observed for different genotypes of chenopodium at mid hills of North-West Himalaya (Prasad *et al.,* 2011) A limited set of accessions is evaluated every year at multi-locations under the AICRP on Under-utilized Crops.

Breeding effort in this crop are preliminary, as a result, local cultivars are largely being grown by the farmers. Cytoplasmic male sterility controlled by a single dominant gene has been detected in quinoa, suggesting the possibility of production of commercial hybrids through planned breeding programmes. Vegetative propagation by shoot tip culture has also been found successful in quinoa indicating the possibility of perpetuation of such hybrids. Although high yield level of some of the lines has been reported no grain type variety has been released for cultivation in India so far. However, a vegetable type, 'Pusa Bathua l' has been recommended for cultivation in the north Indian plains because of its high green leaf yields.

Underutilized Pulses

Horse Gram (*Macrotyloma uniflorum*)

Horse Gram (*Macrotyloma uniflorum* (Lam.) Verde.) Horse gram (*Macrotylorna uniflorulm* (Lam.) Verde. Syn. *Dolichos biflorus* L., *D. uniflorus* Lam.), commonly known as 'Kulthi', 'Gahat' or Madras gram is a low growing, sub-erect annual herb with twining branches, 30-50 cm tall and trifoliate leaves. Pods are small, 3-5 cm long, linear, re-curved, beaked, 5- 7 seeded. The seeds are small, light red, brown, and grey, black or mottled with a shining testa.

Origin

Horse gram is an old world crop, native to the Hindustan Centre. It is a short day annual, cultivated on a variety of soils. It is largely grown in the central and peninsular region, and in the hilly slopes in Himachal Pradesh, Uttarakhand and the North-Eastern states up to an elevation of 1800 m.

Uses

Horse gram is a relatively low quality pulse. Its seeds are parched, boiled or fried and eaten either whole or ground as a meal. Boiled seeds are often fed to cattle and horses. Stems, leaves and husks of horse gram are used as forage. The crop is also grown for green manure and as a cover crop on undulating, eroded hilly slopes and red laterite soils. Horse gram is reported to be used as an astringent, diuretic and tonic. It is recommended for removing kidney stones. Extracted juice of the fresh plant is used for treating diarrhoea; decoction of the pulse is useful in cases of hemorrhage from bowels, leucorrhoea and menstrual derangement in women. Soup is used as a diet in sub-acute cases of enlarged liver and spleen.

Genetic Resource and Breeding

In India, germplasm collection efforts were initiated in the 1970s and a total of 952 accessions were assembled. These include indigenous collections from the peninsular region, hilly regions of the North-east, Himachal Pradesh and Uttarakahnd, and exotic material from Myanmar and Sri Lanka. The germplasm is being maintained at NBPGR Regional Stations at Akola and Thrissur. Germplasm evaluation of horse gram has revealed considerable variation for plant height, pod number, pod length, 100 seed weight, number of branches and seed yield. Screening of genotypes for reaction to pulse beetle (*Collosobruchus chinensis*) reniform nematode (*Rotylenchulus reniformis*) and yellow mosaic virus revealed some resistant types.

Crop improvement work has been largely confined to selection for high yielding desirable types. Single plant selections from locally adapted bulk proved useful and resulted in the development of several varieties. Two types of lectins - a) seed lectin and b) stem and leaf lectin (DB 58) were shown to be encoded by separate genes as revealed by isolating the genes from a library of *Macrotylama uniflorum* nuclear DNA. Evidence was presented for sequence insertions in the nuclear genes of plants that were derived from non-coding regions of the mitochondrial genome. The morphogenetic response of genotype CO-1 was showed differentiation of callus and shoot buds only from cotyledon derived callus on MS medium, supplemented with B-5 vitamin and different concentrations of NAA.

Rice Bean [*Vigna umbellata* (Thunb.) Ohwi and Ohashi Syn. *Phaseolus calcaratus* Roxb.]

Rice bean is an important multipurpose under-utilized grain legume with much higher grain and protein yield potential as compared to traditional pulses such as green gram and black gram. It can thrive in areas where green gram and black gram cannot survive.

Origin

Rice bean is a native of South and South-East Asia. It is also cultivated to a limited extent in various parts of the world. In India its distribution is mainly confined to the tribal regions of North Eastern hills and the Eastern and Western Ghats in peninsular India. In the Eastern and North Eastern states, it is cultivated as a mixed crop under shifting cultivation or in the kitchen garden as a backyard crop. In the north it is cultivated in Himachal Pradesh and Uttarakhand hills. Wild rice bean has been found to grow in the Himalayas. Rice bean possesses resistance against yellow mosaic virus disease not found in other *Vigna* sp. and has good potential as a crop in the mid hills.

Figure 35.5: Rice Bean [*Vigna umbellata* (Thunb.)]

Uses

Rice bean is primarily used as a food and fodder crop in Asia. In other countries it is also used as a green manure and cover crop. The nutritive value of rice bean is exceptionally high. The presence of a high amount of protein and limiting amino acids such as tryptophan and methionine in good quantity rank it as one of the best among the pulses. The whole dried grains are usually eaten boiled, in soups or as a dahl (pulse). Beans and seeds have a very good cooking quality comparable to green gram. The beans are often boiled or cooked alone or with other vegetables and eaten as a dish or as a soup. Young immature pods and leaves are used as a vegetable and are highly popular in the Philippines and Indonesia. Sprouts are also consumed occasionally. Rice bean has been successfully used as a substitute for black gram in the preparation of a number of popular South Indian food items. It is an excellent source of forage because of its high biomass production even under marginal and acidic soil conditions. It nodulates profusely even in virgin soils with no history of legume cultivation. Multiple cuttings can be obtained under irrigated conditions from forage rice bean crop if the first cutting is done carefully. The presence of certain anti-nutritional factors is a limiting factor in the large scale adoption of rice bean.

Cropping System

Longer maturity duration coupled with non-synchronous maturity makes it difficult to fit rice bean in the existing crop rotation programmes in the hills, even though the indeterminate nature is a desirable attribute for a higher fodder production. Separate grain and fodder type varieties should be evolved with the former focusing primarily on the rain-fed mid hill areas while the latter for irrigated conditions in the plains. The following cropping systems involving rice bean are commonly practiced although mixed crops are generally popular.

- ☆ Rice bean- Wheat/Barley/Mustard
- ☆ Rice bean + Grain amaranth/ Wheat
- ☆ Rice bean + Maize/Wheat
- ☆ Rice bean + Other pulses/Wheat

Genetic Resource and Breeding

Assemblage of sizable collections in several countries of South and South-East Asia has been made. In India collection efforts made since 1970s have resulted in the collection of 930 germplasm lines of both indigenous and exotic origin which are being maintained at the NBPGR, Regional Station, Shillong. Wild forms of rice bean have also been collected from the North-Eastern and North-Western Himalayan region. Systematic evaluation of 690 germplasm accessions has been done recently at NBPGR, Regional Station, Bhowali.

Cytological investigations and crossability studies among the species of Vigna (Phaseolus) has concluded that *V. umbellata, V. trilobata, V. radiata* var. *sublobata* and *V. dalzelliana* are closely related forming a distinct group, while *V.radiata* and *V. mungo* form the other. *V. dalzelliana* was considered to be the wild relative of *V. umbellata* as both resemble in plant and seed morphology and have sympatric distribution. The amount of DNA from different species has also been found to differ significantly and has supposed to play a major role in the evolution of the species

Mutation breeding involving indeterminate types with 100 kr gamma rays has resulted in the production of photo insensitive, determinate, early maturing mutants. Detection of early maturing mutants with 65 days maturity was reported at UAS Bangalore. The seed coat colour of rice bean shows wide variation and is not stable. The inheritance of seed coat colour was controlled by two independent genes with three alleles each. Crosses were made between exotic introductions from China, USSR and Nepal and the indigenous collections from Sikkim and Mysore. Early maturing and anthracnose resistant selections were made from their progenies. Early maturing genotypes have also been reported in intervatietal crosses of rice bean at HAU, Haryana.

Considerable breeding material has been generated at GBPUAT, Hill Campus, Ranichauri and PAU, Ludhiana through planned breeding programmes. A number of varieties have been recently released for cultivation. RBL 1 and RBL 6 have been developed by PAU for the northern plains and PRR 1 and PRR 2 developed by GBPUAT, Hill Campus for the hill region of the country. The adzuki bean-rice bean linkage map was developed from an inter—specific cross between adzuki bean and rice bean. In total, 14 linkage groups, each containing more than 4 markers was constructed with one phenotypic, 114 RFLP and 74 RAPA markers. In intervarietal crosses involving genotypes from the hills and plains narrow leaved morphological sergeants were identified in F_2. The narrow leaf cross derivatives were characterized along with their parents using RAPD marker analysis. Out of the 30 primers used 17 were found effective for hybrid analysis showing wide diversity.

Table 35.7: Performance of Frontline Demonstration for PRR 2 in Tehri Garhwal Districts Under AICRP ON UUC

Year	*No. of Villages*	*No. of Farmers*	*Per cent Increase Over Farmers' Practice*
2005-06	05	07	45.75
2006-07	05	15	59.63
2007-08	05	33	75.02
2008-09	04	12	33.60
2009-10	02	10	20.45
2010-11	07	69	64.68

Adzuki Bean (Vigna angularis)

Adzuki bean (*Vigna angularis* (Willd.) Ohwi and Ohashi) is a small bushy annual plant with determinate growth habit known for its multifarious uses. It is commonly grown by the farmers along with rice bean crop under a mixed cropping system. Adzuki bean is a short duration crop as compared to rice bean but having lower yield potential.

Figure 35.6: Adzuki Bean (*Vigna angularis*)

Origin

Adzuki bean has been cultivated for centuries in Japan, Korea, China, India and Manchuria. In India it is grown on a limited scale in the north-eastern states and the north-western hill states. Presently its cultivation has extended to other parts of the world particularly, Southern USA, Hawaii, Latin America and Zaire. It is probably native to India or Japan and has long been established as a crop in China. It is also believed to have originated in China. Unlike other cultivated Vigna species Adzuki bean is adapted to temperate to sub-temperate regions. The nutritive value of adzuki bean grains is comparable to that of rice bean.

Uses

Adzuki bean grains have high nutritional value with a good amount of protein (19.9- 25.3 per cent) and minerals. The seeds are used as human food, either cooked or made into a meal used in soups, cakes and confectionery items. In India it is mostly used as a pulse either whole or split. In Japan it is largely used for human consumption in the form of meal or paste. Sprouted beans are used as a vegetable. The bean may be popped like corn, used as a coffee substitute or as sugar candy. The flour is used for making shampoos and facial creams. The crop is also grown for forage and green manure in Japan and China. Adzuki bean seeds possess medicinal properties and are reported to be used in the treatment of kidney trouble, constipation, boils, abscesses, threatened miscarriage, difficult labour, retained placenta and non-secretion of milk. The leaves are used in the treatment of fever. The sprouts are used to treat threatened abortion caused due to injury.

Genetic Resource and Breeding

A set of 38 active germplasm accessions, mostly introduced, are being maintained at NBPGR, Regional Station, Shimla. 24 accessions are being maintained at NBPGR, New Delhi under long term storage.

Inter-specific hybridization programmes with rice bean have revealed a close relationship with adzuki bean, while it is distantly related with green gram and black gram. Inter-specific hybridization with black gram has been attempted but the seeds obtained did not germinate. Crosses with *V. angularis* and *V. umbellata*, *V. mungo*, *V. acutifolius* and *Phaseolus vulgaris* have been partially successful Genetic improvement work in India on adzuki bean is still preliminary, confined largely to selection with limited hybridization work. Unlike other cultivated *Vigna* species adzuki bean is adapted to temperate to sub-temperate regions. The incidence of *Phoma* disease in recent years has become a serious problem for successful cultivation of adzuki bean particularly during high rainfall years.

Challenges in the Promotion of Underutilized Species

Many workers engaged in the promotion of underutilized crops. Though it is true that the funding for addressing research and hence improvement of these species around the world is still extremely small, it should be also noted that other important factors, similarly determinant in the promotion process, should not be under-estimated. For example, during a participatory conference organized by IPGRI in 1998 in Aleppo, Syria, on "Priority Setting for Underutilized and Neglected Plant Species of the Mediterranean region" 10 major categories of constraints were identified as most relevant for the promotion of underutilized species. Since, some of the constraints were not linked to lack of funds as their solution lies in the strategic mobilization of existing resources to address the real problems hampering the full utilization of the species. Most challenges in the promotion of underutilized species are now well understood and appreciated and have been addressed in numerous works. However, an increasingly key issue in improving the use of underutilized species is the globalisation of the agricultural market. This is part of a much wider phenomenon that is changing dramatically our life and includes urbanization, loss of small commercial activities replaced by large enterprises, the homogeneization of local cultures as a result of the spreading of few models and cultures and improved communication. With specific reference to agriculture, globalisation is bringing about a further specialization in favour of a few crops that might best serve "*Global Village*". Obviously, this trend will lead to much narrower agricultural basket, estimated to hold only some 150 widely commercialized crops (Prescott-Allen and Prescott- Allen 1990), in favour of some commodities which will enter newer areas and displacing local ones. For many supporters of the globalisation process, this is the great opportunity for economic growth, but for many more it is a dealer of illusions, carrying prospects of a wealthy life that cannot be met and hiding the high cost to pay in terms of loss of sustainability, traditions, culture and local economies. Without entering the debate over the issue of globalisation, we would like to respond to the question that is raised in many occasions by those who see no future for underutilized crops in the *Global Village*: why should underutilized species be promoted in an increased global world if these are characterized by regional, national or local importance? The answer is very simple: securing the resource base of underutilized species, particularly in developing countries, is crucial to maintain the "safety net" made of diversified food and natural products that has provided options to address food needs in a sustainable way. Diversification in agricultural systems is indeed an important asset for those fragile social groups who may never be able to afford certain commodities and to whom the more diversified is the portfolio the greater is their self-sustainability and self-reliance in difficult areas where they live, often characterized by conditions not suitable for the cultivation of improved varieties of commodity crops. Generally speaking, plants are a fundamental component of the natural resource available to the poor and the role played by underutilized species is indeed central to reducing poverty and empowering the poor so as to allow poor rural communities to pursue resources-based rather than commodity-based development. But the people who benefit from underutilized species in a globalized world are not just for the poor. The benefits in terms of more balanced diets, diversified income to farmers as well as related sectors of the society, better maintenance of agroecosystems and greater use of marginal lands along with enhanced preservation of cultural identity can be shared by all the humankind.

Securing and Strengthening the Work on Underutilized Crops

Efforts so far have been directed to raise the awareness on the need to work on underutilized species and to start redressing their status of neglect. An area that has not been investigated yet is the analysis of when *a species will evolve* from its status of underuse and neglect *to become a well utilized crop*. This is a very relevant point if we want to make sure our efforts will lead ultimately to the "full promotion stage" beyond which the crop will no longer be considered underutilized. It is also very relevant to ensure that efforts to promote a specific crop have the desired effects with respect to the interests and needs of poor hill farmers, the maintenance of diversity and strengthening sustainability of production. Below are some of the issues that we should consider to develop a monitoring system for assessing the impact of our promotion process:

Conservation Aspects

1. How many accessions should be conserved (*in situ* and *ex situ*) to safeguard of the representative genetic diversity of these species and to provide at the same time the variability needed by breeders and users in general?
2. Given the new opportunities brought about by the gene transfer, how broad should be our efforts in the conservation of the gene pool of the species. Should we include the tertiary gene pool in our collecting efforts?
3. What should be the minimum level of knowledge on the eco-geographical distribution of the taxa and its genetic erosion status?
4. How much local knowledge should be safeguarded?

Utilization Aspects

1. What would be the level at which we would consider the economy of the species self sustainable?
2. What is the research threshold that is the minimum amount of research addressing the economic development of the crop, beyond which we would consider the species "properly addressed"?
3. What would be the minimum information needed on nutritional aspects, processing aspects required for successful industrial applications?

Agronomic Aspects

1. What should be the minimum know-how required for enabling a proper cultivation of the species?
2. What should be the basic information required for understanding the multiplication method and regeneration capacity of the species?
3. What should be the minimum level of information on pests, diseases and other cultivation-related problems?

Policy-Legal Aspects

1. What should be a basic policy framework to enable the proper deployment of the species among the farmers?
2. What should be the minimum level of policy attention required to ensure a sustainable use of these resources (particularly for wild species)?

Conclusion

Underutilized is commonly applied to refer to species whose potential has not been fully realized. Underutilized crops have rightly been termed as the 'Crops for the Future' at the global level signifying their importance for the future. The rapid loss of biodiversity together with habitat loss and the impending climate change has forced the policy makers to look for alternatives to the presently followed practice of monoculture and high resource use. Crop diversification and sustainability of agriculture with eco-friendly technologies such as organic farming and low input agriculture have therefore become key issues in recent times. The importance of the Himalayan region becomes all the more significant as it still contains the treasure of biodiversity of many crops. However, underutilized crops have to be made competitive so that they do not lose out to their more competitive counterparts. They could be made competitive if commensurate investments are made on their collection, conservation and scientific exploitation as is being done for other crops using conventional as well as cutting edge technologies. However, prioritization of crops has to be embarked upon keeping their strategic importance along with innovative utilization of the products with suitable interface with market and industry.

References

Dutta, M. (1998). Breeding grain amaranth for the high altitude Himalayan region: In: *Food Security and Crop Science, Proc. Int. Conf. Hisar,* (Ed.) M S Punia. CCS HAU, Hisar, India, pp. 74-75.

Dutta, M. (2004). Buckwheat improvement in India: Current status and future prospects. In: *Advances in Buckwheat Research, Proceedings of 9th International Symposium on Buckwheat*, Prague, Czech Republic, pp. 302-312.

Dutta, M., Phogat, B.S. and Dhillon, B.S. (2007). Genetic improvement and utilization of major underutilized crops in India. In: *Breeding of Neglected and Underutilized Crops, Spices and Herbs,* (Eds.) S. Ochatt and S.M. Jain. Science Publishers, Enfield (NH) USA, pp. 251-298.

Dutta, M. and Yadav, V.K. (2009). Small millets and underutilized crops in Himalayan agriculture. In: *Agriculture in Uttarakhand Hills,* (Eds.) B.S. Bisht and D.P. Singh. Directorate of Experiment Station GBPUA&T, Pantnagar Publisher, pp. 6-18.

IPGRI (2002). *Neglected and Underutilized Plant Species: Strategic Action Plan of the International Plant Genetic Resources Institute*. International Plant Genetic Resources Institute, Rome, Italy

Joshi, B.D. and Paroda, R.S. (1991). *Buckwheat in India*. NBPGR Shimla, Science Monograph No. 2, Shimla, Himachal Pradesh, India, 117 p.

Padulosi, S., Hammer, K. and Heller, J. (Eds.) (1996). Hulled wheats: promotion of conservation and use of valuable underutilized species. *Proc. of the First International Workshop on Hulled Wheats*, 21-22 July 1995. Castelvecchio Pascoli, Tuscany, Italy. International Plant Genetic Resources Institute, Rome, Italy, 262 pp.

Padulosi, S., Eyzaguirre, P. and Hodgkin, T. (1999). Challenges and strategies in promoting conservation and use of underutilized and neglected species. In: *Perspective on New Crops and New Uses,* (Ed.) J. Janick. *Proceedings of the Fourth National Symposium on New crops and New Uses. Biodiversity and Agricultural Sustainability*, 8-11 November 1998, Phoenix, Arizona, pp. 140-145.

Partap, T., Joshi, B.D. and Galwey, N.W. (1998). Chenopods: *Chenopodium* spp., Promoting the Conservation and Use of Underutilised and Neglected Crops, 22. Institute of Plant Genetics and Crop Plant Research, Gatersleben / International Plant Genetic Resources Institute, Rome, Italy

Prasad, B., Bahuguna, A. and Shukla D.K. (2011). Evaluation of Chenopodium (*Chenopodium spp.*), Genotypes for Yield and Maturity at Mid Hills of North West Himalaya. In: *National Symposium on Advances in Biotechnological Research in Agri-Horticultural Crops for Sustaining Productivity, Quality Improvement and Food Security*. Held at SVBPUA&T, Meerut (U.P.) from September 14-16, Abstr, 61.

Prasad, B., Bahuguna, A. and Shukla, D.K. (2011). Genetic Variability of Some Selected Grain Amaranth (*Amaranthus* spp.) Genotypes for Morphological Traits Under Mid Hills of Uttarakhand. In: *National Symposium on Advances in Biotechnological Research in Agric-Horticultural Crops for Sustaining Productivity, Quality Improvement and Food Security*, Held at SVBPUA&T, Meerut (U.P.) from September 14-16. Abstr, 38-39.

Prescott-Allen and Prescott, A. (1990). How many plants feed the world? *Conserv. Biol.* 4: 365-374.

Vietmeyer, N. (1990). The new crops era. Pp, xviii-xxii *in* Advances in New crops. Proceedings of the First National Symposium on New Crops: Research, Development, Economics. (J. Janick and J. Simon Eds.), Indianapolis, Indiana, 23-26 October 1988. Timber Press, Portland, Oregon.

Wilson, E.O. (1992). *The Diversity of Life*. The Penguin Press, London, 434p.